Essays on
form
in
plants

C W Wardlaw

Emeritus Professor
of Botany in the
University of Manchester

Essays on
form
in
plants

Manchester University Press

Barnes & Noble Inc., New York

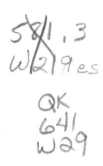

© C W Wardlaw 1968

Published by the
University of Manchester
at the University Press
316–324 Oxford Road
Manchester 13

First published 1968

GB SBN 7190 0331 8

USA
Barnes & Noble, Inc.
105 Fifth Avenue
New York

Printed in Great Britain by
Butler & Tanner Ltd, Frome and London

To the
University of Manchester

This volume is dedicated,
as a sincere if modest
expression of my thanks,
to the University of Manchester
where, for many years, I enjoyed
all that any scholar could desire
in personal freedom, encouragement,
friendship and help freely given

It is also dedicated to my
former staff and research students,
some of whom are now
scattered far and wide,
but all doing their University
and their one-time Professor
much credit

Preface

In this volume I have collected together some thirty essays written during my tenure of the Chairs of Cryptogamic Botany (1940–58) and of Botany (1958–66) in Manchester University. The essays deal with plant morphology and morphogenesis and with such topics as the need for achieving the fullest possible integration in botanical science, especially in the interests of continuity of sound scholarship, which I regard as an essential part of this ancient, liberal and humane subject. The essays, presented in chronological order, illustrate something of the progress in my general experience and thinking which went on as an integral part of my experimental investigations, published in the *Annals of Botany*, the *Philosophical Transactions of the Royal Society*, etc. The essays also indicate some of the initial thoughts and ideas which prepared the way for the writing of my several books. The essays have, I hope, to some extent been welded together by an introductory essay (Part one); and the volume is sent on its way by a free commentary on my part, however erroneous the passage of time may prove this to be, on what I think may be some of the future developments in my subject.

The essays are all as originally published with the exception that some repetitious and irrelevant matters have been eliminated.

I take this opportunity of expressing my thanks to the editors of journals and proprietors of publishing houses for permission to reprint the essays contained in Part two of this volume.

Contents

Part one

Introduction:
The development
of plant morphogenesis
(with elements of a
personal record)

The beginning

Morphology is the science of form in plants and animals: morphogenesis emphasizes the developmental aspect. So morphogenesis is in no sense new, or even recent. Some of the essential features of developing plants are evident and have long been recognized. However, the beginning of what is now regarded as scientific developmental botany goes back to a German observer, Caspar Wolff. In his book *Teoria Generationis*, 1759, he discussed the fundamental nature of the shoot system in plants and he illustrated dissections of the shoot apical growing point, or *punctum vegetationis*. His significant and lasting observations were (i) that he saw nothing in the vegetative plant but leaves and a stem, (ii) that new leaves developed successively from an undifferentiated apical growing point, itself perennially embryonic, and (iii) that the process must be considered to be a true *epigenesis*, i.e. a system with new accretions of organs not previously present as rudiments. In floral studies he showed that the several organs originate from the stem apical growing point in precisely the same way as do foliage leaves, the two kinds of rudiments being indistinguishable. After two hundred years, how very contemporary all this sounds!

For various reasons, discussed in this volume, Wolff's observations did not 'trigger off' an energetic new phase of botanical inquiry. That was to come considerably later when, having cast off the fetters of Goethe's Theory of Metamorphosis, also considered in this volume, Schleiden, in 1842, trenchantly asserted that the highway to new botanical discoveries lay in

the study of *developing plants*. A period of quite remarkable
activity, successful disciplined work, and insight, then followed. Naegeli, by inquiring how new cells are formed, founded
the theory of cell formation and directed attention to (i) the
importance of studying germ cells and growing regions, i.e.
apices; (ii) the remarkable similarities between the processes
involved, whether in the lower cryptogams, the pteridophytes,
or the seed plants; and (iii) the fact that the inception and
further development of organs could be traced back through
cell lineages to the formation of cells in embryonic regions.
Hence, in Naegeli's mature view, the study of development was
not merely one of several ways of approaching the problems of
plant form: it was identical with the study of organic nature.
This period of new, disciplined study is regarded by some as the
real beginning of modern scientific botany. But others, who
recalled the remarkable initial microscopic studies of Hooke,
Malpighi and Grew during the middle and later decades of the
seventeenth century, have regarded the Schleiden–Naegeli
period as a belated renaissance of scientific botany, rather than
as a new beginning. In Grew's work (1672, 1682), for example,
there is unmistakable evidence of systematically thorough treatment, lucidity of thought and exposition and, not least, deep
scientific curiosity and a surprising insight into the pervasive
physical aspects of the conformation and construction of plants,
all being 'contrived and brought about in a mechanical way'.

From such beginnings, the study of development and of
form in plants made great progress at the hands of unusually
gifted observers such as Wilhelm Hofmeister. Most students
are aware of his revealing studies of the life histories of bryophytes and pteridophytes, eventually published in an English
edition in 1862 as the *Higher Cryptogamia*. But Hofmeister was a
man of penetrating insight. He not only observed the constant
changes in size, form and complexity that attended any embryological development: he also carried out physiological experiments and he constantly inquired: How does the observed
form come to be? Here he had in mind the need to formulate
explanations or interpretations, in general terms, incorporating
mathematics and the physical sciences. These studies, which
disclosed a new approach to morphology, were presented in a
considerably less well-known book, issued in 1868, and never

published as an English translation. It bears the significant title: *Allgemeine Morphologie der Gewäche* (or *The General Morphology of Growing Things*). An excellent account of Hofmeister's work, and of relevant antecedent developments, is available in a volume by K. von Goebel on *Wilhelm Hofmeister*. It was published in 1926 to mark the centenary of the birth of this modest but great botanist (English edition translated by H. M. Bower and published by the Ray Society, London).

Hofmeister's approach to causal or general morphology was continued and extended by Sachs, K. von Goebel and others. Sachs, an eminent botanist and physiologist, formulated theories of great importance on correlation and chemical morphogenesis. Indeed, his ideas are still of paramount importance in contemporary investigations of plant morphogenesis. Sachs and Goebel, as also their contemporaries, working in the heyday of Darwinian evolution, were much preoccupied with the phenomena of *organography*, i.e. the relation between form and function, typically *in fully developed organs*. But Goebel was also greatly attracted by observation of developmental processes in growing plants, and by the experiments that could be conducted on them. So, as well as his great books on *Organography of Plants* (three German editions and an English translation), he also had to his credit a smaller but excellent volume: *Einleitung in die experimentelle Morphologie der Pflanzen* (1908). Later Goebel was to draw a distinction between *organography* and *organogenesis*, the latter being concerned with changes in configuration during development. But, in fact, in the later versions of his *Organographie der Pflanzen*, the definition of organography tends to become somewhat obscure.

Comparative and causal morphology

With the sweeping success of *special* or *comparative morphology* as the principal means of validating and expanding the Theory of Evolution as proposed by Darwin, the *general* or *causal* morphology of Hofmeister, although not without distinguished adherents, lost ground as a central theme or discipline in botanical science. In fact, studies of the comparative morphology of fossil and living plants provided students of evolution not only with a central goal but with a rich and satisfying field for exploration for several decades. But even in science, however rigorous the discipline, things change and pass. By the second decade of the present century, the great days of comparative morphology were on the wane for many botanists, especially the younger ones. Doubts were expressed about some of the main conclusions based on comparative studies. Morphology and physiology had drifted apart and the sometimes facile, and usually pseudo-physiological, 'explanations' of the function and adaptive value of organs and tissues, proposed by evolutionary morphologists, were regarded with increasing distrust, not to say distaste, by adherents of the growing science of physiology. Indeed, some comparative morphologists had become critics of their own procedures and findings. A single but critical example may be cited. As comparative studies of organs and tissues widened, it became increasingly evident that more or less closely comparable structural features were present in plants of quite different taxonomic affinity, i.e. that were in no way closely related genetically. In fact, the concept that close morphological similarity, i.e. homology, was indicative of genetical affinity had, in some instances, to be qualified by the introduction of the further concept of *parallel* or *homoplastic* development, appropriately referred to as *homology of organization*.

So, about 1910–20, we approach the beginning of the contemporary period. As some part of the newer generation of botanists began to turn away from comparative morphological studies, it became increasingly appreciated that the problems of causal or general morphology, as envisaged by Hofmeister, still remained and that these were of the very essence of biological inquiry. This view was cogently urged by W. H. Lang in 1915 and later, in 1923, Tansley suggested to those who still had a predilection for the study of plants as morphological entities that they should study the *process of development*, in which physiology would have an essential part. So, for those who still maintained an interest in morphology a new beginning seemed essential. But, at a time of change and disillusionment, the great questions were how and where to begin. If you were interested in causal morphology, what were the 'right' problems to tackle first? Believe me, it was by no means easy to see how to begin in the early 1920s. Some of the special techniques, essential and invaluable, had still to be discovered.

Enter the author
From this point onwards, in presenting this collection of essays, something of a personal record must be introduced, not (I hope my readers will concede) out of egoism, but because it is always interesting to know at first hand how things began. As Aldous Huxley wrote somewhere, it is individuals, and not groups in committee, who do new things.

As a student and assistant of Professor F. O. Bower, a great botanist and a distinguished exponent of the phylogeny of the ferns based on comparative morphology, I was introduced to the wide range of living and fossil species on which the theory of evolution of plants, in its great historical aspect, necessarily rests. But Bower also introduced me to the interest of causation in plant form and structure through the study of size-and-form correlations. At that time, this concept was described as the 'size factor'. Much later, I was to realize that this term, though 'handy', was inappropriate. (More shrewd and critical minds had appreciated this from the outset!) I now realize that I was fortunate in that my initial university studies, which began in 1918, had included mathematics, physics and chemistry, including physical chemistry, and the idea of trying to understand the factors underlying the development of form and structure provided me with a deep, abiding and ever-growing interest. However, it should be said, with due emphasis, that the way in which one should begin one's research was far from clear. More generally, probably because I have a small tincture of critical philosophy in my make-up, I have never thought that the problems of morphogenesis are simple, or that they can be 'explained' in simple ways. But I have always thought that we should not only try to explain, but to aim at formulating comprehensive interpretations.

In those early days, Bower did me a signal service when he put into my hands a copy of D'Arcy Thompson's book *On Growth and Form*. I have never forgotten the delight with which I first read this masterpiece of scholarly exposition and, having a sufficient background of the physical sciences, how eagerly I seized upon, and was never to forget, such wonderful passages as 'Cell and tissue, shell and bone, leaf and flower, are so many portions of matter, and it is in obedience to the laws of physics that particles have to be moved, moulded and conformed. . . .'

In brief, it was borne in upon me that all specific organic forms, however palaeontologically ancient or genetically modern, must have come into existence, and must eventually be interpreted, in relation to the laws of chemistry, physics and mathematics. Many years later (*see* Essay No. 23) I was to have the opportunity of expressing some of these things as a mature scholar. It must be said, however, that although D'Arcy Thompson's book showed how phenomena of form and structure *might* be interpreted, it did not really serve as a guide to the young experimental worker, i.e. how to make an effective beginning. The passage of time has, I think, shown this to be a valid judgment.

During what was still a formative period as far as I was concerned, I read with much interest Professor K. von Goebel's book on *Wilhelm Hofmeister*. Of course, as students who had studied under a great exponent of the Pteridophyta, we were sufficiently well versed in Hofmeister's *Higher Cryptogamia*; but we knew little of his *General Morphology of Growing Things*, i.e. his *Allgemeine Morphologie*. In fact, this is a rather difficult book to read for all but those with a highly competent knowledge of German. That, at least, has been my experience. But Goebel's book opened new doors and one began to appreciate some of the real profundity of Hofmeister's thinking about causation in the inception and development of form.

In 1928, having completed three papers on the 'size factor', and having in the meantime become deeply interested in mycology and plant pathology, I went off to the tropics to work in applied research for twelve years. (The simple fact is that I had a consuming thirst for travel.) As it happened, I still had occasion to do some morphological and anatomical work and, of course, had wonderful opportunities to enlarge my knowledge of plants. To plunge into the immense range of new species in a rich and varied tropical rain forest region is both an exhilarating and humiliating experience. But it was to bear fruit later in days of more mature reflection (Essay No. 24). One result of this experience, which I have expounded with a little due emphasis from time to time, is that the *Oxford Dictionary* is a dependable work of scholarship! For does not it state explicitly that Botany is the *science of plants*, i.e. it is not merely some of the processes in them, or aspects of them abstracted from the totality of specific evolving organisms. Processes and particular aspects, it need hardly be said, must be isolated for special study. That is the way of science. But let us never forget that our understanding of biological processes can only become the fullness of knowledge in the context of the holistic organism. The wilful isolation of special branches from the general corpus has done botanical science little good and less credit. The assiduous reader will encounter a fuller statement of these and related views in some of the essays now presented.

A further and quite unexpected result of my years of varied applied work in the tropics was that I found that I had become detached from my earlier 'indoctrination' in comparative morphology and phylogeny. By this I do not mean that I had forgotten the lessons or the philosophy learnt from a revered professor. Nor did I think they were unimportant. I appreciated this phase of my studies but with a new detachment. In time, I began to see that if comparative morphology was not in itself enough, neither was causal morphology. It was not a question of one or the other: both were essential. My essay on 'Process and Record' indicates my initial groping towards a synthesis of the two disciplines. This was later to be expounded at greater length in my book *Phylogeny and Morphogenesis* (1952).

New beginnings

By the late 1920s and early 1930s, new and important innovations, or practical implementations of earlier ideas, were on their way. The isolation, constitution, growth and morphogenetic effects of specially active biological substances—auxins, hormones, growth-regulating substances (call them what you will)—were attracting ever increasing attention; and by 1942 Dr. J. Needham had published his comprehensive, scholarly book on *Biochemistry and Morphogenesis*. Along other lines, in which factors relating to the inception of form and the manifestation of organization in species were being actively investigated by experimental methods, Dr. C. M. Child had followed up his inspiring little book on *Individuality in Organisms* (1915) by a greatly enlarged work entitled *Patterns and Problems of Development* (1941). Reference must also be made to Dr. J. S. Huxley's useful excursion into mathematics as applied to certain characteristic morphological developments. In his book on *Problems of Relative Growth* (1932), he showed how comparatively small differences in the differential growth rates, established in early embryogenesis, could have remarkable cumulative effects on the morphology of the adult. Meanwhile, the science of genetics was reaching out in many directions and problems of morphogenesis began to have an important place in theories of genic constitution and action.

Thus, when I returned to academic life in 1940, to fill the Chair of Cryptogamic Botany in Manchester University in succession to Professor W. H. Lang, I knew, at least in a general way, what I specially wanted to do, even if I was none too clear as to how I should set about it. For me, it was to be experimental morphology, now generally referred to as morphogenesis —an appropriate term, though I cannot avoid the impression that there are some who use it without an adequate realization that the term is primarily and essentially concerned with *form* and its *inception* or *genesis*. For me, also, it was to be the ferns. I knew them well from earlier phylogenetic studies. What could I contribute on morphogenetic aspects?

Finding my way
It had now become clear to me that, if I was to begin to understand what is involved in the inception and development of form and structure, I must devote myself to the study of meristematic and developing regions, i.e. embryos and the apices of shoots, leaves and roots. Of course, this seems self-evident. But, as a simple truth, it was by no means a standard procedure if you had been brought up among comparative morphologists. The latter, typically, though not exclusively, worked with mature organs and tissues. For example, Professor Bower used to teach us that if you wanted to understand an elaborate stelar configuration, e.g. in a fern shoot, you prepared cross-sections from below upwards, i.e. beginning with the thin axis close to the point of attachment to the gametophyte, with its small protostele, and thereafter you worked progressively upwards into the obconically enlarging axis with its increasingly complex vascular system—solenostele, dictyostele, polystele, etc. Now, this procedure undoubtedly yields essential information: it indicates the progressive elaboration of tissue differentiation during ontogenesis. But it took me quite

a long time to see that, methodologically, it was inadequate. If your aim is to understand how development takes place at each level of stelar complexity, you must examine the nature and constitution of the shoot apex and the differentiation of the stelar tissues in it at that stage. This, evidently, is an essential procedure if our aim is to contribute by direct observation to our knowledge of morphogenesis. I am *not* suggesting that the great classical morphologists and anatomists paid no attention to differentiation at the apices of shoot, leaf and root. Quite the reverse: from the days of Naegeli, Hofmeister and Hanstein, onwards through Sachs, de Bary, von Goebel and Bower, apices had received a great deal of dedicated practical skill and scholarly interpretation. This work is still being continued by interested observers. What I am trying to indicate is that one's training in morphology and anatomy at the hands of exponents of comparative morphology did not prepare one imaginatively or conceptually for experimental studies of morphogenesis.

On due reflection it seemed to me that there were problems in classical morphology, of long standing, that might lend themselves to critical experimental investigation. Thus, whereas some of the prevailing morphological ideas tended to be accepted as being self-evidently valid, others had been the subject of much inconclusive controversy, though little or no thought had been given to devising experiments which might decide the issues, or at least yield new information. It gradually dawned on me that both categories of ideas could be re-examined experimentally. It had long been known that the formation of new organs and the differentiation of the primary tissue systems took place in the shoot apex—perennially embryonic in the vegetative phase. The descriptive treatment was mostly adequate; indeed, much of it was excellent. What impressed one was the number of unanswered questions! How were these developments actually determined and controlled? Were the causal factors of primary morphogenesis present in the apical meristem itself, or did they proceed from the older, differentiated regions of the shoot? Was the whole distal region of the shoot involved in these developments, or did the determining and regulating factors reside in the most distal region of the apex, i.e. in the conspicuous apical cell region as in a fern? Put briefly, to what extent was the shoot apical meristem a self-determining region?

And if so, did the initial differentiation of the vascular tissues take place basipetally? Or, alternatively, since the pattern of differentiating tissues was typically in continuity with that of the mature tissues below, could there be any doubt that acropetally-directed forces, sometimes loosely described as 'influences', were involved? Then there were other propositions, e.g. that, in many pteridophytes, the progressive elaboration of the stele during ontogenesis was a constitutionally-imposed obligatory development, that the stele was a very 'conservative' region, not readily subject to modification and therefore of special value in comparative phyletic studies. There were also controversies about the origin, nature and function of the leaf-gaps in solenostelic and dictyostelic species.

When I returned to these topics, as it were refreshed and detached, I thought that new and relevant information might be obtained by comparatively simple micro-surgical and chemical treatments, provided shoot apical meristems could be 'got at' and that they were not too small and delicate to remain viable after treatment. Already Mr. and Mrs. Snow of Oxford had demonstrated that problems of phyllotaxis in selected dicotyledons could be tackled using such techniques. It is perhaps a curious fact that I had not had occasion to read their papers (published from 1929 onwards) until after my own work was well on its way. While this new phase of work was still in the making, my mind harked back to some of the great experimental achievements in physics and chemistry. In many instances, these had resulted from crucial experiments, i.e. experiments which would have predictable results provided the underlying reasoning and hypotheses had been soundly based; i.e. that correct inferences had been drawn from precise prior observations. Accordingly, from the outset and wherever possible, I tried so to construct my experiments in morphogenesis that they were, in some measure at least, crucial experiments. This has been a rewarding approach. Close anatomical observation of the shoot apex in *Dryopteris dilatata* showed that the incipient vascular tissue of the axis, as seen in serial basipetal transverse sections, formed a continuous ring while the incipient vascular strands of the very young leaf primordia appeared as semicircular masses which become confluent with the periphery of the axial vascular tissue. On proceeding downwards in the axis the leaf-traces, which had now developed as enlarging but still continuous horse-shoe-shaped masses of tissue, began to break up into a number of separate incipient vascular strands. This same tangential enlargement of the leaf-bases was associated with the appearance of small, nascent parenchymatous leaf-gaps in the axial ring of incipient vascular tissue. Still lower down, the differentiating axial vascular system became a dictyostele, with parenchymatous leaf-gaps. With these facts in mind, it thus appeared that the gaps in a shoot dictyostele were a direct consequence of the mode of growth in leaf primordium. Accordingly, if all the very young primordia could be punctured or suppressed over a period of time, the expectation was that the treated region of the axis should contain a solenostele and not a dictyostele. This predictable result was duly obtained as were others where the background observations, ideas and inferences had been *au point*.

As new information and experience of the reactivity of shoot apices under experimental treatments accumulated, I was always afraid that, sooner or later, the whole system of work—the spring—would dry up. But this never did happen. There always seemed to be so much to do, and so many new apices, and the special problems they posed, that my research students and I were continuously occupied with new and exciting ideas. Of course, I appreciated the need for assistance from a fully-trained physiologist and one was duly appointed and became deeply involved in the work, as have many others elsewhere.

The gift of an idea
As my research experience deepened, I began to see that experimental results were not enough: I was constantly trying to edge closer to the essential, or central, problems of morphogenesis. This effort pertained to a rather different realm of experience and scholarship, i.e. as compared with undertaking a discrete and well considered individual problem. It is a common fact of experience that the deeper one tries to go in biology, the more difficult the task becomes, not least in the matter of cogent, explicit exposition. As an analogy, one can do a lot of hard uphill clambering on a steep rocky hill; but one can keep going. But when one is eventually confronted with a smooth, glassy precipice, devoid of evident handholds, the

situation is very different indeed. Similarly, one does not go very far in morphogenesis, whether in plants or animals, before one encounters the glassy-fronted precipice. The origin, or inception, of primary organogenic and histogenic patterns in embryonic regions has long presented this kind of difficulty. Moreover, any theory proposed should be such that it would satisfy the criteria of physics, chemistry, mathematics and biology. But relevant ideas, for me and for others, remained elusive and obscure. Then, one day, light began to break through. I was introduced to the late Dr. A. M. Turing so that he might discuss with me his diffusion reaction theory of morphogenesis, then in the making (*Phil. Trans. Roy. Soc.* 1952). Turing was not only a gifted mathematician, but he had great scientific alacrity and curiosity and an unusual capacity for novel thinking about certain biological phenomena. In particular, the mathematical regularity of the patterns of organs and tissues, newly formed in embryonic regions, held a special fascination for him. They were a kind of living geometry and, being deeply and imaginatively versed in the 'mother of the sciences', he thought that, since there are so many instances of these regular inceptions of pattern, there must be common underlying principles that must surely be based on physical chemistry as applied to complex metabolic systems. So out of this thinking came his new theory. The mathematical treatment and statement of the theory are rather complex—in fact, probably too complex for the average biologist. But the main idea is not difficult to grasp, in that it is based on relatively simple processes in physical chemistry, e.g. rates of diffusion, autocatalytic reactions, etc. The theory states that there is a reaction system in embryonic regions or tissues. If the reacting substances, i.e. the general and special metabolites, are initially homogeneously distributed, the reactions within the system eventually result in a heterogeneous, regular distribution of the reacting substances and their products. For example, in a locus in which an active autocatalytic process had been 'triggered off', the relevant enzyme, its substrate and its by-products will tend to accumulate at the expense of adjacent regions or loci. In the latter, other, complementary reactions will also begin to take place. A chemical pattern will thus begin to be established and will subsequently become stabilized. This pattern underlies and determines the formation of organs in regular, characteristic positions, or the differentiation of regular tissue patterns. Several of the essays here presented relate to Turing's theory.

Not all biologists have accepted Turing's theory but, in my view, if it is not the last word on the subject, as it almost certainly is not, if only because it is a greatly simplified statement about what must be very complex biochemical situations, it belongs to the *kind of thinking* that, with modifications and amplifications, must surely contribute to more penetrating interpretations of the enigmatic events in primary morphogenesis. If it does no more than provoke a search for a more adequate physico-chemical theory of primary pattern in plants and animals, it will have served biology well. For myself, certainly, Turing's idea was a source of inspiration—the kind of idea for which I had been looking for a long time. In more general terms, one might say that the investigator of morphogenesis may recognize that a property, or function, of embryonic regions is that, as loci of reaction systems, they give rise to patterns which are at once genetically specific and also of considerable geometrical commonality. Contiguous organs are evidently physiologically related one to another and the overall development, e.g. of a leafy shoot, is one of equilibrium and harmony. Accordingly, it appears that the reaction system involved must be a unitary or holistic one. So whether Turing is right or wrong in his physico-chemical, mathematical exposition, it is quite inescapable that, in morphogenesis, the formative process must eventually be interpreted, or understood, in terms of the principles of physical chemistry as applied to complex organic reaction systems. In some of the essays in this volume, I have tried to show that there is ample botanical evidence supporting the theory, both in its general and in its more particular applications.

Towards a major theme
While these new ideas were being borne in upon me, the general theme of *Phylogeny and Morphogenesis* was still firmly in my mind, if only because, once there, a topic embracing such wide scholarship does not allow itself to be cast out. Also, like others, I had an acute awareness of the further partitioning of botanical science as a result of the unavoidable contemporary specialization. As a scholar I had a compulsion to do something about it.

For one thing, I began to see that, in morphogenesis as in other
biological disciplines, there was beginning to be such an accumulation of new factual information that we might soon be
in the old, sorry position of not seeing the wood for the trees.
Some major unifying theme, or system of ideas, leading if
possible to the enunciation of new principles, seemed pre-emi-
nently desirable. Needless to say, the central, or centralizing,
idea had been there all the time: morphogenesis is concerned
with both general and specific formal developments in plants
past and present, with their many close similarities and all their
immense diversity. Even the simplest organisms are highly
organized entities, or specific physical systems. Evolution in its
historical aspect is essentially concerned with the progressive
evolution of organizations, both in their general aspects—e.g.
in that all plants are generally alike in some respects and all
vascular plants closely comparable in their main vegetative
and other features—and in their special aspects, in that, over a
vast span of time, large numbers of species have been evolved,
each being a unique physical system. In brief, I became aware,
as others, in some degree at least, had done before me, that
organization should be studied not only as the eventual aim in
morphogenesis but also in relation to the great historic evolutionary progressions. These nascent ideas, which had already
been touched upon in my essay on *Process and Record*, were again
discussed in one of my Prather Lectures in Harvard University
in 1951 and published as an invited paper when the new journal
Phytomorphology was launched in the same year. Much later, and
after a great deal of effort, my book on *Organization and Evolution
in Plants* (1965) was published. In one of the essays, taken from
the book, I have set forth a first attempt at discussing possible
'Principles' of Organization in Plants.

 During recent years I have been encouraged to note that
other botanists have begun to show an enlightened interest in
the topic of organization. Personally, I shall not be surprised if
it becomes a leading theme during the later decades of the
century, especially when biologists, by proper expositions of
their problems, have induced effective numbers of percipient
students of the physical sciences and mathematics to enter *fully*
into the problems of biology.

14 June 1967

Part two

I

Unification of
botanical science

Botany is the science which treats of plants.—*Oxford Dictionary*.

Every speculation about a single phenomenon wrenched from the continuity of life, is playing indeed a thankless part in the present condition of the natural sciences.—SCHLEIDEN, 1838.

An unflinching determination to take the whole evidence into account is the only method of preservation against the fluctuating extremes of fashionable opinion.—A. N. WHITEHEAD, 1926.

An upward outlook is in itself a practical application of any evolutionary view.—F. O. BOWER, 1935.

Any biological phenomenon can be considered from several different points of view, each of which may lead to the formulation of particular and distinctive concepts. In some instances concepts relating to different aspects may overlap: in other instances they may belong fundamentally to different categories. Now, it is a fact that botanical science has developed erratically and spasmodically; botanists have embraced strange and irreconcilable philosophies; they have welcomed innovations and canalized them into fashions or moulded them into new branches of the science. Moreover, the results forthcoming from the several distinctive phases have not invariably been studied with due regard to their mutual relationships, nor have they necessarily been envisaged as contributing directly to a generalized scheme. The piecemeal character of the scientific advance and the diversity of its branches of underlying philosophy have increased notably during the last fifty years; for it has been a period of great, if non-co-ordinated, activity along many seemingly divergent lines of inquiry. Cogent reasons, therefore, exist for the view that the time is at hand when an effort should be made to achieve some closer integration of the science as a whole.

As a result of the growing volume of research on almost every branch of botany and the concurrence of certain contemporary lines of investigation, there is a reasonable hope that certain gaps that have hitherto hindered synthesis may in due course be bridged and that this may lead towards a real unification of outlook. To some, no doubt, this may appear as unjustifiable optimism. The question may indeed be asked why this topic is considered to merit special attention at the present time. It may be argued that botanical science does not in fact lack unity, that it has been adequately unified at various times in the past, or that complete unification is difficult or even impossible. Moreover the nature of the unification envisaged itself requires elucidation: What kind of unification and for what purpose? I have in mind both the conceptual and methodological aspects: the former is concerned with general questions, that is, possessing significance for the plant kingdom as a whole; the latter considers the results of a particular discipline in relation to those of all other relevant disciplines, instead of such results being treated in comparative isolation. It is this latter aspect of unification that I consider of particular importance at the present time.

It seems evident that failure to achieve some progressive method of integration in the near future will be attended by such an accumulation of non-co-ordinated data as to dismay contemporary botanists and bewilder their successors. There is no novelty in this view, but emphatic reiteration seems timely. In the long view, teaching and research are inseparable; failure to collate the main results of contemporary and past work will certainly militate not only against reasonable advance but also impede the proper teaching of the subject.

In the realm of biological science, where a sense of ever-increasing complexity appears to be the chief reward of the most profound investigations, the less complicated structure and mechanism of plant life, as compared with animal life, might be expected to afford a more direct approach to the fundamental problems of organic Nature. Hence an unbiased observer might take it for granted that botanists would automatically take the lead in the formulation of new concepts and the enunciation of broad generalizations. Reference to some recent works on general biology can scarcely be considered to support this view; indeed, a certain neglect of botanical work by adherents of other branches of biology is apparent. It may be that botanists are to blame; for while collectively their science abounds with new and impressive discoveries, it may be argued that these have not been presented in such a way as to be readily accessible to the general reader. The justification for seeking to co-ordinate the various branches should lie not only in its desirability on philosophic grounds but also in its results. Past prophecy as to the future direction of scientific achievement has not enjoyed such a success as to encourage present attempts, but it seems evident that a period of great synthetic development is on its way. This potential development, however, will only become actual if a synthesis, dynamic and progressive, of all new knowledge is steadily maintained in some reasonably accessible form. Thus a contributory channel of specialized research would be seen not only in relation to neighbouring channels but also to the course of the main stream.

Botany and the botanists

'Botany is the science which treats of plants' (*Oxford Dictionary*), and in an extended definition is usually understood to include a consideration of their growth, development and reproduction, the functions of their organs, their origin, systematic affinities, geographical distribution and relation to their environment. Some years ago Professor W. Stiles indicated that in his view there could be no such thing as a *general botanist*: that an investigator may take a sympathetic interest in other branches of the science and realize the bearing of such work on his own, but that the latter, in these days of specialization, must necessarily lie in one particular field. Clearly in such a statement there is ground both for agreement and dissent. In so far as an investigator fails to realize the relation of his work to the science *as a whole*, so may he fail to appreciate the actual and potential development of the subject in his time.

What, then, do we hope to make of 'the science which treats of plants'? What, in particular, is to be the relationship of the specialized branches, for example, plant biochemistry or genetics, to the parent science?

The last five decades have witnessed a great expansion of botanical research in the course of which new aspects, each requiring detailed investigation by means of special techniques, have become distinct, specialized and almost separate branches of biological science. Today, a botanist tends to be labelled systematist, cytologist, geneticist, ecologist, mycologist, morphologist, palaeontologist, physiologist or biochemist, the underlying assumption being that he is that and little else. A further unfortunate consequence of specialization lies in the fact that common ground for discussion becomes more and more difficult to find, and in extreme cases may even lead to the view that it does not, for practical purposes, exist. Although an official cleavage between botanical morphology and physiology was avoided at the British Association meeting of 1894 in Oxford, nevertheless an adequate sense of mutual aid in the common pursuit is still lacking. The morphologist will tell you, with some over-emphasis but not without justification, that when he looks into textbooks of physiology in search of information bearing on his own work, he finds that such books can tell him little that he specially wants to know; and the modern physiologist, though not habitually addicted to helping himself by making the fullest use of morphological observations, rather tends to view the professed morphologist as a relic of a former phase of botanical development and as a less adaptable and less inventive scientific man who is still plodding along in an overworked field; moreover, he, too, may complain that morphological literature fails to provide the information he specially requires. So, too, uneasy relationships exist between other and newer branches of the science. Each new aspect that arouses enthusiasm is soon attended by a profuse outpouring of specialized literature, and this, no doubt, must be accepted as being in the nature of things. But several important consequences should not be overlooked. These include: (1) vast accumulations of reading matter which are such that a worker *in any single branch* has to read constantly to keep up to date; and (2) an increased specialization of outlook, which, if left uncorrected, will make for a progressive disintegration of the science as a whole.

Now, no one would desire that specialization in the various branches of botany should cease. Such close investigation of particular phenomena is of the very nature of the scientific method. Indeed, the more facts these branches can produce, the better for biological science as a whole, provided they are made readily available to readers working in other branches. The crux of the problem, then, is this: How is the contemporary *botanist*, conscious of the need *for achieving a full and coherent account of the plant and its life*, to make the best use of the several contributory branches without wishing to deny to them the fullest freedom to pursue their own aims? For two things are certain: (1) no single human being can now hope to read in detail the literature of the several special branches; (2) no particular branch will relinquish its aims or limit its scope for the sake of the mother-subject. Thus it is not a question of re-uniting the several separate branches, or of making any one the hand-maiden of another, but of being able to synthesize or integrate the facts of these branches in the interests of the central aims of botany.

Before attempting to take the matter further, it may be advantageous to consider briefly some selected aspects of the development of botanical science.

Early developments

It is almost certain that the early systematists, having arranged in orderly fashion the species of plants known to them, must have been conscious of having imparted coherence where none had previously existed. Here it is appropriate to refer to the works of John Ray (1628–1705), of whom Sachs has written that he 'not only knew how to adopt all that was good and true in the works of his predecessors, and to criticize and complete them from his own observations, but could also joyfully acknowledge the services of others and combine their results and his own into a harmonious whole'. The tribute is deserved, for though Ray's *Historia Plantarum* consists essentially of a series of descriptions of all plants then known, the work is prefaced by an account of morphology, anatomy and physiology as then understood. Later, too, Linnaeus, having achieved the completion of his system, must have been conscious of having created a new and desirable unity. Along entirely different lines the German botanist, Caspar Wolff, discussed general questions such as the fundamental nature of the shoot system, and concluded, as stated in his *Teoria Generationis*, 1759, that he saw

c

nothing in the plant but leaves and stem. Later, inspired no doubt by Wolff's generalization, Goethe formulated his theory of metamorphosis in which all the various and diverse append-ages of the shoot in higher plants were regarded as being the metamorphosed products of a single fundamental organ, the 'ideal' leaf. Here, indeed, the nature philosopher had conferred a unity on the objects of his study, but as Schiller pointed out to him, the abstract conceptions which he employed belonged to the realm of ideas rather than of facts.

The scholastic tradition which prevailed during the first half of the nineteenth century died hard, and botanical science made indifferent headway until Schleiden trenchantly preached a new gospel—that the highroad to new discovery lay in the study of development. It remained for his disciple Wilhelm Hof-meister to show what could thus be done. To view his work in proper perspective it is necessary to realize how very little was then known of the Cryptogams and indeed of the life-history of higher plants. The reproduction and embryology of Bryo-phytes and Pteridophytes constituted practically an unexplored field, while precise data relating to the development of the embryo-sac, fertilization and embryology in the higher plants were still being collected—all this rather less than a hundred years ago. The contemporary botanical world may well have been astonished when it first read the curious catalogue title of Hofmeister's paper: 'Comparative Researches on the Ger-mination, Development and Fruit Formation of the Higher Cryptogams (Mosses, Ferns, Equisetaceae, Rhizocarpaceae and Lycopodiaceae) and the Seed-formation of the Conifers' (1851). It must have appeared as if the wrong things, mosses and gymnosperms, had somehow been run together for comparative treatment. But this was no fallible production, nor was the strange association of data an indication of faulty judgment, for the young German botanist was telling the world that mosses, liverworts, lycopods, equiseta, ferns, gymnosperms and phanerogams all shared a common life-cycle, characterized by the same critical events and developmental phases, and by a recurrent alternation of generations. Here, on a substantial basis of observation, was a synthesis which conveyed a sense of unity hitherto unknown. Later, in 1896, following on the dis-covery of chromosome behaviour, Strasburger was in a position to announce the further important generalization of the relation between the chromosome-cycle and the somatic-cycle.

Hofmeister's general morphology

These and other examples which could be quoted afford evidence of the way in which new data and the generalizations which could be based on them not only widened the scope of botanical science but also conferred on it a new sense of coherence. These instances, however, do not convey any adequate idea of the *methodological unification* which I consider to be desirable. For such a discussion the starting-point lies in the works of Hofmeister, for he was not merely concerned with preparing descriptive accounts of changes in form during development, but he also asked himself such questions as: How does the observed form come to be? (see p. 53).

In introducing his new point of view to a botanical world largely given over to the speculative writing inseparable from the then prevailing idealistic morphology, Hofmeister employed little general argument. Instead of this he set about the task of replacing old conceptions by new ones based on personal investigations. A clue to his general attitude is surely given in the title of his book: *General Morphology of Growing Things* (Gewächse). As Von Goebel has said: 'in this book form-relations are presented as conditions of growth. This growth is investigated.' Lastly, to round off our impression of the all-round 'compleat' botanist, it may be mentioned that Hofmeister interested himself in the question of variability in plants and, while demanding further studies of the influence of external factors on the conformation of organisms, he attached himself to the Darwinian theory of descent.

The phyletic period

As a result of Hofmeister's admirably objective inquiries and of his critical search for relationships between physiological activity and the assumption of specific form or pattern, it might have been thought that botanical science had at length been established on a broad and sure foundation, one in which morphology and physiology were seen to be inseparable aspects of the same theme. But in what has been called the phyletic period —that which followed the publication of Darwin's theory of descent—the details of plant structure, together with such facts as could be culled from the fossil record were regarded chiefly as providing materials for comparative studies and for the construction of phylogenetic systems. The sweeping success of Darwin's views must surely have indicated to professed phylogenists that whatever comparative investigations they carried out were bound to 'fit in' somewhere, to contribute in some measure to the wonderful edifice of evolutionary theory. The facts of development and the characteristic features of the adult were thus accepted by them as purely morphological concepts, while physiological and causal aspects, though not entirely neglected, received at best little more than passing attention. But as Professor W. H. Lang pointed out in an important review of the situation in 1915, the problems of general, that is, causal morphology, would remain even if the phyletic history were before us in full. A bad feature of the phyletic period was the tendency of morphologists and anatomists to resort to facile pseudo-physiological arguments regarding the function and adaptive value of structures and organs. Meanwhile physiological research was going on its own way, out of sympathy, or out of touch, with the historic aspirations and interests of morphology.

Organization and phylogeny

Each of the major phases in the development of botanical science has been characterized by a central idea. Thus in the Linnaean period the *beau ideal* was to know and classify as many species as possible and to add to that number by new collections from all the ends of the earth. In the Darwinian period and after, the problem of descent, which included the construction of phylogenetic systems, was the chief aim, use being made of the natural classifications which had been evolved during the preceding descriptive phase. Since the beginning of the present century, the mechanisms underlying physiological and hereditary processes have constituted leading themes. For contemporary workers in both botany and zoology the processes involved in progressive organization during development are providing problems of great importance and interest.

Now it is evident that organization could be studied as a subject *per se*, without reference to origins. But since all contemporary organisms have come from ancestors possessing greater or less family antiquity, and since the fossil record informs us that notable changes in structure have taken place down through the ages, it is clear that, whatever may be the findings from our studies of organization in contemporary organisms, such findings must also in some way be related or referable to the historic or evolutionary aspect. This, of course, is implicit in the once firmly held view that the ontogeny of any organism is a recapitulation of its phylogeny. It is apposite to note here that the modern study of organization differs from phyletic studies in that it is essentially dynamic in outlook.

The comprehensive viewpoint, therefore, will require that the results of contemporary investigations of plant organization be also considered in relation to the fossil record of past biological events. If an adequate understanding of the factors underlying the organization of living plants can be achieved, a fuller interpretation of the events indicated by the fossil record may become possible, though it can never be absolute. Whether the concepts issuing from contemporary studies of organization will support the criteria of comparison which have been used in the construction of phylogenies or will indicate that they lack validity is evidently a matter of the greatest importance and interest.

Entelechy and holism

Since Driesch considered mechanistic conceptions of life to be inadequate he introduced the idea of a controlling or ordering principle—an entelechy—which was independent of physico-chemical laws though these were operative in living systems.

In his important work on *Holism and Evolution*, General Smuts, too, considered that the explanation of living organisms cannot be purely mechanical and that mechanistic concepts have their place and justification only within the wider framework of the integrated unity of the organism. According to him, holism—defined as the 'fundamental factor operative towards the creation of wholes in the universe'—is a *vera causa*, that is, a causal factor with a real existence; in the process of evolution there is a definite and fundamental tendency towards the creation of wholes, the results becoming more marked at progressively higher levels of organic development. Thus, if we take a plant or animal as a type of a whole, 'we notice the fundamental holistic characters as a unity of parts which is so close and intense as to be more than the sum of its parts; which not only gives a particular conformation or structure to the parts, but so relates and determines them in their synthesis that their functions are altered; the synthesis affects and determines the parts, so that they function towards the whole; and the whole and the parts, therefore, reciprocally influence and determine each other, and appear more or less to merge their individual characters: the whole is in the parts and the parts are in the whole, and this synthesis of whole and parts is reflected in the holistic character of the functions of the parts as well as of the whole'.

Whether one agrees or disagrees with the philosophical or biological implications of holism, a valuable service has been rendered to biology by the author's insistence on the essential wholeness of organisms. In the pursuit of researches into particular aspects this integrated unity should not be forgotten.

Contemporary aspects of integration

A survey of certain current biochemical, physiological, genetical and morphological investigations suggests that opportunities for achieving a useful integration of data derived from these several branches do in fact exist. Admittedly the number of instances which may be cited is not great; nevertheless they constitute a beginning which may in due course be notably extended.

The marked increase of interest on the part of biochemists in isolating and determining the chemical composition and physical properties of a number of physiologically active substances is likely to prove of great importance in promoting certain aspects of contemporary botanical research. In some instances these substances, which have been comprehensively described as activators, show remarkable specificity in their action on living tissues. A number of these substances have now been synthesized; an obvious development in biochemistry is to synthesize yet others. A considerable number of physiologically active organic substances of known composition, not so far known to occur in Nature, have also been produced. Physiologists, meanwhile, have been attempting to ascertain the metabolic origin and functional relationships of naturally occurring activators. This exacting branch of plant physiology is one in which substantial progress may be anticipated in due course.

The relation of these developments to the work of the morphologist may now be considered. Morphologists are concerned with the external and internal configuration of plants and regard the facts of embryology, the development of new organs at apical growing points and the attendant differentiation of tissues as integral parts of their work. Hitherto they have laboured under a serious handicap in that they have had very few working hypotheses to account for the mechanism underlying the differentiation of new organs or of new tissues. The observed developments have been regarded as characteristic manifestations of the specific hereditary substance, or in some such generalized fashion, but hypotheses relating to the operation and interaction of individual factors have been inadequate or lacking. Hence the purely descriptive nature of much morphological work and the indefinite nature of many of the conclusions based on it. But today a fascinating prospect of new possibilities lies before us. For example, it is known that certain substances which the biochemist can isolate or synthesize are more or less directly involved in those all-important initial differentiations of organs and tissues, or in the subsequent growth to the adult condition. Those substances, which apparently exercise a specific morphogenetic effect, have been described by Dr. J. Needham as morphogenetic hormones. Whether or not, in the complex of factors operative in the moulding of an organ, a single substance can properly be referred to as 'morphogenetic' cannot be discussed in detail here; the important fact is that specific structural developments follow on the application of certain substances to plant and animal tissues, provided the latter are in a suitable physiological condition.

Physiologists have not only been exploring the many aspects of what may be described as general cellular physiology, they have also been investigating those difficult problems which are concerned with the movement of substances throughout the plant body. It is now known, for example, that auxins, produced at the apical growing point of the shoot, that is, the region of active formation of protoplasm, become distributed throughout the plant. Important developments ensue, for example, the inhibition of buds, promotion of root development and the progressive enlargement of tissues. Now these several developments provide the materials which the morphologist is competent to investigate and of which he wishes to render an account either in terms of comparative morphology or in explanation of how the organization or configuration observed in the adult comes to be. In short, the biochemical, physiological and morphological aspects are seen to be inextricably linked, and conjoint work is essential to any reasonably adequate account of the processes involved.

Any relationship that can be established between the hereditary constitution of an organism and the possession of those metabolites which are significant in the development of its specific morphology will represent an important advance in botanical science. Here the interests of the geneticist, the physiologist, the biochemist and morphologist become confluent. So far the instances which permit of a co-ordination of data are few in number. They are, however, of great interest, not least because they indicate the possibility of such work being extended.

In an investigation of tall and dwarf strains of maize it has been ascertained (a) that the difference between tall and dwarf races is referable to the action of a single pair of genes, (b) that the initial production of auxin, an important factor in the growth-expansion of tissues, is approximately equal at the growing points in the two races, and (c) that the characteristic dwarfness in one race is due to the destruction of a large part of this auxin by an oxidase not present in the tall race. Here we have a genetical observation relating to an important difference in morphological configuration, which in turn can be assigned to analysable differences in metabolic activity. When precise, co-ordinated data of this kind become cumulative, as they probably will in time, the information already gathered by morphologists and systematists may well acquire new significance and provide a new viewpoint from which to consider the central and continuous problem of evolution. Admittedly this is looking far ahead, but the goal is much to be desired.

The relation between genes and developmental processes has only recently begun to receive the attention which the subject so evidently deserves. The lack of co-ordinated development in related branches has no doubt been a contributing factor. At present very little is known about the actual physiological expression of a gene—how and where it exerts its influence: it may affect only one step in the process of development or a chain of processes; or it may be involved systemically in every aspect of development. Indeed, it is held to be improbable that any single formula will be found to unify all observations on the connection between the influence of a gene and its results.

Biological materials in which some well-marked character is known from genetical analysis to be related to a single pair of genes seem likely to prove of great use in attempts to relate genes with developmental processes. In certain annual and biennial forms of *Hyoscyamus niger* it has been shown that if grafts of annual-flowering plants are applied to biennial stocks in their first year, flower-bud development is induced in the latter; and if the vegetative biennial scions are grafted on to annual shoots they are induced to become flowering shoots. The evidence suggests that a substance (or substances), directly or indirectly operative in flower-bud development, has passed from the tissues of the annual to those of the biennial. Now this factor, which is associated in the hereditary constitution with a single pair of genes, is productive of changes which are of profound interest to both the physiologist and the morphologist. Other instances illustrating the same community of interest relate to such characters as the shape and size of leaves, total growth, branching, difference in chlorophyll content, flower size and shape and sterility. In each, the indications are that gene-dependent diffusible substances are involved; in each the materials are such as to call for detailed physiological investigations and are of the kind on which classical systematic and morphological studies have been based.

Here, perhaps, it is appropriate to utter a word of caution. The subject-matter has been treated in such a way as to illustrate how a plurality of distinct branches of botanical science can be focused, to their mutual advantage, on the same phenomenon. But it would be a mistake to assume that the actual operational relationships involved are simple. Behind every change and development lies all the complexity inevitably associated with the multifarious operations of a metabolic system. Hence, while there is a need for simplification so that main issues and essential relationships are not obscured, the innate complexity of the processes involved should not be under-estimated. Experience suggests that it is unlikely that any so-called specific morphogenetic substance is as direct in its mould-ing activity as the words would appear to imply. It is safer to assume that every morphological development is the result of the interaction of many factors.

To summarize briefly. The guiding lines are these: the heredi-tary constitution of a species, itself a small fragment of the evolutionary picture, is conceived as being subdivisible into genes; these are involved in all developmental processes; they find expression in the action of chemical substances, which in conjunction with temporal, spatial and physical factors are directly or indirectly operative in morphogenetic processes and culminate in the production of the distinctive organization of the adult.

Tissue culture
An all-pervading consideration in biology is the physiological mechanism relating to the assumption of form, that is, to the origin of the ontogenetic growth pattern and its development to the adult status. How are we to approach this very difficult problem, and how are we to test such hypotheses as may be constructed? What, for example, are the factors leading to the institution of an apical meristem? How is it maintained in its active formative capacity? To what can we attribute the orderly development of leaf and bud primordia? What factors determine their bilateral and radial symmetry respectively and the char-acteristic shapes into which they are moulded during develop-ment? And the overriding 'wholeness' of the organism, what of that? These are questions on which we scarcely know how to make a beginning, for the problems are of a manifold com-plexity. We note that they involve the origination of cells (cytogenesis) and of organs (organogenesis)—in fact, all that is connoted by the word morphogenesis; but they also involve a great deal more.

One possible line of approach, on which a beginning if no more has been made, is by means of tissue culture. The ad-vances made during the last thirty years in mycological and bacterial cultural methods have played an important part, and media of precise composition can be prepared on which the tissue of certain plants can be maintained in a state of growth without differentiation for an indefinite period. Here then are the means by which it may be possible to determine experimentally, on materials of known genetic constitution maintained under controlled conditions, the direct or indirect action of many factors considered to be operative in morphogenesis. Whether this hope is vain or whether it will, in fact, be realized, remains to be seen.

Conclusion

In this essay it has been quite impossible to refer to all aspects of the unification of botanical science. Studies of plant distribution, for example, which are of interest not only to the systematist, geneticist and ecologist but also to the morphologist and physiologist, have of necessity been omitted from the discussion. So, too, with taxonomy and other important branches. I have attempted to illustrate a point of view, not to cover the field.

To bring out particular points, I have referred to both conceptual and methodological aspects of unification. There are important instances where the two may be closely related. For example, the phylogenist makes use of certain morphological criteria of comparison. The validity of these criteria depends, among other things, on the accuracy of our knowledge of morphogenesis and this, on a further analysis, is seen in practice to require the conjoint work of the biochemist, physiologist and morphologist.

From the multiple-aspect-study of organization during development we not only derive some impression, however inadequate, of the organism as a whole; we are brought face-to-face with the undeniable wholeness of the organism. A case has been made out for the view that the over-emphasis of any single aspect, while the whole is not kept in proper perspective, will almost certainly lead to the fabrication of unstable theoretical superstructures, destined to crumble because they have not been based on the fundamental reality of organic wholeness. This is a matter which concerns all botanists, though each, according to his capacity, must perform his detailed work in a particular field. But whatever that field may be, he will at one time or another be concerned with some aspect of the distinctive growth-pattern of the organism which he is investigating; this, it need scarcely be said, is of paramount interest to the morphologist at large. An interesting contrast that has been drawn between the 'substance-minded' and the 'relation-minded' man is relevant to the present discussion. 'The substance-minded type of thinking,' says A. H. Hersch, 1941, 'is unquestionably the older, both in the individual and the race. It has all the tenacity of original sin. In morphology it has given us representative particles, preformation, the transmission of acquired characters, and such morpho-chemical hybrids as bristle-producing, facet-forming substances, and so on. The morphologist, when substance-minded, thinks of the developmental pattern in terms of the visible structural characteristics from stage to stage. In short, he thinks in terms of a series of pictures. But when relation-minded, the morphologist recognizes that the pattern at any moment is the expression of the events which produce it, and attempts to gain a knowledge of the durations and rates, and relative durations and relative rates of the component processes in the developmental nexus. Consequently, instead of thinking in terms of a series of pictures, the relation-minded morphologist tends to think in terms of the non-picturable. If the problem of the developmental pattern is similar to the problems of the more exact sciences, then no doubt in time a system of equations will be developed to facilitate our thinking about it.'

There is the modern outlook on *one* aspect of morphology. While it is evident that certain comparative studies and all fossil studies will continue to conform to the older pattern, the new point of view suggests great possibilities for further exploration. The feasibility of pursuing these investigations to a successful conclusion will in large measure be determined by the existence of the tools to do the job. Some of these are already at hand. Here I have in mind certain major biological works recently published or reissued, for example, D'Arcy Thompson's *Growth and Form* (2nd edn.), Needham's *Biochemistry and Morphogenesis* and Child's *Problems of Pattern and Development*. Each tends to emphasize a particular aspect, but taken together they afford both the morphologist and physiologist a working knowledge of the several biochemical, physical, physiological, temporal and spatial factors which, at one or another stage of development, may be operative in moulding the distinctive form of the organism.

The publication of a major work, such as any one of those mentioned above, or of a first-class textbook, is an event of rather occasional occurrence, and depends on particular individuals who possess the experience, capacity and urge to attempt a synthesis. Now, the point of view conveyed in Hofmeister's general morphology, with appropriate modernization and thereafter subject to progressive integrated development, would appear to represent a desirable central aim in botanical science with which few would disagree. With this as a focal point, it is cogent to inquire how we are to make the best use of the data of each of the special branches, having in mind the volume of such literature, the present tendency of individual workers towards intensive research in a restricted field, the fact that this may involve disability to broader vision, and the finite mental capacity of human beings. It is undeniable that the proper comprehension of the subject as a whole is suffering from the inevitable and progressive increase in specialization. How do we propose to deal with this situation?

I claim no originality in raising this general question and offer no solution at this stage. It is evident that underlying the numerous symposia, conferences, joint-meetings and so on, that from time to time have been convened, the same or a not dissimilar point of view has obtained; but a more definite policy needs to be framed, continuously pursued and kept to the fore in our biological deliberations. The question of *how* this is to be done is for botanists collectively to decide. The time for doing so is at hand if a great opportunity is not to pass unheeded.

In conclusion, I wish to express my gratitude to colleagues for suggestions and much helpful criticism; but for the opinions expressed responsibility lies wholly with myself.

2

An experimental treatment of the apical meristem in ferns

In *Dryopteris dilatata* an attempt has been made to observe the effect of isolating the apical meristem from the adjacent lateral tissues, while maintaining continuity with the pith parenchyma below. The first step was to ascertain the practicability of the operation. The procedure involves an operation roughly comparable with 'ringing' in an older region of the shoot, in that the incipient vascular tissue is severed. The difficulty lies in the fact that the treatment has to be applied to a minute region—the meristem—the tissue of which is soft, delicate and readily injured. Thereafter it remained to be seen whether a meristem so treated would prove capable of further growth. The successful outcome of such an experiment seemed likely to throw light on (1) morphogenetic processes at the shoot apex; (2) translocation of nutrients to the shoot apex; that is, whether this takes place by way of the undifferentiated vascular tissue or by upward diffusion over the whole cross-sectional area of the shoot; (3) the factors influencing the differentiation of the vascular system; that is, whether the effective stimuli proceed basipetally from the apex or acropetally from the older preformed parts; and (4) the physiological dominance normally exercised by the shoot apex over lateral buds or bud-primordia.

For these experiments the common broad-shield fern *Dryopteris dilatata* was selected, plants of all sizes being obtainable in abundance. Previous experience has indicated that this species possesses very considerable vitality under experimental treatment.

An apex of *D. dilatata*, as seen in longitudinal median

section, may, for convenience, be considered to include three regions in basipetal sequence: (1) the apical meristem which consists of a single layer of superficial cells, including the apical cell, of distinctive appearance and chemical reaction; (2) the region of initial differentiation, in which leaf primordia, buds and scales originate superficially from cells within or on the margin of the apical meristem and in which stele, cortex and pith can first be distinguished internally; and (3) the region of subsequent differentiation which merges downwards with the fully matured tissue systems. It should be noted that the incipient vascular tissue can be traced upwards to the region of its inception immediately below the apical meristem. Since the principal initial developments take place in region (2), it is there that experimental treatments must be applied.

The first step is to have unimpeded access to the apical meristem. At first sight this does not appear to be a very feasible undertaking. Indeed, because of the large number of rolled leaves and leaf primordia in the terminal bud and the dense investment of scales, the apices of ferns have not generally commended themselves as providing favourable materials for experimental investigations. Nevertheless, it is a fact of experience that there is little difficulty in removing the scales and leaf primordia from stout shoots of *D. dilatata* or *D. filix-mas* and thereby laying bare the apical cone. These operations are carried out by means of small pointed scalpels and forceps, the material being observed under a binocular microscope. A downward view of a small shoot apex is illustrated diagrammatically in fig. 1.

To isolate the apical meristem from the adjacent lateral tissues, and to sever the incipient vascular tissue, the following procedure was adopted. The terminal region of a stout shoot, trimmed to a length of about 2–3 cm. and defoliated as already described, was observed under a binocular microscope at a magnification of fifty times. By means of a small, thin-bladed scalpel four vertical cuts were made in such a way as to isolate the apical meristem, on a little square island of tissue, as indicated in fig. 1. In longitudinal section it will be seen that the isolated apical meristem, together with a thin layer of incipient vascular tissue, is seated distally on a plug of developing medullary parenchyma (figs. 2 and 3).

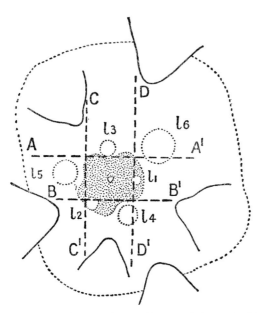

1 Downward view of a small apex of *Dryopteris dilatata* showing the apical meristem (stippled) and a succession of young leaf primordia (l_1, l_2, etc.). The system of vertical incisions by which the apical meristem is isolated from adjacent lateral tissues is indicated. ($\times 54$.)

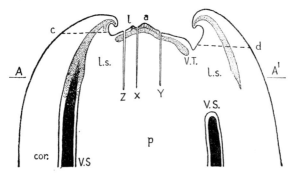

2 Longitudinal median section (diagrammatic) of a shoot apex of *D. dilatata*, showing the apical meristem (*a*) and lateral segments (*l.s.*) isolated by intersecting longitudinal incisions at *x* and *y*, and by two other incisions at right angles. An older leaf primordium (*l*) has also been isolated by the vertical incisions at *x* and *z*. *c* and *d* indicate the levels at which leaf primordia would be pruned off in experimental shoots. *p*, pith; *cor*, cortex; *v.s.* vascular strand; *v.t.* incipient vascular tissue. AA^1, level in shoot at which the differentiation of phloem and xylem begins to be apparent. ($\times 6$.)

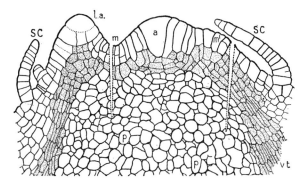

3 Longitudinal median section of a small apex of *D. dilatata* showing the details of an isolated meristem, *m–m¹*; apical meristem, with large apical cell, *a*; *sc*, scale; *l.a.*, leaf apex; *v.t.*, vascular tissue in initial phase of differentiation; *p*, pith. The vertical incisions are shown: the isolated meristem is continuous with a thin sheet of incipient vascular tissue and medullary patenchyma. (× 100.)

The system of incisions described above is obviously only one of many that may be applied. For example, the vertical cuts may be made on the margin of the apical meristem, within the margin, or outside it, so as to include one or more leaf primordia, and so on. Again, leaf primordia of different ages may be isolated on separate plugs. I have found that isolating by four vertical cuts is simple and least damaging to the apical meristem, but other patterns, for example, by three cuts, are also possible. Radial cuts diverging from the margin of the apical meristem have also been made with interesting results. The procedure can be further varied by cutting more or less deeply into the tissue. In passing, it may be noted that the small size of the meristem precludes the use of a cork borer. The experimental and control plants were planted in moist peat in a cool greenhouse.

The isolated meristem is capable of growth. It develops into a short axis and in the course of four weeks may develop three or four new leaf primordia. In the course of three months, one specimen had developed a compact terminal bud in which fourteen leaf primordia were counted. In relation to the severance of the vascular tissue, the lower region of the axis consists entirely of medullary parenchyma. Higher up, a solenostele is

present (fig. 4). Still higher up, in the region of leaf development, the vascular system becomes dictyostelic (fig. 5). The
new leaves in the materials so far examined show normal
phyllotaxis and this is apparently in continuity with that of the
parent shoot. Whereas lateral buds develop rapidly on the
surrounding segments of the parent shoot, the new terminal
axis is devoid of buds.

The lateral segments of the shoot are characterized by the
development of very large axillary solenostelic buds (fig. 6);
they are thus in marked contrast to buds on the normal shoot,
in which a small non-medullated protostele is present. Individual undifferentiated meristeles, isolated in lateral plugs as
a result of the system of incisions adopted, also show some
remarkable and hitherto undescribed developments, such,
for example, as the development of solenostelic structure
(fig. 7).

Individual leaf primordia, isolated on plugs, undergo limited
development only, and show a tendency to be thrust aside by
more strongly growing axillary buds. In contrast to the normal
development, in which the foliar conducting system consists of
six to eight separate vascular strands disposed horse-shoe fashion,
the vascular system of an isolated leaf primordium is solenostelic.

Thus, by using the technique described, many new and
interesting observations have been made. These will be fully

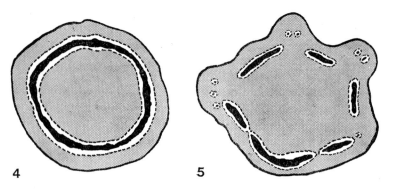

4   **5**

4 Solenostele present in the short axis developed from an isolated apical
meristem. (×16.) **5** Dictyostele present in the terminal foliar region of an
axis developed from an isolated meristem. (×9.) Cortex and pith, stippled;
endodermis, broken lines; xylem, black.

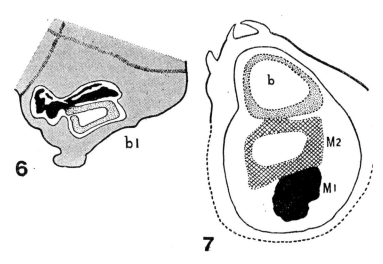

6 Transverse section of a lateral segment of an incised shoot, showing the large solenostele of a bud (b_1) conjoined with the meristele of the parent shoot. **7** Transverse section of a lateral segment of an incised shoot showing the solenostele of a bud (b) conjoined with a meristele which has become medullated (M_2); the compact mass of tracheides (M_1) pertains to a second meristele of the parent shoot.

described and discussed elsewhere in due course. The answers to some of the questions raised in the opening paragraph are evident from the data submitted.

3
The shoot apex
in pteridophytes

Nearly two hundred years have elapsed since Kaspar Friedrich Wolff (1759) stated that growth in plants proceeds from an undifferentiated apical growing point or *punctum vegetationis*. To the botanist of today who is interested in the progressive organization which takes place during development, the general phenomenon indicated by Wolff is of great importance, and, be it said, still far from any comprehensive elucidation. As Sachs (1890) has pointed out, the influence of Wolff's work was not due to its actual correctness but 'to the thoughtfulness of his observations, and to the earnest desire which inspired them to search out the true nature of vegetable cell structure and to explain it on physical and philosophical grounds'. Wolff's discovery had, indeed, far-reaching implications: for if new parts or organs originated successively at the apical growing point, itself perennially embryonic, then the growth of the terminal bud could not simply be regarded as a process of unfolding of pre-existing rudiments, but must be considered to exemplify true epigenesis. Such a conception brings us very close to the starting point of contemporary investigations. It is a matter of botanical history that Wolff's observation did not immediately lead to experimental and analytical investigations. But in time the apical meristem began to receive some of the attention which was its due. The present purpose is to survey some of the developments which have made for a fuller understanding of the shoot apex in pteridophytes, to inquire into the point of view and the aims which inspired different investigations, to consider the contemporary state of relevant knowledge, and to indicate, if possible, opportunities for future work. The study of the apical meristem, though rooted in classic morphology, can no longer be considered the perquisite of the professed morphologist: to the physiologist, the biochemist and the geneticist the apex also provides materials for fruitful research. Indeed, only by integrating the data of the several branches can we hope to arrive at an adequate understanding of this complex but all-important region.

This article has in the main been restricted to the pteridophyte shoot apex. Limitation seems desirable if only because the total subject-matter of apices is so great and diverse as to require monographic treatment (see Schüepp, 1926). In the living pteridophytes we have a coherent and unmistakably related group, well represented by fossil correlatives of great antiquity; not only do they afford a very considerable diversity of external form and internal structural pattern, but these developments are uncomplicated by secondary thickening. Thus the pteridophytes provide favourable material for the investigation of primary developmental processes. Since pteridophyte apices differ in many respects from those of seed plants it seems valid to treat them separately until such fundamental features as they may share in common are more fully understood. Comparisons between the two branches may on occasion, however, serve a useful purpose.

Since the purpose of a review such as this is to prepare the way for further investigations it is relevant to indicate why such work seems desirable. One of the aims of botanical science is to give a coherent account of the processes whereby the individual organism acquires its distinctive appearance and specific character. In vascular plants the shoot apex is the seat of all primary formative activities; all lateral appendages, leaves, scales and buds (excepting those which are adventitious) originate at or near the apex; so, too, do all primary tissue systems. Thus, if we would understand how the external form and internal structure of any vascular plant come to be what they are, we must investigate morphogenetic processes at the shoot apex throughout the development of the individual. The studies envisaged will be essentially dynamic in outlook, the observed developments in the species under consideration being interpreted as the expression of a specific system during growth under a particular set of conditions.

The data obtained should also have another interest for us. It is evident that organization could be studied as a subject *per se*, without reference to origins. But since all contemporary organisms have come from ancestors possessing greater or less family antiquity, and since the fossil record informs us that notable changes in structure have taken place down through the ages, it is clear that, whatever may be the findings from our studies of organization in contemporary organisms, such findings must also in some way be related to the historic or evolutionary aspect. If a more adequate understanding of the organization of living plants can be achieved, a fuller inter-pretation of the events indicated by the fossil record may become possible, though it can never be absolute. Whether the concepts issuing from contemporary studies will support the criteria of comparison which have been used in the construction of phylo-genies or will indicate that they lack validity, is evidently a matter of the greatest importance and interest.

That the situation at the plant apex is complex must con-stantly be borne in mind. Any kind of investigation, whether by morphological observation, experimental treatment, physio-logical measurements, or biochemical analysis, which advances our knowledge of this important region, merits careful attention.

Early developments

Wolff's observations on the *growing point* were necessarily limited by the contemporary state of microscopy and anatomical science. Detailed knowledge of apices begins with the prompt-ings of Schleiden and the researches of Unger, Von Mohl, Naegeli and Hofmeister. It was Schleiden (1842) who insisted that a knowledge of the history of development is the foundation of all insight into morphology. Von Mohl restored the study of anatomy; while Naegeli (1844–6), by addressing himself to the questions of how cells are formed in growing vegetative organs and how far the processes are the same in the lower cryptogams and in vascular plants, founded and elaborated the theory of cell formation. Naegeli's conception of development is given in Sachs's *History* (p. 194) as follows: 'Since in nature everything is in movement and every phenomenon is transitory, presenting itself to us in organic life especially as the history of development, all due regard must be paid to their condition of constant mobility in forming scientific conceptions. The history of development is not merely to be treated generally as one of various means of investigation but as identical with investigation into organic nature.' The lower cryptogams were made the chief subjects of his initial investigations of apices, but later his work was extended to the higher cryptogams and phanerogams, i.e. he proceeded from the simpler to the more complex and difficult types of organization. Closely connected with this methodological innovation Naegeli made the new doctrine of the cell the starting point of morphology, i.e. the initial development of organs and their subsequent growth were both referred back to the formation of the separate cells. In many of the lower cryptogams he made the remarkable observation that every organ has a single cell at its apex, that all succeeding cells are formed by the division of this one cell according to fixed laws, and that all cellular tissue can be traced to an apical cell. These observations, amplified by Hofmeister (1862) and others, have served as a basis for many developments in botanical science. To Naegeli, Sachs has paid the tribute that his was the first attempt 'to apply mechanico-physical considerations to the explanation of the phenomenon of organic life', i.e. he referred the growth and inner structure of organized bodies to chemical and mechanical processes.

Anatomical investigations

As a result of these early investigations, soon to be followed by many others, descriptions and illustrations of the apices of all classes of plants became available for more comprehensive consideration. It was seen that while apical growth was a very general feature, the underlying cellular arrangements were exceedingly diverse. Much controversy ensued regarding the interpretation of this inherent diversity (Koch, 1891; Schüepp, 1926), and this eventually led to the full treatment of the subject to be found in such standard texts as Sachs's *Lectures*, De Bary's *Comparative Anatomy* and Haberlandt's *Physiological Plant Anatomy*.

Three major interpretations of growth and cellular arrangement at the shoot apex have been advanced. In Naegeli's theory all organs and tissues were referred ultimately to the divisions

of a single apical cell.* This view was well supported by reference to the vascular cryptogams, and accordingly it seemed
feasible to assume that the theory held good for phanerogams
also. Hofmeister (1857, 1862, 1863) showed that in many of the
vascular cryptogams shoot growth does proceed from a single
apical cell, but he adopted a conservative attitude and pre-
dicted difficulties for Naegeli's generalization when applied to
all classes of plants (1857). Nevertheless the theory gained adherents: Dingler (1882, 1886), Korschelt (1884) and Douliot
(1890, 1891), for example, attempted to show that the shoot
apex of gymnosperms terminated in a single tetrahedral or
prismatic apical cell; so, too, for the angiosperms, though they
received rather less attention. Hofmeister (1857), Korschelt
(1884), Pringsheim (1869) and Schwendener (1879) concluded
that in certain gymnosperms four equivalent apical initial cells
were present and that the single 'three-sided' apical cell was
exceptional in this group.

In the histogen theory of Hanstein (1868), based on a study
of angiosperms, the shoot apex consists, not of a single apical
cell, but of a central core of irregularly arranged cells surrounded by a variable number of cell layers. Each layer, as also
the central core, originates from an initial cell or from a group
of initial cells. As these initials are not usually distinctive in size
or in the sequence of their divisions, the configuration of the
shoot is not directly affected by their shape or segmentations.
In brief, the behaviour of individual cells is subordinate to the
general distribution of growth in the apical mass of tissue—a
view which has found increasing support. As Foster (1939) has
pointed out, the chief defect of Hanstein's theory lies in his
attempt to assign specific destinies or 'prospective values' to the
various regions of the meristem, e.g. that the *dermatogen*, or
surface layer, produces only epidermis, the *periblem*, or intermediate layer, only cortex, and the *plerome*, or inner core, pith
and procambial strands. There is the further difficulty that in
some apices the several layers are not distinguishable, while in
others the formative activities of plerome and periblem are not
as Hanstein laid down. For the present purpose the important
point is that Hanstein's views have little application to the

* Naegeli was aware that in certain classes of plants a characteristic and
well-defined structural development might be achieved by other means.

vascular cryptogams where, almost invariably, the entire apical
development proceeds from one or more well-defined and
usually distinctive apical initials. Thus it was implicit in Hanstein's theory that there are fundamental differences in the
organization of the apices of pteridophytes and phanerogams.
This conception was naturally opposed by those, like Naegeli,
who supported the apical cell theory and adhered to the contemporary phylogenetic view that the phanerogams had originated from the pteridophytes. Thus Naegeli argued that as the
shoot of phanerogams had evolved from the shoot of vascular
cryptogams the apical growth of the former must be a continuation of the process of terminal growth proceeding from a single
apical cell. The experience of comparative morphology, however, made it inconceivable that the histogen layer type of
growth could have been derived from the apical cell type.
Since, in the phylogenetic view, the gymnosperms were considered to occupy a position intermediate between pteridophytes and angiosperms, support seemed to be given to Naegeli's
views by those investigations which ascribed a single apical cell
to the gymnosperm shoot apex. A survey of the shoot apex in
several groups of the gymnosperms by Strasburger (1872),
however, disclosed no evidence in favour of Naegeli's theory, but
rather supported, with modifications, Hanstein's views; but it is
interesting that he considered the gymnosperm type of meristem
to be derived phylogenetically from the *Lycopodium* type (see
also Schüepp, 1926).

The third interpretation, known as the tunica-corpus theory,
comes originally from Schmidt (1924) and was extended and
amplified by Buder (1928) and his collaborators. Schmidt recognized only two tissue zones in the apex, namely the *tunica* which
consists of one or many peripheral layers of cells, and the *corpus*
or central tissue lying within, the demarcation of the two zones
being indicative of two contrasted modes of growth and cell
division. Thus surface growth, accompanied by anticlinal divisions, predominates in the tunica, whereas a bulk, mass or
volume increase, with irregular divisions in different planes, is
characteristic of the corpus. Expect during the formation of
leaf or bud primordia, the tunica thus remains a discrete and
self-perpetuating layer. The formation of primordia is causally
related to the relative growth development of the two formative

regions. According to Foster (1939): 'The maintenance of a proper balance between surface and volume growth leads to constant adjustments which find their expression in the rhythmical growth of the shoot apex into "maximal" and "minimal" areas with the resultant production of surface folds or leaf primordia.' Since this theory has received wide support from investigators of the apices of phanerogams it will be a matter of interest to consider to what extent it is supported by data from the pteridophytes. In Foster's view the concept of tunica 'is sufficiently plastic to include those types of shoot apices in which only a single discrete surface layer is maintained,' while the general theory serves to focus attention upon the dynamic rather than on the purely formal aspect of the shoot apex.

The apical meristem in pteridophytes

As this subject has received full treatment by Bower (1890–1, 1908, 1923, 1935) and by Schüepp (1926), only a brief outline need be given here. In the four classes of living pteridophytes many different conditions are to be observed at the shoot apex. Thus the Equisetaceae provide a classical example of tissue formation proceeding from the orderly segmentation of a single inverted tetrahedral cell (the so-called 'three-sided' apical cell). As Schoute (1902) has pointed out, Hanstein's generalization does not apply to these plants; nor, for that matter, does the tunica-corpus theory of Schmidt. In the Psilotales (*Psilotum* and *Tmesipteris*) the terminal region of the shoot tends to be broadly dome-shaped rather than conical. A single apical cell is usually to be observed at the tip, but its segmentation is somewhat irregular; indeed, it may be difficult to refer the development of the meristem to the division of a single initial. In this respect these plants resemble certain eusporangiate ferns. In the ferns the considerable diversity in the primary segmentation has been made the basis of far-reaching phylogenetic comparisons (Goebel, 1880–1; Campbell, 1890; Bower, 1890–1). In eusporangiate ferns two to four initial cells can usually be recognized, whereas in leptosporangiate ferns a single apical initial is the rule. It is a matter of great interest, particularly to the student of evolution, that there is no sharp line of demarcation between the two types. The Osmundaceae, for example, show an intermediate condition.

The living Lycopodiales also show diversity in their apical meristems. In the genus *Lycopodium*, e.g. *L. Selago*, a group of three initial cells is present in the flat apical region. By growth and division these cells give rise to the whole of the shoot. The apical cells are small and prismatic in form and closely resemble the adjacent superficial cells which clothe the terminal region of the shoot. Immediately below the initial cells the distal region of the plerome cylinder can be observed. In a recent investigation of some lycopod apices Hartel (1938) has concluded that the data are in conformity with the tunica-corpus theory. Her investigations show that all the tissues of the shoot apex originate from a single tier of initials. These divide by both anticlinal and periclinal walls. Her conclusion, that the apex corresponds morphologically to the 'corpus' region of the angiosperm apex, is a purely morphological conception: the phenomenon clearly needs further consideration in terms of the process of growth.

In the genus *Selaginella* both the shoot and rhizophore apices may show a definite apical cell. As Schüepp (1926) has pointed out, the genus is of particular interest in that it affords evidence of growth proceeding from an apical cell as well as growth from a group of initials, the graded transformation being observable in the individual development; e.g. in *S. Poulteri*, as investigated by Bruchmann (1909), the young shoot apex shows a three-sided apical cell, whereas older branches may show several initials. Bruchmann (1909, 1910) also found in *S. Lyallii* and *S. Preissiana*, as in *Lycopodium*, that the apex is crowned by a group of similar initial cells. In *Selaginella Wallichii* Strasburger (1872) found that two initial cells are present: these can be seen in a horizontal section of the dorsiventral shoot, whereas only one will be seen in the longitudinal plane at right angles. Similar observations have been made by Williams (1931) on *S. grandis*. Wand (1914) has given an analysis of the nature of the apical complex for a number of species: this shows that growth may proceed, in different instances, from a three- or four-sided apical cell, from a group of initials, or from a divided apical cell. It is of interest that in the primitive radially symmetrical species, *S. spinulosa*, more than one initial is present at the apex of the sporeling (Bruchmann, 1909), whereas a single apical cell is established as a distinctive feature at an early stage in the embryogeny of dorsiventral species. In *Isoetes* Hegelmaier (1872, 1874), Farmer (1891) and Lang (1915*b*) are in agreement that no single initial cell is present at the apex, the condition being closely comparable with that of *Lycopodium*.

Sachs and the individuality of the apical cell

The foregoing observations not only direct attention to the individuality of the apical cell (or small group of initial cells) in all classes of pteridophytes but also suggest that it may possess a distinctive physiological importance. This view has had both adherents and opponents. Haberlandt (1914) follows Naegeli in the view that an apical cell is of primary importance in that all the tissues of an organ can be genetically derived from it. It is, in fact, a persistent initial cell. The segment formed at each division is a new formation, both from the morphological and physiological point of view, while the other product of the division retains all the features of the original apical cell unchanged. This conception, which appears to be in accord with the facts of observation and experiment, has not passed without challenge. Thus, from geometrical considerations, Sachs (1878, 1879, 1887) noted that the systematic arrangement of cells at the meristem could be referred to the intersection at right angles of cleavage planes—orthogonal trajectories—and was led to the curious view that the apical cell is no more than a hiatus in the system of intersecting surfaces.* Thus for him the apical cell was devoid of physiological significance. As Haberlandt remarks: 'This negative conception altogether obscures the cell character of the apical initial and *a fortiori* ignores its special physiological activity in its role of primordial meristematic unit.' Moreover, as an examination of any fern apex will disclose, the apical cell

---

* 'Since, as Naegeli showed so long ago as 1845, all the tissues of a root or shoot may be derived progressively and in genetic sequence from the apical cell, wall for wall, the opinion arose gradually that the whole process of growth in the growing point is ruled by the apical cell itself, which, by means of its segments, adds stone upon stone to the structure like a builder. . . . So far as regards the importance of the apical cell as the ruler of the whole growth in the growing point, however, I showed that the cell represents merely a break in the constructive system of the growing point—i.e. the apical cell is that spot in the embryonic tissue in which neither anticlines and periclines nor radial longitudinal walls have as yet been formed' (Sachs, *Physiology of Plants*, pp. 457–8).

does not stand alone but is the principal component of a distinctive and fundamentally important tissue, the apical meristem (Wardlaw, 1943, 1944).

Sachs's conception of the apical meristem was supported by Schwendener (1880), who saw in the formation of the series of trajectories the operation of the same fundamental mechanical principle which also determines the direction of the rows of micellae in cell walls and in starch grains. Later investigators, too, have been attracted to the geometrical-mechanical explanation of the meristematic tissue pattern (Bower, 1923). There is no doubt that Sachs and Schwendener called attention to a very important aspect of the cellular construction of the apical region. Greater precision was given to their conception by Errera (1886), who, following Hofmeister (1863), ascribed to the incipient cell wall the properties of a semi-liquid film and deduced that in its configuration it must conform to the law of minimal areas. This law (which we owe to Errera) states that the incipient partition wall of a dividing cell tends to be such that its area is the least possible by which the given space content can be enclosed. For a full discussion of cell-wall dispositions the reader is referred to D'Arcy Thompson's *Growth and Form* (1942).

Sachs's dictum that the meristematic region should be studied as a whole deserves special attention. It is a necessary condition wherever the apex is studied from a dynamic point of view. As D'Arcy Thompson has pointed out, we are dealing not merely with material continuity—the continuity of the protoplasmic threads across the barriers imposed by the cell walls—but with a continuity of the forces of a comprehensive field of force. 'And such a continuous field of force, somehow shaping the whole organism, independently of the number, magnitude and form of the individual cells, which enter like a froth into its fabric, seems to me certainly and obviously to exist. As Whitman says, "the fact that physiological unity is not broken by cell boundaries is confirmed in so many ways that it must be accepted as one of the fundamental truths of biology" ' (p. 345). This conception is one which may be expected to have a particular application to such processes as the diffusion of metabolic substances to and from the apical meristem.

Sachs's comprehensive outlook on the nature of the meristematic region as a whole is of the greatest value and might well

serve as a guide to contemporary investigators. Thus he pointed
out that, with a few exceptions, the directions in which new
cell walls appear during growth depend on the internal distribution of growth as well as on the external form of the growing
organ; that the arrangements of cells in young growing parts
are quite definite and 'in the highest degree characteristic'; that
the directions of cell divisions are in no sense accidental but are
'in conformity to law, the true meaning of which, however, is
difficult to decipher'; that the daughter cells which arise by the
division of a mother cell are equal in volume to one another, the
new cell wall being at right angles to those already present; that
the 'mode of cell division depends only upon the increase in
volume and the configuration of a growing organ, and not upon
its morphological or physiological significance'; that the phylogenetic nature of an organ cannot be determined or ascertained
by the nature of the cell divisions which take place during its
development; that the system of segmentation or pattern observed in a developing organ is an immediate expression of the
processes of growth taking place therein; that the entire mass
of embryonic tissue at the apex grows as a whole and that
'definite geometrical and mechanical relationships exist between
the whole arrangement of cells and the outward form of the
growing organ'; and the 'distribution of the processes of growth
in the interior of organs which possess like external form may
be very different'; that organs considered to be peculiar to
certain plants—e.g. the hypophysis in phanerogamic embryos—
may be no more than the expression of the general law of cell
division in that particular instance; that the production of
apical cells, e.g. in the embryogeny of pteridophytes, is a necessary consequence of the law of cell division in a growing organ;
that cell division is slowest in immediate proximity to the apical
cell; that cell divisions are *not* 'an essential cause of growth'; and
that growing points with and without an apical cell or cells are
not essentially different—'the presence or absence of an apical
cell appears as an interesting but quite secondary point in the
growth of the growing point'.

Apical segmentation, organogenesis and histogenesis
To morphologists, interested in the precise details of apical
segmentation, it was a natural step to consider how such seg-

mentation was related to the inception of new primordia and to the initial differentiation of the several tissue systems. To those who entertained views of a strictly formal and rigid kind it seemed reasonable to expect that a close and obligate relationship would be found; and they therefore tended to generalize from those instances in which, as a fact, new organs or tissues do arise in a consistent and regular manner in relation to the segmentation of the apical cell. But a wide survey of pteridophytes showed that other views were also possible.

It has been seen that the Equisetaceae provide a classical example of the orderly segmentation of a single three-sided apical cell. The toothed leaf sheaths, each of which is a collection of fused, reduced leaves, arise in acropetal succession by the lateral outgrowth of the apical cone. The number of leaf teeth varies in stems of different diameter and bears no constant relation to the three rows of segments proceeding from the division of the apical initial. Thus Bower (1908, 1935) has pointed out that in this instance, as in many others taken from pteridophytes, apical segmentation does not dominate, or determine in any precise manner, the number and position of the lateral members, i.e. segmentation does not determine organogeny. There is also the further point that whereas successive whorls of leaf teeth alternate, segmentation at the apex remains uniform. Each of the three segments produced from the apical cell divides first by an anticlinal wall, giving six cells, and each of these by periclinal walls. The inner column of cells so produced on further division gives rise to the central pith; the outer series to the stele and cortex. Thus the vascular tissue, as distinct from the cortex, bears no constant relation to the initial segmentations at the apex. In ferns and lycopods, on the other hand, the inner column gives rise to the tissues of the stele. In Bower's view such facts 'are important as showing that the apical segmentation and development of parts or of tissues are not necessarily related to one another in vascular plants at large'. Since, in the Equisetaceae, all organs and tissues can be referred to the segmentation of a single apical cell, Hanstein's generalization on the layered constitution of the shoot apex, as Schoute (1902) has indicated, is inapplicable.

That there is no obligatory correspondence between segmentation at the apex and the origin of the several lateral organs, or of the tissue systems in the ferns, has been shown by researches covering a wide field. Among occasional instances of such correspondence Kny (1875) has demonstrated that in *Ceratopteris* one leaf originates from each segment of the apical cell, while Klein (1884) states that one leaf is produced from each of the two dorsal segments of the rhizome of *Polypodium vulgare*, though none arises from the ventral segment. In the majority of ferns, however, such relationships cannot be demonstrated.

In many ferns the limits of the stele are definitely related to the initial segmentations of the apical cell. For example, the stele in monostelic types originates from the inmost cells of the successive apical segments. According to Conard (1908), however, each apical segment in *Dennstaedtia* divides into three parts; the inner giving rise to the pith, the middle to the vascular tissue, and the outer to the cortex. Schoute's opinion that such a system may be considered as showing a correspondence with Hanstein's meristematic layers seems inadmissible to the present writer. Bower has summarized the position for the ferns as follows: that the genesis of both external organs and internal tissues appears to be independent causally of segmentation, although the two at times may coincide, and that apical segmentation and 'morphological definition, whether external or internal, are distinct processes, each of which is determined by the apical region as a whole, and not by its segments', i.e. 'segmentation and organogeny are proved to be two independent propositions'. A similar conclusion emerges from a comparable analysis of the living Lycopodiales. In the foregoing observations we see a modern vindication of De Bary's well-known aphorism: 'The plant forms (or fashions) cells, not cells the plant.'

Apices of eusporangiate and leptosporangiate ferns

At this point it is convenient to consider how the data relating to pteridophyte apices have been used in comparative morphology. Notwithstanding the warnings which had been given by Sachs and De Bary as to the probable unsoundness of using the facts of apical segmentation in phyletic comparisons, morphologists interested in the problems of descent came to the conclusion that such comparisons were not only of great value in their work but were legitimate and valid. The situation is indeed a curious one. For while it was recognized that apical segmentation may not be closely related either to the external form or to the internal pattern, nevertheless a survey of the relevant data seemed to indicate that a knowledge of apical segmentation was of real value in the phyletic grouping of the ferns. As Bower (1923, p. 105) says: 'It is on segmentation that the generally accepted distinction of Leptosporangiate from Eusporangiate Ferns is based.'

It has already been seen that in some ferns one initial cell is present at the shoot apex, in others several initial cells are present, and in yet others a fluctuating intermediate condition is the rule. In the group of ferns distinguished by Goebel (1880–1) as *leptosporangiate*, the less massive and more precise plan of construction—that proceeding from the segmentation of a single apical cell—is seen not only in the development of the shoot but also in the development of leaves, roots and sporangia. By contrast the more bulky or robust and less precise plan of construction of the *eusporangiate* ferns is based on the segmentation of a group of initial cells, such groups being recognizable at all the meristems of the plant and in the sporangial rudiments. Here, then, within an unmistakably coherent and natural group of organisms, we have apparently two different constructional systems. Linking the eusporangiate and leptosporangiate ferns, however, there are intermediate forms, and these, on a detailed analysis, are found to show a fluctuating and intermediate condition in the construction of their meristems. It is scarcely to be regarded as surprising that such differences appeared to offer a basis for comparison and phyletic seriation. One end of the series would be occupied by those ferns in which segmentation proceeds from a single apical initial: the other end by those with a group of initials. We have already seen that observers like Sachs, working towards a causal explanation of development at the apex, considered that the cellular pattern there is determined by the bulk of the organ and has no other significance. But, as Bower (1923) has indicated by many telling examples, the relation between bulk or massiveness and apical segmentation is in no sense simple and direct. The manner of segmentation is not, in fact, directly dependent on actual size. Both in the largest and smallest leptosporangiate ferns growth may be shown to proceed from a single apical cell. Hereditary factors must, it seems, be

E

considered. That there is some relation between massiveness and mode of segmentation has long been known, but, as Bower points out, it is not 'simple or direct in the individual. It makes its appearance in the race.' In other words, hereditary differences in constitution are involved, separating those ferns with a massive plan of construction from those with a more delicate and specialized plan, these differences extending to all parts of the plants. Which end of the series is to be regarded, in the evolutionary sense, as comprising the more primitive group of organisms? Different answers have been given to this question. Thus, some of the earlier writers suggested that the Hymeno-phyllaceae (filmy ferns) must rank as the most primitive ferns because of their resemblance in a number of vegetative characters to the mosses. This view is no longer regarded as tenable, the filmy character of these ferns being now related to their specialized hygrophytic habitat. In 1890 Campbell argued that the eusporangiate ferns were the more primitive, a view which soon received ample support from the growing volume of fossil evidence (Bower, 1890–1). Thus the ferns of the Palaeozoic period were found to be prevailingly eusporangiate, lepto-sporangiate ferns being conspicuous by their absence. Such evidence appeared therefore to establish the fact that the euspor-angiate type, both in ferns and other pteridophytes, was of prior existence and represented the primitive condition, the more delicate structure and exact mode of segmentation in the leptosporangiate ferns being a derivative and specialized condi-tion. Between these two lay intermediate types exemplified by the Osmundaceae and here, too, the fossil evidence was in conformity with the view set out. Such comparisons can also be extended to the Lycopodiales and to other classes of pterido-phytes.

The available information may thus be read as supporting the view that the eusporangiate was the prior condition and that the course of evolution has been in the direction of a progressive specialization towards a more exact system of segmentation. The factors which lie behind this change are wrapped in obscurity. In the absence of the fossil evidence—and in this instance it is also incomplete—the selection of one or other of the two contrasting types as the more primitive would have been purely arbitrary, and indeed reference to the literature shows clearly how radically opinion has veered from one extreme to the other (Bower, 1889, 1890–1). And, notwithstanding his views on fern apices, Campbell (1940) regards the apex of *Selaginella* with a single initial as more primitive than the apex of *Lycopodium* with three initials. Why, in the ferns, should a bulky apex of several initials be more primitive? What is the essential inwardness of this situation? All that can be said at the moment is that the total available evidence suggests that a certain seriation is to be recognized as a true reading of the facts. But, when we understand more fully the causal factors involved in apical segmentation, it may well be that prevailing notions of phylogeny will need reconsideration.

Causal morphology

Hofmeister (1868) was not merely concerned with preparing descriptive accounts of changes in form during development: he also asked himself how the observed forms came to be; what processes of growth underlie the observed structural developments; what external and internal factors determine morphogenetic processes and specific structural organization. The substance of such investigations he described as *general morphology* (allgemeine Morphologie). Not only did he consider the problems of form from the physiological aspect, he also undertook physiological investigations. A clue to his general attitude is clearly indicated in the title of his book: *General Morphology of Growing Things*, and in this work, as Goebel (1926) says, 'form-relations are presented as conditions of growth'. Later, as we have seen, Sachs did much to stimulate interest in the problems of causality at the apical meristem: the developments taking place there were to be interpreted in terms of the process of growth; the methods of investigation involved a consideration of the relevant spatial, mechanical and physiological factors. But it is a matter of history that during the latter part of the nineteenth century interest in these problems was more or less completely side-tracked by the sweeping interest in phyletic studies. We may therefore pass on to the year 1915 to Lang's important address on 'Phyletic and Causal Morphology'. By the latter term he meant the general morphology of Hofmeister —the investigation of how growing things come to acquire their characteristic configuration; such investigations being in part physiological. Lang pointed out that even if the phyletic history of plants were before us in full, the problems of causal morphology would still remain. And if we ask what these problems are, as they relate to the organization of the leafy shoot, it is at once seen that growth and development at the apical meristem are involved. For example, Lang pointed out that although, as a result of comparative studies, a great deal of accurate information was available regarding the structure of the fully differentiated conducting system in pteridophytes and other vascular plants, little was known of such structure in the light of apical development. 'A gap in our knowledge usually comes between the apical meristem itself and the region with a fully developed vascular system. It is in this intermediate region that the real differentiation takes place and the arrangement of the first vascular tracts is then modified by unequal extension of the various parts. The apical differentiation requires separate study for each grade of complexity of the vascular system even in the same plant.' A recent paper by Williams (1938) shows how little has been done in the investigation of problems of causality in pteridophytes. How can we account for the shoot type of organization, for phyllotaxis, for the development of buds in specific positions, for the dichotomous or monopodial branching of shoots, for the differentiation and distribution of vascular tissue? If we would attempt to answer any of these questions, clearly we must concentrate on the region of continued embryogeny—the apical meristem. The prime difficulty lies in the fact that the situation there is exceedingly complex: nor is it easy to separate out individual aspects for special investigation.

In the experimental investigation of meristematic activity several courses are open. Thus attention may be directed to embryos, to the shoot apex, to the apices of buds, or to bud primordia; in time it may also be possible to make use of tissue cultures (White, 1944). Because of their mode of occurrence and very small size, the manipulation of pteridophyte embryos presents great difficulties. For the most part, therefore, use must be made of shoot apices and of buds and bud primordia. If the apex of a leptosporangiate fern is examined in longitudinal median section it will be found to include the following regions in basipetal sequence: (i) the apical meristem, consisting of a single layer of superficial cells, including the apical cell, of distinctive appearance and chemical reaction; (ii) near the basiscopic margin of this layer, the primordia of leaves, buds and scales originate from meristematic cells; immediately below the apical meristem vascular tissue in the initial phase of differentiation can be distinguished (Wardlaw, 1944); (iii) the region of subsequent differentiation and expansion which merges downwards with the fully matured tissue systems. Since the initial differentiation of primordia and tissues takes place in regions (i) and (ii), it is there that experimental treatments must be applied. Although these regions are of very brief extent, nevertheless, in ferns such as *Dryopteris* spp. experience has shown that the apical region of adult plants can be laid bare and manipulated in various ways. Even in ferns with sunken apices (e.g. the Osmundaceae) the shoot apex can be exposed by careful dissection. Such exposed apices, if suitably protected, do not readily become moribund or decayed as a result of the inroads of pathogens: on the contrary, they may be considered to show a notable degree of vitality (Wardlaw, 1944a).

Phyllotaxis and bud position
It is not proposed to discuss in detail the difficult question of the phenomenon of phyllotaxis. Attention may, however, be directed to one or two aspects where the ferns provide data of critical importance. Since each leaf primordium in leptosporangiate ferns originates from a single prism-shaped cell of the apical meristem,* phyllotaxis should be related to growth processes in that region. The leaf initial cell is at first indistinguishable from adjacent meristematic cells but it soon enlarges, divides by a somewhat oblique anticlinal wall, and thereafter the young primordium is easily recognizable. Studies by Priestley (1928) and others of phyllotaxy in dicotyledons support the generalization reached by Hofmeister (1867) that the most recently formed leaf primordium occurs at a point on the apical cone as far away as possible from the previously formed primordia. Priestley's rendering of this idea is that new primordia 'may be expected to appear in succession upon the opposite flanks of the growing point because each growing fold is a competing growth centre. The next centre of vigorous growth must be established as far as possible from the one preceding it.' Hof-

* Several cells are now known to be involved.

meister's account, as applied to the ferns, may be considered adequate, provided cognizance is taken of the fact that the new primordium originates just within the basiscopic margin of the apical meristem, i.e. at a point as far removed from the apical cell as possible. Priestley's physiological argument would appear to apply to the development of a primordium, once it had been formed, rather than to its actual inception. Other physiological explanations suggest themselves, e.g. that the apical cell and the recently formed leaf primordia give rise to diffusion fields or shells, and that the new primordium arises at the point of intersection of these several fields, but data are lacking. What seems beyond dispute is that any adequate account of the development of a phyllotactic system will call for a consideration of spatial, metabolic and mechanical factors.

Priestley (1928) and others have attributed the development of leaves in dicotyledons to the fact that growth is more rapid in the outer than in the central region of the shoot. Hence folds will tend to develop on the outside of the axial cylinder, in a specific phyllotactic arrangement, such folds being in fact the leaf primordia. While this interpretation may be correct for dicotyledons, it is not adequate as an account of leaf development in leptosporangiate ferns. In them the segmentation of a single superficial cell and its products is involved; there is no evidence of the development of a superficial fold of tissue. Nevertheless, in its megaphyllous character, phyllotaxis, symmetry and development, the leafy shoot of the ferns closely resembles that of the dicotyledons. A knowledge of the fern leaf may therefore be regarded as essential to any general theory of leaf development.

The positional relationships and development of buds in pteridophytes yield many points of interest. Many ferns show frequent dichotomy at the apex: unequal development of the shanks may result in the production of what appear to be lateral buds (Stenzel, 1861; Bower, 1923). In the adult shoot, buds may occur in axillary positions or in extra-axillary positions related to the leaf base. There is therefore a question as to what relationship, if any, these buds bear to the apex of the main shoot. In the main there is evidence of a unity of origin of meristematic tissue throughout the plant. Thus, the lateral buds in the Ophioglossaceae arise from primary meristematic tissue

(Lang, 1913–15). In *Matteuccia struthiopteris* and *Onoclea sensibilis*, in relation to the character of the growth development at the apex, parts of the apical meristem become detached and persist in definite positions along the shoot; these have been described as detached meristems (Wardlaw, 1943). On the removal of the shoot apex they give rise to a bud or group of buds. Some details relating to the origin of detached meristems may be given here. As growth proceeds, a majority of the meristematic cells on the basiscopic margin of the apical meristem become successively transformed into epidermis and cortical parenchyma. From a consideration of the facts of shoot development from the apex backwards it is apparent that the tangential growth and enlargement of the cortical and epidermal tissues must be maximal opposite the widest part of the foliar gaps in the dictyostelic system and minimal at the points of conjunction of shoot meristeles. It is only in proximity to the latter positions that detached meristems occur (Wardlaw, 1943). It has further been shown that if these buds are excised the underlying cortical tissue can also give rise to buds (Wardlaw, 1943a, 1944b). Whether these are positions of minimal mechanical stress or of minimal nutrition, or whether other factors are involved, such as the localized distribution of growth-inhibiting substances, must at present remain uncertain. These observations suggest a definite relationship between the origination of buds (or persistence of bud rudiments) and the position they occupy. Fern buds on normal uninjured shoots are usually characterized by a slow rate of growth. In species of *Dryopteris*, for example, there is little evidence of bud development until the associated leaf primordium is in the third or fourth whorl removed from the apex. Thus while a bud may eventually be axillary to one leaf or basal to another, its position during development may be essentially *interfoliar*, i.e. buds may be regarded as developing from those regions of the apical meristem which have not been used up in the development of leaf primordia. According to the nature of the growth development more or less notable departures from the initial spatial relationships of organs at the apex may result, e.g. in *Dryopteris filix-mas*, the quite remarkable displacement of buds from an initial axillary position to an abaxial position high up on the petiole base (Wardlaw, 1943a). Such observations suggest that, in respect of their initial positional relationship, the shoot buds of ferns (as distinct from the arrested shanks of dichotomies) may all prove to be fundamentally alike, the final position being determined by the specific distribution of growth in shoot and leaf base. Thus in order to understand why a bud develops and why it eventually occupies a certain position, the integrated growth activities at the apex must be considered.

Initial differentiation of the vascular system

In pteridophytes the initial differentiation of vascular tissue is apparent immediately below the apical meristem (Wardlaw, 1944); the subsequent differentiation and distribution of vascular tissue take place within the relatively limited growing region. The factors involved in these important developments are as yet little understood. A particular interest should attach to a comparison of the conclusions which may be reached in causal investigations with those which emerged and were considered valid during the phyletic period, i.e. when the facts of vascular construction were regarded as being of special importance in comparative studies. During that period morphology and physiology were pursued almost as separate branches of the science. Investigations of stelar structure were almost entirely morphologial in their inception and outlook and were largely based on the anatomical investigation of mature tissue systems. The inadequacies of the interpretations undoubtedly issue from the limitations of the method used. It now seems evident that an adequate account of the differentiation and distribution of tissues can only be given if the development of the shoot as a whole is investigated from the apex backwards at successive stages during the ontogeny (Wardlaw, 1944).

In 1915 Lang pointed out that a deeper insight into the nature of the stele may be obtained by regarding it as 'the resultant of a number of factors, as part of the manifestation of the system of relations in development'. He further remarked: 'Possible influences that have at various times been suggested are functional stimuli, the inductive influence of the older preformed parts on the developing region, and formative stimuli of unknown nature proceeding from the developing region.' Of these possibilities the third commends itself as deserving special attention. It is clearly cogent to inquire what relationship, if any, may be considered to exist between the activity of the apical meristem and the initial differentiation of vascular tissue. Growth at the apical meristem involves as a first consideration an increase in the amount of protoplasm, i.e. the meristem is a region of active protein synthesis. From the complex of bio-chemical processes involved, enzymes and other substances, e.g. auxins, possessing important physiological properties, are known to result. These will tend to diffuse or otherwise move from the region of highest concentration—the meristematic cells in which they are produced—into the adjacent cells, and these, as also those at a distance, are liable therefore to undergo changes in their physiological and structural properties.

It has been tentatively suggested that the initial differentia-tion of vascular tissue is attributable to the diffusion of a sub-stance (or substances) from the meristematic cells (Wardlaw, 1944). Since there is also a contemporaneous movement of nutrients of various kinds towards the apex from below, the region of interaction extending downwards from the region of initial differentiation is characterized by further histological changes, i.e. those constituting the phase of subsequent differ-entiation. A further source of interaction in the developing region (which is known to be characterized by a high rate of respiration) may relate to the gradients of moisture, oxygen and carbon dioxide concentrations between the centre of the shoot and the external atmosphere. A working hypothesis regarding the initial differentiation of the vascular tissue in pteridophytes has been stated as follows. Wherever the apical meristem of a shoot, bud, leaf or root is in a state of active growth, of such a nature that the distinctive character of the meristematic cells is maintained, the initial differentiation of vascular tissue will be observable immediately below the apex and in the path of substances diffusing from it, one or more of these substances being causally involved in that process. What the diffusing substance (or substances) may be and how it works is not known. The relationship under consideration should also hold good for leaves and roots. As the differentiation of the vascular system of a leaf primordium is approximately contemporaneous with the differentiation of the adjacent shoot stele, a coherent and unified vascular system results. The leaf trace, in fact, 'passes' backwards *into the shoot*, not *out from the shoot* as is sometimes described in anatomical investigations based on adult regions. This is a point to which Lang (1915a) has called attention. A considerable body of evidence from many sources can be adduced in support of the hypothesis outlined above (Wardlaw, 1944; Holloway, 1939; Williams, 1933). In those instances where the apical growth, whether of shoot or leaf, is not of such a nature that the distinctive character of the meristematic cells is maintained, stelar tissue is not differentiated. Again, in the young embryos of some lycopods and of certain eusporangiate ferns (Ophioglossaceae, Marattiaceae), the delay in the organization of an active shoot apex is attended by a delay in the differentiation of cauline vascular tissue (cf. Campbell, 1921).

In pteridophytes the relative importance of the shoot and of the leaf in contributing to the vascular system has in the past been considered in some detail. Some writers have maintained that the shoot stele is essentially of axial origin, others that it is a composite structure built up from the decurrent vascular strands from the leaves (Campbell, 1921). Evidence has been adduced in support of both points of view, and there is no doubt that, in different instances, differences in the relative development of shoot and leaf vascular tissue do occur (cf. Bower, 1923, p. 139). The hypothesis regarding the initial differentiation of vascular tissue presented above, however, is not only equally applicable to both sets of data but serves to unify them. Active shoot growth is usually productive of an undeniable cauline stele: on the other hand, active leaf development and relative inactivity at the shoot apex may be productive, at least for a time, of a vascular system which will appear to be a composite structure consisting largely of decurrent leaf traces with little or no cauline component.

Avery (1940) has made reference to a number of investigations which suggest that substances such as indoleacetic acid may profoundly affect the transport of materials, and may lead to the accumulation or mobilization of nitrogen and carbohydrates in proximity to the point of application. Now substances such as auxins probably owe their origin to the decomposition of proteins; for example, Thimann & Skoog (1934) have shown that whereas auxin is actively produced by growing buds, none is produced by dormant buds. It is conceivable that a close relationship exists between (a) the maintenance of a terminal meristem in an actively formative condition, (b) the production of activating substances during protein metabolism, (c) the initial differentiation of vascular tissue, and (d) the accelerated movement of nutrients to the formative region.

Stelar morphology

The method of investigating fern steles during the last fifty years or more has consisted mainly in the examination of transverse sections of the shoot from the base upwards, or more locally through one or two nodes only. This now seems inadequate: what is required is an investigation, in plants of different ages, of development in the growing region, not only of the stele but of all tissues. In other words, the development of the shoot as a whole should be investigated from the apex backwards at successive stages during the ontogeny (Wardlaw, 1944).

In the ferns, the shoot stele, whether protostelic, solenostelic or dictyostelic, consists initially, i.e. just below the apical meristem, of a solid or hollow, uninterrupted, subconical mass of tissue. During the subsequent differentiation, notable changes take place and internal features of considerable complexity may be produced. Such changes include the development of pith, leaf gaps and perforations, and the interruption or disruption of the continuous vascular cylinder. The classical accounts of vascular structure in leptosporangiate ferns show clearly that the interruption of the conducting cylinder of the shoot is associated with the insertion of the leaf traces, i.e. the vascular strands of the petiole or leaf base. In other words, the vascular cylinder which otherwise would be continuous, is interrupted by non-vascular tissue, usually parenchyma. In shoots where the leaf gaps overlap, the vascular system becomes an open meshwork or dictyostele, each individual strand being described as a meristele. Various tentative physiological explanations have been advanced to account for the development of these leaf gaps (Tansley, 1907). A considerable volume of literature also deals with the question as to what extent the vascular tissue of the shoot is of truly cauline origin, or alternatively, is a composite structure largely composed of decurrent leaf traces.

At the apex of a dictyostelic fern such as *Dryopteris dilatata* no leaf gaps are present at the points of conjunction of the very youngest leaf traces. As the leaf primordium enlarges, however, the distribution of growth is such that its initial single vascular trace becomes subdivided and greatly distended, particularly in the tangential direction. This is attended by the development of a leaf gap—a region of parenchymatous tissue—in the hitherto unbroken vascular cylinder of the shoot. These facts suggest that leaf gaps develop in those regions of the shoot stele which are subject to mechanical stress as a result of the enlargement of the vascular systems of the leaf bases; they also indicate that groups of cells, initially differentiated as vascular tissue and potentially capable of developing into phloem, xylem, etc., have been transformed during growth into parenchyma (Wardlaw, 1945).

That a direct relationship exists between leaf development and the internal morphology of the shoot can be tested by experimental means. If the hypothesis is correct, then, by destroying a succession of very young leaf primordia, the formation of the corresponding leaf gaps should be prevented, and the vascular system of the shoot should develop, not as a dictyostele, but as a continuous solenostele. When the simple process of laying bare an apex of *D. filix-mas* or *D. dilitata*, as already described, was carried out and all new primordia destroyed, a conclusive result was obtained. Transverse sections from the apex downwards of shoots thus treated showed every stage between complete solenostele and the normal dictyostelic condition in the older region of the shoot (Wardlaw, 1944*a*).

In the ontogenetic development the first appearance of the pith marks a critical phase in the morphology of the stele. Bower (1930) has suggested that, in the individual species, the size factor may be implicated. Whatever may be the factors involved, metabolic or mechanical or both, the development of pith is a result of differentiation at the apex. It is there that it should be studied and not merely, as is so often the case, in the mature shoot. The several parenchymatous regions, cortex, leaf gap and pith, constitute notable components of the internal pattern of the shoot. They become increasingly conspicuous during the ontogenetic development and progressively less conspicuous when the shoot diminishes in size, as in 'starvation' experiments. The variety in the internal morphology of the leafy

shoot system of ferns is largely referable to the relative development of these three regions during growth at the apex (Wardlaw, 1945).

The size factor and related problems
Data submitted by Bower (1921, 1930) for all classes of pteridophytes indicate that actual size is a factor in stelar morphology.
Thus in the obconical development of the individual, from the
sporeling to the adult, the progressive increase in the dimen-
sions of the stele, is, as a rule, accompanied by an increase in
the complexity of its outline, or by a complication of the pattern
of its constituent tissues.

The particular pattern to be observed in any transverse
section of a fern shoot is determined by the relative proportions
and positions of the pith, the vascular tissue and the cortex. The
pith, for example, may be only slightly or very extensively
developed and the whole cross-sectional pattern thereby correspondingly modified. In many instances medullation is directly related to the complexity of the pattern which is developed.
Now, vascular tissue, pith, leaf gap, and cortical parenchyma
are all differentiated more or less contemporaneously in the
growing region. The rates of growth of these tissues are not
only initially different: they also change during development.
Hence the internal pattern by which the mature leaf-shoot
system is characterized can only be properly understood if the
apical region as a whole is investigated as a dynamic system.

It has been argued (Bower, 1923) that medullation in the
pteridophyte stele represents a 'change of destination' of elements
initially and potentially tracheidal, and that phyletically the
origin of the pith is to be sought in a loss of conducting function
in the central core of tracheids. A somewhat different view of
these histological developments is taken here; for even if it be
accepted that phyletically the pith represents modified tracheidal
tissue, the process of medullation at the apex of the individual
solenostelic or dictyostelic shoot must still be accounted for in
terms of causality. The facts, indeed, justify the view that the
development of the pith is a direct expression of the process of
growth. So, too, there is justification for the view that all initial
developments at the apex should be interpreted in terms of
growth and not of adult functional activities.

The effect of differences in the rates of growth of adjacent tissues on the internal pattern is well seen in the genus *Selaginella*. In the region of initial differentiation, just below the shoot meristem, the cells of the inner cortex and of the stele are of approximately equal size as seen in transverse sections. The rate of enlargement of the cells of the inner cortex greatly exceeds that of the cells of the central stele, and there is no compensating increase in the number of stelar cells. Hence a disruption takes place along the plane of junction of the two tissues and characteristic lacunae appear in which the stelar cylinder or ribbon is suspended by a trabecular endodermis (Wardlaw, 1925).

The development of internal pattern is thus seen to be referable to growth processes in the apical or subapical region. So, too, with regard to the operation of the size factor Bower (1921) has remarked: 'The behaviour of meristems remains now, as it has always been, the greatest enigma of the plant body, and not the least of the questions which it raises is this: How does the meristem forecast in its embryonic tissues those proportions of surface to bulk which will be necessary when the tissues still embryonic shall have matured to their full size?' The idea which prompted this question is seen in another form in a later work (1930) and again in 1937 when he writes: 'the centre and focus of the whole problem lies, not in the details of the adult state, though these may suggest what are the functional requirements for success, but in that imperfectly known and still problematical region of initiative, the Growing Point. It is here, rather than in the matured parts, that the key to causality should be sought, and the Size-Relation tested by comparative measurement.'

Movement of nutrients to the apical meristem
The supply of nutrients to the apical meristem, though a problem of the greatest importance, is almost completely unexplored. In the subterminal region of a shoot of *Dryopteris* it can be seen that although both cortex and pith are already well developed, the vascular tissue is still in the initial phase of differentiation, i.e. phloem and xylem cannot be distinguished. It is therefore cogent to inquire if this incipient vascular tissue is the means whereby metabolites are translocated and distributed, centrifugally and centripetally to cortex and pith respect-

ively and to the apical meristem; or whether, alternatively, the growing region is supplied by upward diffusion over the whole cross-sectional area. To answer this question would be to effect a notable advance in our knowledge of formative processes at the apex. On analysis, the situation seems likely to prove one of great complexity. Together with the distribution of metabolites must be considered the incidence of mechanical factors in the growing region.

Activating and inhibiting substances

As in flowering plants the removal of the physiological dominance of the terminal meristem in pteridophytes usually results in the active development of such dormant or inhibited buds, or bud rudiments, as may be present further down the shoot. There is little doubt that the inhibition of lateral buds is due to the basipetal movement of substances formed at the growing apex. Space does not permit of a detailed consideration of relevant data, but reference may be made in passing to investigations on the development of buds by Lang (1913–15) on *Botrychium lunaria*, Rostowzew (1892) and Goebel (1928) on *Ophioglossum vulgatum*, Goebel (1900) and Sahni (1917) on *Diplazium* (*Asplenium*) *esculentum* and *Platycerium* spp., Lang (1924) on *Osmunda regalis*, Wardlaw (1943*a*) on species of *Onoclea*, *Matteuccia* and *Dryopteris*, Williams (1933) on *Lycopodium selago*, Solms-Laubach (1902) on *Isoetes* spp., and Ludwigs (1911) and Praeger (1934) on *Equisetum* spp.

In species of *Selaginella* the angle meristems, which are present at every branching of the shoot, normally grow into rhizophores, i.e. root-bearing organs. If, however, the adjacent shoot apices are removed, the angle meristem grows out as a leafy shoot when such materials are used as cuttings. Williams (1937) has shown for *S. Martensii* and *S. Lobbii* that if the cut upper surfaces of shoots, from which the apices have been removed, are smeared with 3-indoleacetic acid in lanoline, the angle meristem grows out into a rhizophore. The view that a substance diffuses backwards from the apex of the normal untreated shoot and controls the morphogenetic processes in the angle meristem is supported by this experimental evidence. I have shown that if rhizomes of *Matteuccia struthiopteris* and *Onoclea sensibilis* are decapitated and the exposed surface smeared with 3-indoleacetic acid in lanoline the development of buds from detached meristems does not take place (unpublished results).

The relation of nutrient supply and of activating substances to morphogenetic processes requires investigation. Little has so far been done, but certain morphological observations indicate the nature of some of the problems and the materials which may be suitable for investigation. In this connection reference may be made to studies of such phenomena as leaf regression in ferns by Goebel (1900, 1908), Lang (1924) and Wardlaw (1945*a*); the transformation of potential sporophylls in ferns into sterile foliage leaves by Goebel (1900); and the growth of fertile cones of *Selaginella* and *Equisetum* into vegetative shoots by Goebel (1900, 1928), Ludwigs (1911) and Williams (1938). Variations in habit and external form are also, in many instances, referable to growth processes at the apex. (See Mekel (1933) and Wardlaw (1943) on *Matteuccia struthiopteris*, Ludwigs (1911) on *Equisetum*, and Goebel (1928) and Williams (1938) on *Selaginella* and *Lycopodium*.)

Essentials of the shoot type of organization

It may be anticipated that the investigation of the apex from the standpoint of causality will eventually lead to new concepts regarding the fundamental nature of the shoot type of organization. Current views are broadly speaking of two kinds, i.e. phytonic and axial (or strobilar). In the former the existence of the shoot or axis as an independent member is more or less explicitly denied, the plant being envisaged as a construction of phytons or segments of which the leaf bases, or extensions thereof, are the fundamental units. Although in some instances such theories may have a limited application or a certain value for purposes of description, they impress us today as being artificial, rigid and divorced from the facts of physiology and embryology. Supporters of axial theories regard the shoot 'as a phyletically pre-existing axis or stem from which the leaves may have arisen by enation' (Lang, 1915*a*). In his later period Sachs held that the shoot, including both leaves and axis, is the real unit. A similar view is adopted by Bower (1935, p. 546): he observes that 'axis and leaves act together as a physiological whole, and are so initiated in the embryology; also, in evolutionary history, as based on comparison of early fossils, such as the Psilophytales. The shoot-unit of Sachs is the natural, that is the developmental and evolutionary unit.' Bower (1922) has pointed out that the young vascular plant consists essentially of a simple spindle or axis, with a distinction of apex and base. Given such a unit, it is held that plant bodies, progressively more elaborate in construction, may result from dichotomy, and this in turn 'may pass over to monopodial branching, thus producing lateral appendages' (Bower, 1935). Such lateral appendages may include (a) lateral branches in which the shoot structure is repeated, and (b) megaphylls which are thus regarded as being of cladode origin. Other lateral appendages may originate 'by enation of parts from [shoot] surfaces previously not tenanted' (Bower, 1935, p. 540): such appendages would be exemplified by the microphylls of *Lycopodium* and the scales of ferns. It is evident that in these instances Bower regards the fundamental unit as being the axis or spindle, not the leafy shoot of Sachs.

Now, in these views, the concepts involved are essentially morphological and are largely non-physiological in character. They are the outcome of comparative investigations of living and fossil plants and have been elaborated against a background which was predominantly evolutionary in outlook. Only in occasional instances have the causal aspects been considered in any detail.

Every growing plant may be regarded as an organic individual. Such an individual, in the view of Child (1915, 1941), may be considered as a system of relations between a physical substratum or structure—the specific hereditary substance— and chemical reactions. Metabolism is thus considered to be *the* effective factor, and physiological gradients part of the essential mechanism. Such a point of view may be of considerable value to the botanist. Whether embryonic development or growth at the shoot apex is under consideration, investigation of the following aspects seems likely to yield more precise data relating to morphogenetic processes than we have hitherto possessed: (i) the character and position of the embryonic or meristematic tissue, (ii) the sources of nutrition supplied to the embryo or meristem, its nature and path of translocation, (iii) the effect of different nutrient and other substances on

F

meristematic and differentiating tissues, and (iv) the relation of tissue differentiation and development to the final structural organization observed.

When reduced to the simplest terms there must be, for the development of a shoot, (*a*) a stimulus to growth, (*b*) cells capable of meristematic activity, and (*c*) a source of nutrition in close proximity. In other words, it is not necessary to begin with an axial unit. In many instances the establishment of polarity may be directly related to the path of diffusion of nutrients, i.e. the young shoot may be alined along the axis of diffusion. In the majority of pteridophytes the polarized growth development of embryonic or meristematic tissue yields a structure recognizable as an axis, spindle or shoot. The successive stages of this development are usually very constant and characteristic for the individual species, but by modifying the environmental conditions the development of other types of structure may be induced. For example, a detached meristem of *Matteuccia struthiopteris* or *Onoclea sensibilis* may yield, not a single shoot, which is the rule under 'normal conditions', but a plurality of shoots, or a coralloid outgrowth, i.e., a tissue mass of indeterminate structure. In contradistinction to classical morphological views, therefore, the shoot is here regarded not as a unit of construction but rather as an expression of meristematic activity under certain environmental conditions. In this connection it may perhaps be anticipated that notable advances in the analysis of morphogenetic processes may result from the use of tissue cultures (White, 1944). So far pteridophyte tissue cultures have not been prepared.

In leptosporangiate ferns the apical meristem can be precisely specified and distinguished from adjacent tissues. Along the basiscopic margin of the apical meristem some of the meristematic cells develop into scales. Leaf and bud primordia also originate from one or more of the prismatic cells which constitute part of the apical meristem. These several developments are dependent on supplies of nutrients from below and probably on 'morphogenetic', i.e. activating, substances, produced locally. A possible hypothesis is that the type of lateral organ developed is in some way related to its position on the meristem at the time of its inception, to the balance of nutrients supplied and to the presence in critical concentration of particular metabolites. Now in the phyletic view the pteridophyte shoot or axis is accepted as a pre-existent morphological unit; shoot branching has been referred to dichotomy with equal and unequal development of the resulting shanks; megaphylls are considered to be of cladode nature and are ultimately referable to branching of the shoot with dorsiventral development of the lateral branch; while microphylls are 'enations', i.e. shoot outgrowths occupying superficial areas not previously tenanted. It may be that such views do represent in broad outline the course of evolution of these several members. Nevertheless, in any leptosporangiate fern species we observe that branches, megaphylls and microphylls all originate from the same fundamental tissue: they are formed at the apex and are derived from meristematic cells. In brief, we are led to a conception of the apical meristem as a morphologically plastic region, of specialized metabolism, and of regulated formative processes, capable of various developments within the limitations imposed by the potentiality for development of the specific hereditary substance.

4
The Metamorphosis
of Plants

Goethe's essay on 'The Metamorphosis of Plants', first published in 1790, provides a theme of recurrent interest. Yet it would probably be not untrue to say that while the majority of contemporary botanists are familiar, in a general way, with the underlying idea of this work, few have studied the original edition or had access to the English translations. Indeed, the latter are not readily accessible to the ordinary reader. By preparing a new and critical translation, Dr. Arber has rendered a signal service to botanists. But more than that, she has rendered a service to botany, for the translation is preceded by an introduction which is a model of its kind. To those who have occupied themselves with the history of botany, particularly that relating to the last two hundred years, the introduction will indeed prove all too short. For the author has much to say that is interesting and important about the genesis and development of Goethe's idea, its intrinsic merit, its place in botanical science and, more generally, in the philosophy of biology. The aphoristic terseness and sureness of touch with which these matters are set out make it difficult to do more than emphasize the value of the new translation and introductory essay.

Students of plant morphology are familiar with the general idea underlying Goethe's theory of metamorphosis, namely, that all the external parts of the shoot are regarded as being due to

The Metamorphosis of Plants (1790) and Tobler's Ode to Nature (1782). By Agnes Arber. *Chronica Botanica*, Vol. 10, No. 2. Pp. 63–126 + pl. 23–26. (Waltham, Mass: Chronica Botanica Co.; London: Wm. Dawson and Sons, 1946.)

the transformation of a single organ, that organ—an ideal leaf—being itself an abstraction. Or, in the words of the new translation . . . 'the laws of transmutation according to which she (Nature) produces one part from another, and sets before us the most varied forms through modification of a single organ . . . the process by which one and the same organ presents itself to our eyes under protean forms, has been called the *Metamorphosis of Plants*'. Contrary to a view widely held, Goethe was apparently not acquainted with the earlier related work of Kaspar Wolff ('Teoria Generationis') published in 1759, when he wrote the 'Metamorphose'. The view now before us is that he was an independent observer, a philosopher who looked closely at plants, and who was imbued with the idea of developing some general conception, or nexus of ideas, to cover the diversity of form which he saw everywhere in Nature, as well as in the individual plant. His method of presenting his views was not that of the man of science, but, as Dr. Arber points out, essentially that of a man of letters. The ideas in the 'Metamorphose', which are set out in an easy, familiar and somewhat tentative fashion, on close examination prove to be rather elusive. Here Dr. Arber supports other critics in the view that the difficulty of grasping Goethe's ideas of metamorphosis is largely due to the fact that he did not always succeed in grasping them firmly himself. Nevertheless, that he was preoccupied with morphological developments of a most important kind cannot be denied; moreover, he was interested in the underlying mechanism, he tried to formulate general ideas admitting of synthesis; and he produced an essay, which if not good science, still provokes thought. There is, of course, always a danger of reading into a work of this kind considerably more than the author intended. Nevertheless, after reading some passages in the 'Metamorphose', it is interesting, if idle, to speculate on the contribution which Goethe might have made to biological theory had he been alive today.

Dr. Arber has not only concerned herself with the text of the 'Metamorphose': she has also made use of much additional matter from Goethe's correspondence and the comments of his contemporaries. Hence she has been able to present as critical an estimate of his contribution to botany as we are likely to get.

Thus she emphasizes that Goethe's great service to morphology —we owe the word to him—was his recognition that its basis must be essentially comparative. On the difficult question of Goethe's scientific status, she remarks that ... 'This question still remains fraught with difficulty, for the catholicity of his mind, and the kaleidoscopic character of his activity, defy neat labelling. As a botanist, he began with a simple utilitarian interest in plants; he passed through a brief period in which he studied the multiplicity of the plant world from the standpoint of the descriptive naturalist; this was succeeded by a phase in which his mind was entirely possessed by comparative morphology, a subject to which the value of his contribution, and the inspiration which later workers have derived from it, are undeniable; and, finally, by a transition natural to his mental growth, he reached a stage in which his morphological thought reached out to the reconciliation of the antithesis between the senses and the intellect, an antithesis with which traditional science does not attempt to cope. It has been suggested by a literary critic that Goethe was "a great poet who grew out of poetry". Approaching him, as we have done here, through the medium of his plant studies, we may perhaps offer the comparable conclusion that Goethe was a great biologist, who, in the long run, overstepped the bounds of science.'

The publication under review also contains the original and a translation of the rhapsody on Nature, attributed to Goethe— 'Nature: Aphoristic'—a translation of which by T. H. Huxley opened the first issue of *Nature* in 1869.

By this new work of scholarship, Dr. Arber has again placed a wide circle of botanists in her debt.

5
Process and record:
aspects of botanical science

It is surely a strange and surprising thing that Morphology—
the study of Form in plants and animals—which Charles Darwin
regarded as 'the most interesting department of natural history',
indeed, 'its very soul', today stands in danger of suffering eclipse
by failing to appeal to the younger generation of botanists.
Without some knowledge of the external form and internal
structure of plants there can be no approach to the most comprehensive theme in Biology—the process known as Evolution
or Descent with Modification. What the botanist knows of this
process is necessarily based on the comparative study of plants.
It is therefore a matter for surprise that, among botanists,
interest in this branch of the science has undergone a marked
decline in recent decades. Not only that, but many of the major
conclusions regarding the evolution of plants have been
seriously challenged. What the botanists of the Darwinian
and post-Darwinian period tried to do was to show the relationship of the various classes of plants during the course of their
descent from common ancestors. They tried to reconstruct the
family or genealogical tree of plant life and, indeed, they
considered that their efforts had met with a very considerable
measure of success. But today many botanists see the matter
in a different light. It has been asserted that the search for
common ancestors is a hopeless quest, the genealogical tree an
illusory vision. But if the subject possesses the merit which Darwin claimed for it, there is clearly need for inquiry into the
present state of affairs. The record of plant development during
remote past ages, together with the comparative study of living

forms, would appear to provide a theme of perennial interest to men of philosophic mind. Nevertheless, it has been said that the whole of this branch of botany—the investigation of the phylogeny of plants—leaves the majority of the younger botanists cold.

Some contemporary botanists, no longer attracted by comparative morphology, and even holding the pronouncements of the professed phylogenist in contempt, see a promising and little-worked field in the study of the organization which becomes apparent during the *process of development* of the individual; in other words in the study of *Morphogenesis*. The aim of this study is to explain, in terms of mathematics, physics and chemistry, how, at each stage in the development of the individual, the distinctive form comes to be what it is. In other words it attempts to answer the question: How, during the growth of a plant, is the characteristic form, or succession of forms, produced, and what are the factors involved?

It now seems evident that the post-Darwinian botanist, intent on comparative studies with a view to the construction of a phylogenetic system, or 'family tree', had an insufficient knowledge of the process of development. But the contemporary student, in attempting to make good that deficiency, is liable to fall into a not dissimilar neglect of the historical aspect. Both studies have distinctive and important contributions to make to the common theme. Each is related to the other though the nature of that relationship may often be obscure. In brief, the wider vision requires consideration of both Morphogenesis and Phylogenesis—i.e. of both Princess and Record. My theme on this occasion, then, is to show how our views on these two aspects have developed, and how they may be related to each other.

Botanical trends during the nineteenth century

The two important trends in botanical science with which we are concerned here can be discerned during the middle decades of the nineteenth century. One of these related to the process of development both in the lower and higher plants, the other to the natural classification of plants and, after Darwin, to their genealogical relationships. Let us briefly consider the distinctive features of these two aspects.

Problems of development

[References are made to the work of Schleiden, Naegeli, Hofmeister, etc. (see pp. 1, 20, 21, 53).]
It might be thought that as a result of Hofmeister's investigations and his critical search for the relationship between physiological activity and the assumption of specific form or pattern, botanical science had at length been established on a broad and sure foundation—one in which morphology and physiology were seen to be inseparable aspects of the same theme. It might further be anticipated that these investigations of the process of development would have made rapid progress in the hands of his successors. This, however, was not the case as we shall presently see.

Evolution and the phylogenists

In the development of botanical science the initial phase of collecting, cataloguing and grouping of plants (usually on an arbitrary basis) was followed by attempts to construct *natural systems* of classification. The aim of these systems was to indicate the gradations of natural affinity among organisms. Plants which bore a general resemblance to each other, and which shared a number of the more important characters in common, were held to belong to the same circle of affinity and, by implication, shared the same hereditary constitution. Thus groups of genera were associated together into families, or natural orders, and these, as also the larger subdivisions, became defined with a considerable degree of certainty and precision, just as, at an earlier stage, Linnaeus had fixed the boundaries of genera and species. By the middle of the nineteenth century these natural systems had reached a high degree of completeness. It may perhaps strike the modern biologist as not a little strange that the founders of these natural classifications were nevertheless adherents of the dogma of the Fixity of Species. With the publication of Darwin's *Origin of Species* in 1859, these natural classifications were thus at hand, ready for use in the construction of phylogenetic systems, i.e. systems showing the *natural affinities, or genetical relationships, of organisms during descent.*

In the period—sometimes described as the Phyletic Period— that followed Darwin's enunciation of the Theory of Descent with Modification, the details of the form and structure of

living plants, together with such facts as could be gleaned
from the fossil record, were regarded chiefly as providing
materials for comparative studies and for the construction of
phylogenetic systems; in everyday language, for reconstructing
the 'family' or 'genealogical tree'. There was, in brief, a very
marked swing away from the causal outlook which had characterized Hofmeister's later studies. Botanists were no longer
preoccupied with the question: How does the observed form
come to be, in terms of physical, physiological and other factors?
but, what family relationship is indicated by the observed
form or structure and what light does this information throw
on *the course of evolution*? The *general* or *causal morphology* of
Hofmeister was, in fact, displaced by the *special comparative
morphology* of the students of evolution. Now, it is evident that
there can be no study or interpretation of evolution in plants
or animals without having recourse to comparative morphology.
Evidence of evolution is seen in the extensive changes in the
form and structure of plants during the passage of geological
time and in the diversity of form and structure in living plants.
As a result of the studies which were undertaken in the phyletic
interest, our knowledge of the range of plant structure was
enormously extended. A vast new realm, so worthy of exploration, must indeed have opened out before the comparative
morphologist. In particular, the ferns and their allies, the
progenitors of which reach back to the Upper Silurian and
Early Devonian Periods, were found, more than any other class
of plants, to provide the raw materials for the study of evolution. Comparative or Formal Morphology, now provided with
an integrating thesis of the most far-reaching possibilities and
interest, had come into its own. In these studies, the conclusions
from detailed and comprehensive investigations of the structure
of living organisms were checked against the evidence of the
fossil record; and, bit by bit, a coherent if still speculative and
necessarily incomplete account of the genetical relation of
plants during descent was sketched.

The sweeping success of Darwin's views among biologists is
understandable. It would be surprising had it been otherwise.
The raw materials were at hand in the more or less complete
natural classifications which had been devised in the previous
fifty years. The dogma of the Fixity of Species, so long and

tenaciously held, was ripe for destruction. Darwin's full and masterly outline of the theory of *Descent with Modification* must evidently have provided a key which would not only unlock many doors but would enhance the value of the treasure that lay within awaiting discovery. To the student of phylogeny, it must surely have appeared that to whatever group of plants his researches were directed, fruitful results were assured. Since Nature was a unity, all findings were bound to 'fit in' somewhere; all destined to contribute in some measure to the wonderful edifice of evolutionary theory. The highroad to achievement lay in the study of comparative morphology, which, in fact, became a restricted and well-defined discipline. The facts of development and the characteristic features of the adult, as ascertained by the observation of form and structure, were used in the formulation of what we should today regard as purely morphological concepts, physiological and causal aspects receiving at best little more than passing attention. The relation between form and function, which Sachs and later Goebel described as Organography, was, of course, of importance to those who maintained the Darwinian position and assumed the all-pervasiveness of adaption in plant life. On the other hand, the factors relating to the formation of organs and the differentiation of tissue systems for the most part remained unexplored.

Decline of comparative morphology

Several decades passed before the methods or conclusions of the phyletic morphologists were seriously challenged. But gradually doubts began to accumulate. A bad feature of the phyletic period was that plant morphology and physiology were largely pursued as separate disciplines. Morphologists and anatomists, moreover, tended to resort to facile, pseudophysiological assumptions: these were typically not tested by experiment. Then other difficulties became evident. We have seen that both natural and phylogenetic classifications necessarily rested on a basis of comparison. Indeed, the search for criteria of comparison was a major part of the morphologist's task. Similarity of form and structure in the principal organs and tissues was accepted as an indication of genetic affinity or blood relationship; in other words, organisms which showed the same underlying form and structure, and which passed through comparable stages during development, were held to be related and to share a common origin. But, as the survey widened, it was realized that similar structural features might occur in plants which, as judged by other criteria, were in no way related. In short, it became evident that *parallel* or *homoplastic* development was of frequent occurrence both in the plant and animal kingdoms. As information of this kind accumulated, faith in the validity of the previously accepted criteria of comparison declined. This eventually led to the view that instead of there having been one family tree, or one main line of descent, as had generally been assumed by biologists of the Darwinian period, it was probable that there had been several, even many, parallel lines of descent from primitive ancestors. The family tree, in fact, had become a thicket. Later, because of the recurring difficulty of linking one main group of plants with another by means of common ancestors, the ultra-cautious and the pessimistic were to hold that the 'genealogical tree' had been reduced to a bundle of sticks; and that the search for common ancestors was a hopeless quest.

A protagonist of the extreme polyphyletic view, Church (1919) asserted that land vegetation developed from transmigrant algae which had already reached a high degree of differentiation and elaboration in the sea. The chief phyla or families of land plants already had marine precursors before the period of the transmigration. Church even suggested that Pteridophyte classes such as the Lycopods and Ferns represent independent lines which have run distinct, though in many respects parallel courses from the earliest times, even perhaps from the original flagellate ancestors. So, too, with other classes of plants. In this extreme view the monophyletic genealogical tree would be replaced by a series of parallel lines of descent, these arising from a group of unicellular organisms of great antiquity.

Briefly summarized, this is the situation: we do not know how the Mosses and Liverworts originated from more primitive organisms such as the Algae, or how, in turn, to relate the Mosses and Liverworts to the Ferns and their allies, which represent the next level in structural organization; we do not know how to link the Pteridophytes with the Gymnosperms, or the latter with the Flowering Plants. And finally, among Flowering Plants, we do not know, at least, with any precision or assurance, how one great cohert or natural group is related to another. Nevertheless, the broad fact of evolutionary change throughout geological time is not in doubt. Moreover, within individual major classes, such as the Ferns, of which we have knowledge going back to Carboniferous and Devonian times, there is unmistakable evidence of a broad and sustained progressive development from ancient and primitive to modern and derivative types. In brief, whereas we can indicate the course of evolution in a single phyletic line, i.e. a single branch of the hypothetical genealogical tree, difficulties arise when we try to relate this branch to neighbouring branches and to the main trunk. In this connection D'Arcy Thompson (1942) has pointed out that a 'principle of discontinuity' appears to be inherent in all our classifications. Within any natural group there may be evidence of the continuous process of evolution, but between groups considered to be allied, more often than not discontinuities are evident. Yet that a natural affinity is involved seems scarcely to be doubted. As D'Arcy Thompson says, 'there are gaps between the groups but we can see, so to speak, across the gap'. But, in other instances, 'the breach is so wide that we cannot see across the intervening gap at all'. Unless much new fossil evidence comes to light, it now seems that the bridging of the several gaps—if they can be bridged— is unlikely to result from the methods of comparative morphology alone.

Renewed study of process of development

We may now inquire what precisely is meant by 'process of development' which seems to provide the contemporary botanist with opportunities for fruitful investigations. I understand it to mean the investigation of the factors which determine the characteristic external form and internal structure of an organism, and the integrated 'wholeness', or organization, which it shows throughout development. Eventually studies of process would also include a consideration of development in its wider evolutionary sense and of the forces which bring it about. It will be seen that the aim of those who would investigate the process of development is very similar to that which inspired

Hofmeister's *General Morphology of Growing Things* nearly eighty years ago. No justification is needed for these studies: as Professor Lang pointed out many years ago, even if the phyletic history of plants were before us in full, the problems of causal morphology would still remain; while it is evident that the relevant investigations are of the very essence of scientific inquiry.

In studies of Flowering Plants and Ferns, particular consideration will be given to the factors which determine the leafy-shoot type of organization, the development and shape of leaves, buds, hairs and scales, roots, and reproductive organs, and the differentiation of the more or less complicated tissue systems within. These considerations prompt yet a further inquiry, namely: Should these researches be pursued to a successful conclusion, to what further use can the data be put? To this question we shall return later.

If the advances which have been and are now being made in the application of physics and mathematics to biology can be sustained; and if the data of certain aspects of physiology and genetics can be integrated with those of experimental morphology, substantial progress in the study of development may be anticipated. Already, in such notable works as Professor D'Arcy Thompson's *Growth and Form,* we find numerous fascinating examples which show how a knowledge of the laws of mathematics and physics contribute to an understanding of the development of form and structure in plants and animals. 'Cell and tissue, shell and bone, leaf and flower, are so many portions of matter, and it is in obedience to the laws of physics that their particles have been moved, moulded and conformed' (p. 10). True, in a majority of the instances cited, we still await experimental proof: nevertheless there seems to be little doubt that many aspects of form and structure—which, as we have seen, are of great interest and value to the comparative morphologist—are largely explicable in terms of physical and mathematical laws.

From another angle evidence is accumulating that certain morphogenetic processes are due to the action of specific biochemical agents, i.e. particular metabolites, or substances used in growth. Biochemistry, in fact, is everywhere becoming of increasing importance in botanical investigations. As Dr. Joseph

Needham has said (in *Biochemistry and Morphogenesis*), 'Form is not the perquisite of the morphologist. It exists as the essential characteristic of the whole realm of organic chemistry . . . there can be no sharp distinction between morphology and biochemistry.' In biochemistry, as in biology, we are concerned with two basic considerations, energy and organization. Nevertheless, it is no easy matter to show precisely how biochemistry and morphology are related. We now know that a close and possibly obligate relationship exists between the presence of certain activating substances—morphogenetic hormones—and the formation and differentiation of organs and tissue systems. But to explain why an organ or tissue acquires its characteristic configuration because a certain chemical substance is present is no easy matter. It seems probable that even the simplest biological form or pattern is due to the action of several factors, including the biochemical factor. In the exploration of this aspect we have still a long way to go but a beginning has been made.

It is also pertinent to note that recent investigations by plant physiologists of particular phenomena such as photoperiodism —the effect of exposure of growing plants to different periods of light and darkness—are adding to our knowledge of the factors which determine form and structure; while their studies of growth have an essential place in the investigations under consideration.

Genes and development
All the foregoing considerations, however, only take us part of the way. Every living organism possesses a number of more or less well defined external and internal features, or characters, which become apparent during development; such as the size and shape of the leaf, the colour and construction of the flower, and so on. These characters are specific for each individual species and are heritable. Fundamentally they are controlled by determining agents or factors—the genes—which are located in the cell nucleus. There are good reasons for the view that these genes are particulate in nature, that they occupy definite positions in the chromosomes, and that they retain their individuality throughout the cycle of development of the individual plant or animal. For us, the basic questions are these:

How do genes operate during morphogenetic processes? How and when does each gene work so as to contribute to the orderly development characteristic of any particular species? Such questions have only recently begun to receive attention. Indeed, a relatively new branch of biological science, described as Physiological Genetics, is beginning to gain adherents. The methods of this branch of biology would, as I understand the situation, include the physiological and morphological investigation, during development, of plant materials of known genetical constitution. It would, in fact, be the meeting place of the geneticist, the biochemist, the physiologist, the mathematician, the physicist, and the morphologist—in short, it would represent a desirable return to the central aims of Botanical Science. So far, the instances that permit of a coordination of the data of these several branches are few in number. At present very little is known about the nature of the gene and its primary action, how and where it exerts its effect, whether it affects only one step in the developmental process, whether it affects a succession of steps, or whether it is involved in every aspect of development. In broad outline, the aim of Physiological Genetics will be nothing less than to establish the chain of causation between particular genes and the ultimate appearance of characters in the adult organism. In practice this will require comprehensive studies of the action of biochemical substances known to be gene-controlled, together with the other factors which affect morphogenetic processes. These investigations cannot fail to prove of considerable complexity. Nevertheless, much may be achieved in the course of the next few decades.

As to the fundamental nature of the gene, and the related problem of genic mutation, suggestions of a most exciting and far-reaching kind have been made. It is thought that the gene may consist of a very large organic molecule—it has been described as an aperiodic solid—and that when a genic mutation takes place, either spontaneously or by induction, the change is a quantum change of the kind studied by atomic physicists. As a result of mutation the properties and biochemical action of the gene may be modified to a greater or less extent, and this may be reflected in changes in the eventual form and structure of the adult.

Towards a synthesis

Perhaps at this point it may be permissible to speculate on possible developments in the next few decades, bearing in mind the fallibility of all such speculation and the probability that the performance is likely to fall far short of the expectation. If it can be shown that evolutionary changes, or some aspects of evolutionary change, are due to the cumulative effect of genic mutation in conjunction with the selective action of the environment, and if the rôle of genes in morphogenetic processes can be more adequately explored, a comprehensive biological synthesis should become possible. Here it is proper to note that some of the hereditary factors may be located in the cytoplasm. But, generally, the hereditary constitution of a species, itself a small fragment of the evolutionary picture, is conceived as being sub-divisible into genes. These genes, large organic molecules, of which the composition and structure may one day be known, are involved in all developmental processes; they find expression in chemical reactions which, together with the other factors at work during growth, directly or indirectly determine form and structure, and culminate in the distinctive appearance of the adult. These large organic molecules are apparently subject to mutational changes of a kind that are of present interest to the organic chemist and molecular physicist, and may one day be more fully investigated by them. If now, in a particular plant species, mutations appear or are induced in one or more genes or in the cytoplasm, modifications in the whole chain of reactions during development may result; and an adult configuration may be produced which differs in some respect from that of the original species. In a manner that is not possible in the present state of knowledge, it may eventually be possible to explain—or at least to give some approximate indication of—what is happening at critical stages during development. The basic assumption on which the biologist works is that all the characters which we see in plants can ultimately be referred back to the hereditary constitution of the organism, to the action of mathematical and physical factors, and to factors in the environment. In practice it is impossible to separate the action of the genes from the action of the many other factors which also take part in determining the development of form and structure. Nevertheless, by observation and experiment, it

G

may in time be possible to indicate more specifically the particular kinds of morphological development that result from the action of particular factors. Some aspects of form and structure, as we have seen, are due to mathematical and physical factors and, in a sense, are extrinsic to the hereditary or genetic constitution. But others, those which we regard as being gene-controlled, are inherent or intrinsic. Ultimately, it is the developments which are gene-controlled, or are controlled by the specific hereditary substance, that are of paramount interest to the student of evolution. Hence in the course of the next few decades—for Science, like Art, is long—it may be possible to indicate in broad outline the chain of causation which begins with the mutating gene and culminates in the appearance of new morphological characters in the adult mutant plants. Morphological studies of the parent and mutant forms would be an essential part of such investigations.

The comprehensive investigations which I have indicated would necessarily relate to plants now living. But if a clearer understanding of the factors that determine form and structure in living plants can be achieved, a fuller and more adequate interpretation of the developments indicated by the fossil record may become possible, though it can never be absolute. It may even be possible to fill in some of the numerous phyletic gaps: indeed, in some instances they may not prove to be gaps in the sense hitherto understood. How the new knowledge will affect our views on the process of evolution remains to be seen: the topic is clearly one of the greatest importance and interest to biologists.

6
Phyllotaxis and organogenesis in ferns

From experimental studies of phyllotaxis in flowering plants, M. and R. Snow[1] concluded that the next leaf primordium to be formed will arise in the 'next available space' on the apical meristem, that is, the first space or gap between existing primordia that attains a certain minimum width and which is situated a certain distance below the extreme tip of the shoot. They showed that the position of the new primordium within a gap can be modified by experimental treatment and that it is determined by those primordia which abut on the gap and not by *all* the primordia of the top cycle. In comparable experiments with *Dryopteris dilatata* Druce, I obtained evidence which substantially supports these findings.[2] In ferns, an evident feature is that the newly formed primordium occupies only a small part of the interfoliar meristematic area. There is evidence that tensile stresses are induced in the fern apical meristem by the existing leaf primordia, new primordia typically arising in regions of minimal stress. Accordingly, it has been suggested that these stresses may determine, or define, the space in which the next primordium can be formed; but that other factors are probably responsible for leaf formation.[2]

In studying organ formation at the fern apex we have to account for the very curious fact that whereas leaf primordia are formed, shoot buds are not, although the rudiments of both lateral organs are histologically identical.[3] Now why should one group of cells of the apical meristem, those in a presumptive leaf position, be able to grow out and form a primordium—a leaf primordium—whereas an adjacent but closely comparable group, those in a presumptive bud position, remain inhibited in respect or organ formation? It seems clear that any adequate account of organogenesis at the fern apex must not only include leaf formation but must also include the phenomenon of bud inhibition.

A unifying conception of morphogenetic processes that is emerging from contemporary biological investigations involves a recognition of the existence of *growth-centres*. It is held that each growth-centre, by its metabolism, gives rise to a physiological field within which a new growth-centre cannot originate: but a new centre may arise in meristematic cells outside the field. In 1913, Schoute[4] put forward the view that in the process of leaf formation the leaf-centre is determined first, the primordium being organized around this centre. He also held that a specific substance is produced by leaf growth-centres which inhibits the inception of others in the immediate vicinity. Hence new primordia only arise between older ones when the space on the meristem has become sufficiently large. The new centre is typically situated at the point of intersection of equal circles drawn around the two adjacent primordia. In Schoute's hypothesis the inhibition of new growth-centres by the shoot apex was also postulated: hence the position of a new primordium well down the side of the apical cone. The substance proceeding from the apical cell was considered to be somewhat different from that produced by the leaf-centres. The conception of growth-centres and their relation to organogenesis have recently been given support by Richards,[5] and Bünning.[6] In the course of studies in which the apical meristem of *Dryopteris dilatata* was incised in various ways, certain developments gave a strong indication that Schoute's ideas, suitably modified, might provide the basis for a comprehensive conception of morphogenesis and the regulated development to be seen at the shoot apex in ferns.

Hypotheses now put forward, and sustained by a considerable body of experimental data, are as follows: (1) The apical cell of the shoot (together with its adjacent segments), as also each young leaf primordium, constitutes a growth-centre with a surrounding physiological field; new growth-centres can only originate in regions of the apical meristem which lie outside the existing physiological fields. (2) No fundamental metabolic differences exist between shoot and leaf apices; but there may

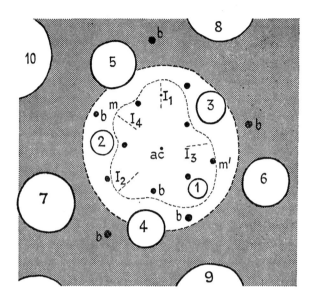

1 Diagrammatic representation of the apex of *D. dilatata* as seen from above. 1, 2, 3, etc., leaf primordia in order of increasing age: I_1, I_2, etc., the next primordia to be formed, in the order of their appearance; *ac*, apical cell; the circular broken line indicates approximately the base of the conical apical region.

be differences in the extent (or intensity) of their fields. (3) The physiological field round a growth-centre is established at an early stage; that is, a field develops round a leaf primordium contemporaneously with or very soon after its inception.

The application of these ideas to the fern apex is illustrated in figs. 1 and 2. The positions of the existing leaf primordia (P) (1, 2, 3, etc., in order of increasing age) and of primordia yet to be formed (I_1, I_2, I_3, in the order of their appearance) are shown. Approximate bud positions (b) are indicated in fig. 1. These are situated some distance above the axils of the leaf primordia; they may also be described as occupying lateral or *interfoliar* positions. It will be seen that in respect of their distance from the apical cell, some bud positions are closely comparable with those of the more recently formed leaf primordia. Nevertheless, bud rudiments remain inhibited whereas leaf primordia are formed. Figure 2 illustrates the

growth-centre hypothesis as it may apply to the fern apex,
physiological fields being shown in relation to the apical cell and
the adjacent leaf primordia (P_1–P_5). Needless to say, this
rendering is simplified and arbitrary: it will at once be evident
that many different representations are possible. But, as investigations progress, it may be possible to indicate with some
degree of precision the shapes, sizes and physiological properties
of the several fields. For the time being, the assumptions are
made that each leaf primordium develops its effective field at an
early stage and that the field of the apical cell extends well
down the flanks of the cone. As shown in the diagram, the next
area of the apical meristem—the 'first available space' of M.
and R. Snow—which is free from the inhibitory effects of
adjacent growth-centres and in which a primordium could
develop, is the position normally occupied by I_1; the next space
to become available after that is that normally occupied by I_2;
and so on. Bud positions, on this hypothesis, are subject to

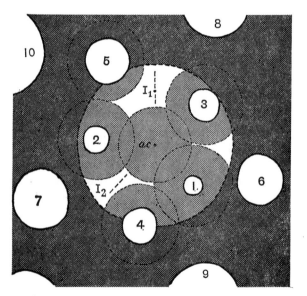

2 Diagram as in fig. 1, the shoot apex and each leaf primordium of the top
cycle being regarded as a growth-centre, with a surrounding physiological
field in which new growth-centres cannot arise. I_1 is the first space on the
apical meristem which is free from this inhibition and therefore becomes the
site of a leaf primordium.

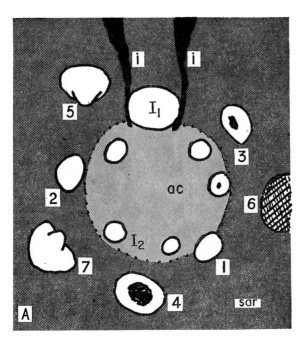

3a An apex as seen from above, in which the I_1 site was isolated from P_3 and P_5 by deep radial incisions (i, i). The new primordium is larger than any of the others in the apical cell region (tracing from photograph).

inhibition by the young leaf primordia, that is, the inhibition of buds *at the apex* is not solely due to the shoot apex.

The hypotheses indicated above can be subjected to experimental tests: it should be possible, by the partial isolation of particular regions, to observe whether or not inhibitional effects are, in fact, associated with the inception and growth of primordia. Some typical experiments and the inferences drawn from them are briefly indicated below.

When the apices of the leaf primordia of the top cycle (P_1-P_5) were punctured or incised, the new primordia I_1-I_4 were formed in their normal positions. From this it is inferred that fields are established contemporaneously with, or soon after, the inception of primordia.

When the position of the next primordium to be formed, that is, I_1 (situated between P_3 and P_5), is isolated by two radial incisions, the primordium formed at I_1 soon becomes larger than the older primordia, P_1, P_2 and P_3 (fig. 3). It is inferred that the inhibitive or regulative effects of P_3 and P_5, the adjacent primordia, have been precluded by the incisions. Comparable results were obtained when P_1, P_2 or P_3 were similarly isolated. (In these and related experiments the effect of wound hormones has still to be evaluated; the evidence suggests that such effects are localized near the wounds.)

When radial incisions are made through the positions I_1 and I_2, the next primordium to be formed, I_3, arose in its normal position; but I_4 was displaced towards the I_1 incision, and I_5 towards the I_2 incision. Moreover, I_4 was soon larger than I_3, and I_5 as large as I_3. These displacements of I_4 and I_5 and their large relative size are considered to result from the absence of physiological fields at I_1 and I_2.

In the normal development, leaf primordia are formed on the sides of flanks of the apical cone. As the growth-rate is considerably greater on the abaxial than the adaxial side and as the latter may also be subject to inhibitive effects proceeding from the shoot apex the dorsiventral symmetry of foliar members

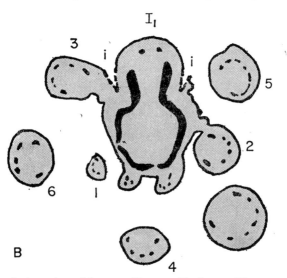

3b A section of the apex illustrated in fig. 3a. The new primordium which has appeared at I_1 has grown rapidly and has become considerably larger than some of the older primordia.

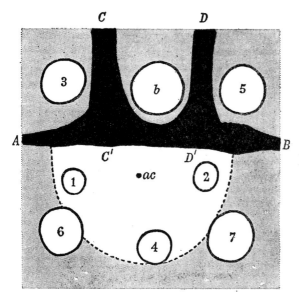

4 An apex in which the I_1 position was isolated from the shoot apex by a tangential incision, A–B, and from P_3 and P_5 by radial incisions, CC', DD'. A large solenostelic bud (b) has been formed in the presumptive leaf position.

seems a natural consequence. Buds, on the other hand, typically arise in interfoliar positions in the sub-apical or adult regions of the shoot. When the position I_1 was isolated from the apical meristem by a tangential incision, or by a tangential and two radial incisions, a large solenostelic *bud* developed in a majority of the experimental materials; that is, a bud was induced in a presumptive leaf position (fig. 4). In such experiments inhibitive effects from the shoot apex on I_1 will have been precluded; there also appears to be a levelling up of the growth-rates on the abaxial and adaxial sides. (Very occasionally, a leaf primordium was formed at I_1; in some instances a leaf and a bud appeared to have been formed contemporaneously.)

Perhaps the most interesting experiments were those based on the following considerations. In ferns, buds, which may be described as axillary, are actually situated on the shoot a little above the leaf axil. They are more aptly described as being interfoliar. Bud positions occur on the meristem at about the same level as the top cycle of leaf primordia (fig. 1). Bud formation in

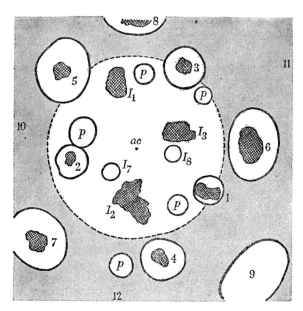

5 The apices of P_1–P_5 and positions I_1–I_3 have been punctured (cross-hatched). Interfoliar leaf primordia (p) (and axillary to the older primordia P_8–P_{12}) have been formed in what are normally bud positions. No new primordia have arisen in the I_4, I_5 or I_6 positions, but two new primordia, above the cycle P_1–P_5 and labelled I_7 and I_8, are in approximately normal positions for these primordia. *ac*, apical cell.

these positions is, however, inhibited in the normal development. (These inhibited bud rudiments, that is, groups of meristematic cells, become distributed in characteristic positions along the shoot during growth. They have been described as detached meristems; and will give rise to shoot buds when the apex of the shoot is destroyed or inactivated.) On the hypotheses under consideration, the inhibition of buds at the apical meristem is attributed partly to the shoot apex and partly to the adjacent leaf primordia. It was, therefore, argued that if (*a*) the apices of P_1–P_5 were punctured, and (*b*) positions I_1–I_4 were punctured or incised so that the development of their physiological fields would be precluded, conditions would have been established which would admit of the formation of lateral organs in the normal bud positions at the apex. In a considerable series of experiments along these lines, *leaf primordia* have been induced in positions between the existing leaf primordia, that is, in what are normally bud positions. These *experimentally induced leaf primordia* were typically basal or lateral to P_1–P_5 and axillary to the older primordia P_6–P_{10} (fig. 5). The leaf primordia which were formed on the further growth of the apex were in approximately normal phyllotactic sequence with the induced primordia.

It is concluded that these experimental data are compatible with the following views: (1) the positions in which lateral members arise on the apical meristem, and the regulated development of the leafy shoot, can be referred to growth-centres and their physiological fields; (2) the apical meristem is totipotent, the essential differences in the morphology of leaves and buds being due to the positions in which they are formed; that is, leaves and lateral shoots, viewed as distinctive organs, are not specifically pre-determined either by the hereditary constitution of the organism or by the pre-existence in the race or organs of different fundamental categories; (3) with certain qualifications, the shoot apex is a self-determining region.

7
'The Natural Philosophy of Plant Form'

This is a treatise on plant morphology, written by an expert. It deals in particular with the parts, or organs, of flowering plants, the relation of these parts to one another, and the kinds of interpretation that have at one time or another been put on them.

In a short review it will not be possible to do justice to this interesting and important book. Into some two hundred pages there is compressed a survey of relevant aspects of the history of botany, a considerable body of botanical fact (illustrated), a full statement of the conclusions at which the author has arrived after a lifetime of work as a plant morphologist and, not least, a great deal of thought. It is necessary to read this book carefully and, to get the full good of it, sympathetically. It is quite easy to see that to some contemporary botanists, preoccupied as they are with specialized advanced lines of research, the book may make no appeal at all, or even rouse antipathy. Theirs, however, will be the loss: to any botanist a careful perusal of this work will bring its own reward, for it offers accurate and refreshing glimpses of the development of botanical science and thought from ancient times, it conveys a definite point of view, it is provocative of thought and ideas and, not least, it is a work of scholarship and philosophy of a kind that botanical science has not been overburdened with in recent years.

As the title indicates, the author is concerned with the form or configuration of plants, what it is, how it is to be comprehended or conceived, how botanists, of different periods, have regarded

Review of *The Natural Philosophy of Plant Form*, by Agnes Arber, M.A., D.Sc., F.R.S., F.L.S. [Pp. xiv + 247, with frontispiece and 46 figures.] Cambridge Univ. Press. *Science Progress*, 1950.

the problems of plant form, and how a student with the necessary philosophic outlook and equipment would regard the several ideas and conceptions which have emerged. Morphology, the science of form, is not necessarily a narrow and restricted discipline: on the contrary Dr. Arber, following the best of the ancients, from Aristotle onwards (whose ideas, in different chapters, she discusses in some detail), looks on *form* in its most extensive connotation, in the fullest and most comprehensive sense that can be attributed to it, and by so doing she shows that it affords a means of connecting many aspects of botanical thought into a coherent whole. She emphasizes the value of concentrating, with proper detachment, on the problems of form itself, not, for example, solely for comparative purposes as did so many of the post-Darwinian phyletic botanists. In this connection she shows the value of a reconsideration of the findings and ideas of some of the pioneer workers, for, as she points out, the number of occasions on which the science of plant morphology has received the impetus of quite new ideas has, after all, been limited; therefore we stand to profit by a close re-reading of those writers who were capable of generating new conceptions.

The earlier chapters, devoted to the meaning and content of plant morphology—and by this is intended the fullness of meaning and content—and to the work of selected biological writers from Aristotle to Goethe and de Candolle, are the means of placing before the reader the general outlook and attitude of mind which the reviewer has tried to suggest. The author then goes on to consider the concept of organization type, the partial-shoot theory of the leaf, and the mechanism of plant morphology and its iterpretation. Readers who are unfamiliar with Dr. Arber's scientific papers may well be a little startled when they see a chapter with the admittedly curious heading: 'The Urge to Whole-shoothood in the Leaf.' They may even be shaken by 'The Partial-shoot Theory of the Leaf'. But once over the initial shock they will find that the author has not merely safeguarded her position on philosophic and logical grounds, but that she has interesting facts to support the views which she advances. Sachs stated that 'the morphological conceptions of stem and leaf are correlative' and held that the expressions *stem* and *leaf* denote 'only certain relationships of the parts of a whole—the shoot', Dr. Arber has endorsed this view, and has pointed to the fact that even the most elaborate plant systems can be analysed as a 'root-and-shoot complex', these organs being 'in some sort, primary units', the shoot concept having a higher degree of validity than that of stem and leaf. In an interesting review of notions of plant organization she refers to various earlier writings—in particular, those of Aristotle, Spinosa, Schultz-Schultzenstein—in which 'self maintenance' is held to be an important principle—the *law of biological maintenance* according to Bertalanffy. In flowering plants, with their continued apical growth, the urge to self-continuance is seen in the repetitive production of branches each resembling the parent shoot: or, as Dr. Arber says, 'The whole plant may be said to consist of a series of shoot generations, together with a series of root generations'.

Dr. Arber has attempted to develop the hypothesis of the shoot as a unit in such a way as to give an adequate account of the relation of leaf and shoot and of root and shoot. In so doing she has criticized the acceptance by Troll of root, stem and leaf as three unanalysable categories but supports the idea of Kant and de Candolle that 'the leaf is a partial-shoot, arising laterally from a parent whole-shoot'.

In discussing the most evident difference between a leaf and a stem, i.e. the dorsiventral symmetry of the one as compared with the radial symmetry of the other, Dr. Arber points out that the difference may be one of degree rather than of kind, many radial shoots being not completely radial in their symmetry and with a trend towards dorsiventrality, just as some leaves may tend towards radial symmetry. She also cites evidence of various kinds which can be interpreted as upholding the view that leaves show a tendency towards whole-shoot characters.

Her view that a leaf is a partial-shoot tending to whole-shoot characters has no phylogenetic implications, for it is concerned, not with the historic process of the evolution of the leaf, but with what it is as it is formed in the individual plants.

In the final chapter Dr. Arber deals in masterly fashion with the difficult problem of 'cause' in morphology, and the reader is led into those regions of thought where natural philosophy and metaphysics converge—regions in which the contemporary investigator would do well to pause and ponder.

8
Organization in plants

An organism is an organization, i.e. an organized body or structure. (By way of example, Locke is cited in the *Oxford Dictionary* as follows: 'That being then one Plant, which has such an Organization of parts in one coherent Body.') Organization, then, is something that relates to the essential wholeness of an individual plant, and, by extension to any of its component parts. When we speak of a plant as an organization, evidence of that integrated unity is apparent not only in the adult form but at all stages of development from the zygote or spore. Moreover, the orderly succession of phases during which the organization becomes manifest, are distinctive and specific, so that, from quite an early stage, the plant can be recognized as belonging to a particular genus or even species. A central problem (perhaps *the* central problem) in biology is to account for this specificity and constancy of form and structure. What conceptions have we that do justice to the remarkable phenomenon of organization? How is it that the developmental pattern is reproduced, generation after generation, with such fidelity, each species according to its kind?

Contemplation of this problem led Driesch (1908) to conclude that purely mechanistic conceptions of life do not adequately account for the distinctive organization that characterizes the development of any particular species. He, therefore, introduced the idea of a controlling or ordering principle—an entelechy—which was independent of physico-chemical laws, though these were operative in living systems. Smuts (1921), too, thought that purely mechanical explanations fail to do justice to the harmonious, unified construction of living organisms; certainly, he admitted, mechanistic concepts do have their place and justification, but only within the wider framework of the integrated unity of the organism. Hence his theory of Holism, holism being defined as the fundamental factor which works towards the creation of wholes in the universe. Smuts regarded it as a causal factor with a real existence; and the higher the level of evolutionary development of the organism, the greater will be the manifestation of its state of integrated wholeness or organization.

Conceptions such as those of Driesch and Smuts are undoubtedly of value in reminding us of this universal phenomenon in living organisms. Whether, in their present form, they are either valid or necessary is clearly a matter that deserved careful consideration; for a full and adequate comprehension or organization is the central aim in studies of morphogenesis and morphology.

In the life of any organism, be it simple or complex, the inescapable but always amazing and enigmatic feature is the orderly manner in which it develops, each and every stage or phase testifying to the fact that growth and the assumption of form do not take place in a haphazard manner but, on the contrary, are evidently harmoniously regulated processes in which the essential unity and equilibrium of the whole is maintained. Plants at large appear to exemplify a great many kinds of organization, e.g. a gasteromycete as compared with a seaweed or conifer. Yet, in major classes which, on grounds of comparison, are not held to be closely related, comparable organizational features may be observed: for example, a great many plants including algae, bryophytes and higher plants, show polarity from the outset, have an apical growing point and an axial development. The Vegetable Kingdom, in fact, yields so many remarkable examples of what Goebel (1900) and Lang (1915) have referred to as *homologies of organization* that we are led to inquire whether this seemingly endless diversity of form and structure cannot, in fact, be referred to a comparatively small number of 'organizational types'.

Observations of this kind have affected botanical ideas in two ways. In the post-Darwinian Phyletic Period, comparable developments in different groups were accepted as an indication of hereditary relationship; and it was confidently asserted that all the systematic groups of the Plant Kingdom could be assembled into a monophyletic 'tree' or pedigree. Later, when the prevalence of closely parallel developments in widely separated groups had been incontrovertibly demonstrated, it was recognized that a study of the factors determining these homologies of organization was a major task, not only for the morphologist, but for botanists in general. Thus, whether our interest be in phyletic or causal morphology, the study of organization must rank as one of our principal aims. Our approach should be objective and cautious, with a full measure of that suspension of judgement which T. H. Huxley enjoined; and since the organization of a species is the summation of all that has gone to the making of it, embracing every aspect of its constitution and development, so must the approach of the botanist to the problem be a multi-aspect one, and he must draw as required on all the branches of his science. It may be that the day will come, as Hersch (1941) has suggested, when we shall think of a growing organism, not as a visible structural pattern or configuration, changing from stage to stage, but in terms of the non-picturable, i.e. a system of equations expressing all the rates, relationships and events which produce the visible structure. However, some time is likely to elapse before the morphologist becomes 'relation-minded', in the rarified degree implied above. In the meantime, in conjunction with the physiologist, the geneticist, the physicist and the mathematician, the morphologist has a rich and varied field in the study of morphogenesis. These comprehensive studies should lead to, and culminate in, the formulation of conceptions of organization which do justice, not only to what we observe in the individual species, but to those homologies of organization which are so prevalent in the Plant Kingdom as a whole.

Theories of organization

Various 'explanations' or partial explanations, of organization in plants have at one time or another been attempted. To some extent all of these accounts are of a multi-aspect nature, though often one particular aspect tends to be emphasized. Accounts of organization are of the following kinds: (*a*) morphological, (*b*) physiological, (*c*) physical and mathematical, (*d*) genetical and (*e*) epigenetic, multi-aspect or integrative. Considerations of space do not permit of a full review of these several accounts, but a few comments may be offered; these will be restricted to vascular plants.

(a) *Morphological Theories*   Under this heading should be included Göethe's theory of metamorphosis, Sach's theory of the fundamental categories of parts, phytonic and axial theories of shoot construction, and the telome theory of Zimmermann. These all relate to the organization of the leafy shoot. In Göethe's theory, the shoot is little more than an axis on which the really important organs, the leaves, in their several manifestations, are disposed. Sachs's fundamental categories included caulome, phyllome, trichome and rhizome: these were taken for granted but were held to be subject to modification in various ways. Sporangia were not included among the fundamental categories: they were held to be the result of a transformation of one or other of the categories. In phytonic views, the existence of the shoot as an independent member is more or less explicitly denied, the plant being envisaged as a construction of phytons, or foliar units, the extended bases of which become conjoined to form the stem or axis. By contrast, supporters of axial theories hold that the shoot was 'phyletically a pre-existing axis or stem from which the leaves may have arisen by enation' (Lang, 1915). In his later writings Sachs, followed by Bower (1935), held that the leafy-shoot was the real unit, axis and leaves acting together as a whole. Yet Bower also held that the prototype of vascular plants consisted of a simple, vascularized axis, without leaves or roots, and that the plant body became progressively more elaborate in construction as a result of dichotomy; and this may in turn have passed over into monopodial branching, thus giving lateral appendages. Megaphylls were held to be of cladode origin, whereas microphylls were no more than enations or small outgrowths from a shoot surface 'not previously tenanted'.

These several theories or conceptions were almost entirely morphological both in inception and outlook. In each, the so-called fundamental unit, whether axis or phyton, was taken for granted, as a unit ready-made. To the contemporary experimental morphologist, whose aim is to understand morphogenetic processes and the factors which determine organization in plants, these theories not only seem artificial, but they take for granted precisely those features which stand most in need of investigation by every method at our command.

The hypothesis of Arber (1950), that the leaf is a partial shoot, is also in essence a morphological theory, but space does not permit of consideration of it here.

(b) *Physiological theories*   Sachs's theory of chemical correlation has reached its heyday in contemporary botanical and biochemical studies. A knowledge of the formation, transport, and action of growth-regulating substances and other special metabolites, the inception of growth centres and physiological fields, the setting up of diffusion gradients, and other physiological processes, all contribute to our understanding of the growth and morphological development of plants. But do the facts of physiology and biochemistry really tell us how the distinctive form and structure and the specific organization are brought about?

No botanist now doubts that certain specific organic substances, e.g. auxin, are of great importance in morphogenetic processes: a vast fund of observation testifies to the truth of this statement, while the general processes of growth are involved in every morphogenetic situation. Still, we may ask if biochemical concepts alone will enable us to give an adequate account of organization. Sinnott (1946), while admitting to the full the importance of physiological and biochemical data, has urged caution in an all-out acceptance of any biochemical theory of organization. As a basis for research its value is not in doubt; but as a *full* explanation of the facts it proves insufficient on close analysis: a knowledge of the successive biochemical changes during development will not explain the assumption of form or the underlying organization that pervades the whole of organic development. The reaction induced by a growth-regulating substance depends very largely on the state of the affected cell or tissue, i.e. on its competence to react. Hence Boysen Jensen has referred to such substances as 'realizers'. The actual morphogenetic effect produced is, as it seems, due to the underlying organization, not to the activating substance. Nevertheless, it is evident that a knowledge of the specific effects of growth substances, together with theories of the inception of growth centres and their surrounding physiological fields (Weiss, 1939; Wardlaw, 1949), and the gradient concept or organization as expounded by Child (1941) and Prat (1945), make an important contribution to our understanding of the growth, configuration and regulated development of vascular plants.

(c) *Physical and Mathematical Theories*   The best known of these theories have been fully discussed in D'Arcy Thompson's *Growth and Form* (1917, 1942). Already in the nineteenth century, Naegeli, Hofmeister and Sachs had instituted causal inquiries and had attempted to interpret form and structure in terms of physical and mathematical laws. With many fascinating examples, D'Arcy Thompson has shown how a knowledge of physics and mathematics may contribute to our understanding of the development of form and structure. The growth of every organism is characterized by an accumulation of material arranged in a particular way; and it is 'in obedience to the laws of physics that their particles have been moved, moulded and conformed'. In this view, the problems of form in plants are mathematical and physical problems: the shape of any portion of matter, and changes in its shape, are due to the forces acting on it. These forces include gravity, cohesion, surface tension, mechanical pressure, molecular diffusion, and chemical, electrical and thermal forces. The phenomenon of differential growth, the size-structure correlation, and cell division by minimal surfaces, may be cited as examples where mathematical conceptions can be used in the analysis of organic form. Some theories of phyllotaxis are essentially mathematical in conception, while Vöchting's theory of cell differentiation relates directly to spatial considerations.

Mention should here be made of the comprehensive electro-dynamic theory of life as expounded by Northrop and Burr (1937). They point out that the problem of organization has always been a crucial difficulty in bringing biology and physics into some acceptable relationship. In their view the introduction of qualitative and purely biological concepts has obscured the relationship between the physico-chemical constituents and their organization in living things. Biological notions, they say, have failed and they have proposed the theory that living organisms are complex electrical fields with a definite pattern of potential distribution as a whole.

(d) *Genetical theories*   According to modern genetical theory, hereditary factors pervade, determine and control each and every phase and aspect of development. The units are the genes —giant protein molecules, or molecular aggregates, capable of self-reproduction and of mutation—which are arranged in a particular manner in the chromosomes, while other kinds of genes may be present in the cytoplasm; they are combined and redistributed at syngamy and meiosis and their mode of action is chemical. It is now known that certain important metabolic substances, e.g. auxins and enzymes, or their precursors, are gene-controlled. The action of certain genes may be evoked at particular times by the physiological condition of the cell, this having been determined by the action of other genes under the particular environmental conditions. Writers such as Sewell Wright (1945), Mather (1948), and Waddington (1948), have suggested schemes showing how morphogenetic processes may be determined and controlled by genic action. The details are complex, and, in the present state of knowledge, necessarily highly speculative. Yet the need for some such conceptions becomes evident if we ask the negative question: Can we think of any morphogenetic process in which genetical factors are not in some way involved? A very little reflection will show that, in every consideration of morphogenesis and specific organization, assumptions concerning the hereditary constitution are unavoidable.

Physiological genetics ranks as one of the most recent branches of botany and many of our views on genic action are necessarily speculative. And just as a knowledge of the chemical components can tell us little about the organization which is produced, so also the action of genes must be taken in conjunction with factors of other kinds. Since the aim in physiological genetics is to combine two basic aspects, the action of units of the genetical system and the biochemistry of growing cells, we may expect that, in time, adherents of this discipline will make important contributions to our knowledge of organization.

(e) *Epigenetic, Multi-aspect and Integrative Theories* It may seem a little old-fashioned to speak of an epigenetic theory of plant organization, but the term is a useful one. Epigenesis connotes the progressive development of an organism on a mechanistic basis, each stage being determined by those that have gone before; i.e. any suggestion of development according to a plan is inadmissible. Broadly speaking, epigenetic theories differ from the morphological theories set out under (*a*) above, in that all conceptions based on preformation are ruled out, as are also those in which any unit of construction—phyton or axis—is taken for granted. In an epigenetic theory nothing is taken for granted but the hereditary constitution of the zygote, spore or bud rudiment and its capacity for growth. Thus, if we begin with the fertilized ovum of a fern or a flowering plant, the aim is nothing less than to give an account, stage by stage, of the formation of the adult plant with all its characteristic features. Factors in the genetical constitution determine and limit the potentialities for development of the zygote or spore, and these, together with the many other factors which become incident, account for the changing form and structure during the individual development until finally the full expression of the hereditary constitution in the particular environment becomes manifest. Some such view was indicated two hundred years ago when Kaspar Wolff first announced his discovery of the apical growing point—a view which is probably widely accepted by experimental botanists at the present time.

More than most other writers, Needham (1942) has shown how comprehensive must be the system of ideas relating to the organization of plants and animals. In his view, the universe may be perceived as a series of levels of organization and complexity, ranging from the sub-atomic level to that of living organisms. The laws which apply to one level cannot be expected to apply to other levels: each, in fact, requires its own appropriate concepts. There must, therefore, be a simultaneous pursuit of knowledge at all the levels of complexity, and an attempt made to ascertain to what extent the data of one level are related to those of another. The final aim must be to bring the organization perceived at the several levels into some integrated relationship.

An integrative conception of morphogenesis and organization

In the scheme set out below the writer has tried to give his conception of the main features of the morphogenetic processes which culminate in the organization of the adult vascular plant. It is recognized that organization connotes the relation of the constituent organic particles to one another, of the cells and organs to the whole, and of the whole to the parts.

In this view, while genetical factors enter into every phase of development, the organization is the result of the action and interaction of many factors, both intrinsic and extrinsic, and the whole process is essentially one of epigenesis. The sheer

Generalized scheme of morphogenesis

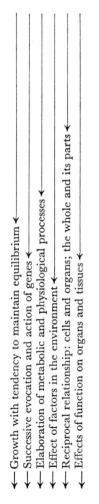

— Growth with tendency to maintain equilibrium →
— Successive evocation and action of genes →
— Elaboration of metabolic and physiological processes →
— Effect of factors in the environment →
— Reciprocal relationship: cells and organs; the whole and its parts →
— Effects of function on organs and tissues →

a Zygote or spore with sub-microscopic organization and a specific genetical constitution (comprising nuclear and cytoplasmic genes) which controls metabolic processes in the cytoplasmic substrate.

b Under suitable environmental conditions gene-controlled enzymes, auxins, etc., become available and active; nutrients are taken up, or mobilized from reserves in the zygote; growth begins.

c The polarity of the enlarging embryo is determined by inherent factors, or by factors in the environment, the initial partition wall being at right angles to the axis.

d On further growth, the dispositions of the succeeding partition walls are in general conformity with the principle of cell division by minimal areas.

c^1d^1 Concomitantly with **c** and **d,** a differential utilization of metabolites at the apical and basal poles is established, the former becoming the locus of active protein metabolism and meristematic activity.

e Increase in size brings changes in spatial relationships; there is a decrease in the ratio of surface to volume, a separation of parts and the distinction of superficial and inner tissues.

e^1 Concomitantly with **c, d** and **e,** biophysical and physical factors become incident and have a large share in determining form and structure; also, growth centres and their physiological fields are established and diffusion gradients set up; and these have a large share in determining the integrated, regulated or harmonious development of the organism as a whole.

f The distal apex continues as a self-determining morphogenetic region, the tissues to which it gives rise becoming the mature, rigid tissues of the axis.

g The *Organization*—the regulated inception and development of form and structure—characteristic of the species becomes manifest as development proceeds.

regularity with which the developmental pattern is reproduced, stage by stage, is perhaps the most remarkable phenomenon of growth and the most difficult to understand. At least a partial explanation would appear to be that not only one, but several kinds of factors, are simultaneously contributing to this orderly development. Thus, geneticists have suggested how a regulated morphogenetic process could result from the orderly sequence in which genetical factors are evoked and become active. Physiologists have now obtained much experimental evidence relating to the inception of growth centres and their physiological fields, to the setting up of gradients and to the phenomenon of correlation, each of which contributes to the orderly, regulated development of an organism. To the physicist, all growth, involving increase in size and change in form are, in the first instance, physical phenomena, and as such must take place in conformity with the laws of physics and mathematics. Like all physical systems an organism, during its growth, will constantly tend towards equilibrium. Not least, factors in the environment exercise both immediate morphogenetic effects, i.e. during the individual development, and selective effects during the development of the race. Each and all of these kinds of factors are exercising something in the nature of a controlling or regulative effect throughout the individual development; and with this in mind, although the growth of an organism to its characteristic size, form and structure, is indeed one of the great mysteries, we begin to see that it would be an even greater mystery if the organism failed to do so.

Homologies of organization
Any plant organization is the result of a complex nexus of factors. Where two plants, which are in no way closely related, show closely comparable morphological or structural developments, we say that they exemplify parallel development or homology of organization. These homologies of organization are surprisingly numerous in the Plant Kingdom. Moreover, they are often very striking: e.g. the spindle-like embryo, with distinction of apex and base, in practically all classes of plants; growth by an apical growing point and axial development; alternation of generations; sexuality with a trend towards oögamous reproduction, and so on. To what factors can such

homologies of organization be attributed? When homologies occur in two organisms which are evidently related, the simple answer to this question would be that there are genes or groups of genes common to the two species. But where no such close systematic relationship exists, is the parallel development to be attributed to the evolution of parallel genetic systems, or to the evolution of comparable but different genetical systems which produce closely comparable morphological effects? It is evident that such conceptions raise new problems, e.g. as to why such parallelism should occur; they also have the effect of transferring the problem from the morphological to the genetical sphere rather than of affording a solution of it. Certainly, these are questions to which answers should be sought. It has been said that in organic evolution the changes which occur are those which, on a basis of probability, are most likely to occur. In some such fashion it may be possible to account for the evolution of parallel or convergent genetical systems.

It is, however, conceivable that some homologies of organization are due more to extrinsic than to genetical factors. As we have seen, many important factors in morphogenesis are essentially extrinsic in character, though always the substrate on which they work is a specific living substance. In vascular plants the determination of polarity, axial development, the inception of growth centres and gradients, the development of the histological pattern, the distribution of mechanical tissue, etc. and the incidence of yet other factors consequent on increase in size and changing spatial relationships, are phenomena which undoubtedly have a genetical aspect or basis, but in which the action of extrinsic factors is predominant. To some observers it may well appear that it is to extrinsic factors and relationships, rather than to genetical factors, that some homologies of organization are due. In other words, although genic action pervades every phase of development, it may be less important in determining the more evident morphological developments than is sometimes assumed. But these are weighty matters calling for a due suspension of judgement: much more needs to be done before a statement of definitive conclusions should be attempted. Nevertheless, the tentative ideas set out above are of value in that they stimulate thought and prepare the way for the assembling of new evidence.

9
Comparative morphogenesis in pteridophytes and spermatophytes by experimental methods

The contemporary interest in morphogenesis has resulted in new investigations into the constitution, growth and formative activity of the apical meristem in all classes of plants. Thus the past two decades have seen the publication of many papers in which the shoot apex has been re-examined and elegantly illustrated by modern techniques. It can be said with truth that many of these studies have been largely anatomical and histological in inception and outlook. An important result has been to call attention once again to a fact that has long been known and much discussed, namely, that the shoot apex in different classes of vascular plants shows great histological diversity. What greater contrast could there be than that shown by the apex of a leptosporangiate fern, with its conspicuous apical cell and large prism-shaped superficial meristematic cells, the bulky and complex apex of a cycad, and the precisely zoned but relatively small apex of many flowering plants? The more the anatomist has widened his knowledge of apical construction, the more complex his problem—to bring all shoot apices into some general scheme—has become. He finds, for example, that descriptive terms that seem appropriate for the apex of a fern are inadequate for the zoned or layered apex of a flowering plant; common ground for a common terminology has yet to be found. Yet, as formative regions, all shoot apices have evidently much in common, i.e. they occupy a distal position and give rise to an axis with regularly arranged lateral members. That all shoot apices have essential features in common can

Paper contributed to the Seventh International Botanical Congress, Stockholm, July 1950, and printed in *The New Phytologist*, Vol. 50, No. 1, May 1951.

scarcely be doubted. The problem is thus to make meaningful and to reduce to a few general concepts the diverse histological patterns which they show. Nor is it only in the histology of the shoot tip that such problems are encountered; comparable problems present themselves when we try to give a generalized account of the inception and development of the vascular system in the several classes of vascular plants. The details are seemingly so very different in different plants. Thus nearly two hundred years after Wolff first recognized the *punctum vegetationis*, botanists have still to give a full and adequate *general* account of the configuration and morphogenetic activity of that enigmatic but all-important region.

The relevant histological data for a large number of plants have now been ascertained. Among vascular plants at large we can observe those with: (i) a single, conspicuous, superficial apical or initial cell; (ii) several superficial but rather less conspicuous initial cells; (iii) inconspicuous but definitely superficial initial cells; (iv) a weakly zoned or layered construction; (v) a fluctuating layered construction; (vi) a highly definite layered construction; and (vii) a radiating construction apparently originating from a central, subsurface group of mother cells. Our task is to understand this diversity of construction, to discover the factors that determine the configuration in each type and its characteristic morphogenetic activity. *These problems cannot be solved by the methods of the anatomist alone.* To have exact anatomical knowledge of the apex is an essential preliminary step in the study of morphogenesis, and this undoubtedly depends on the work of the anatomist. But the comparative investigation of apices by strictly anatomical methods has definite limits which are soon reached. Other methods must be sought. Given the anatomical data which have now been obtained, it appears to the writer that new advances may be made along several lines including (i) experimental morphology, (ii) physiological genetics, and (iii) the use of new concepts. Some examples of each may now be briefly considered.

Experimental morphology

In a number of investigations, the same experimental treatments were applied to differently constituted apices to see if closely comparable or divergent organogenic developments would ensue.

When the shoot apex in a eusporangiate fern, *Angiopteris evecta*, and a leptosporangiate fern, *Dryopteris dilatata*, was isolated by vertical incisions, with concomitant severing of the incipient vascular tissue (or prestelar tissue, or early procambium), the apex continued to grow and gave rise to a short, vascularized, leafy shoot in which the normal anatomical pattern was soon reconstituted (Wardlaw, 1947, 1950). When the same technique was applied to the apex in *Primula* spp., which has a two-layered tunica, a similar result was obtained. In *Primula*, as in *Dryopteris*, the vascular tissue in the experimental region consisted of an uninterrupted column; and in both plants the cross-sectional outline of the vascular 'ring' was in conformity with the outline of the plug of tissue on which the apex had been isolated; i.e. a triangular or a rectangular stele could be induced at will (Wardlaw, 1950). Moreover, in *Primula* as in *Dryopteris*, the leaves formed on the new axis were in approximately normal phyllotactic sequence with those already present at the beginning of the experiment. Lastly, in both species, there was clear evidence that the inception of the new vascular tissue was due to a basipetal influence, e.g. an activating substance, proceeding from the active terminal meristem. Comparable results have been demonstrated by Ball (1948) in similarly isolated apical meristems of *Lupinus albus* L.

In the normal adult shoot of *Dryopteris dilatata*, or *D. filix-mas*, the vascular system consists of an open meshwork or dictyostele, wide parenchymatous gaps being present at each leaf-insertion. When all the very young leaf primordia were removed over a period of time, the experimental region of the shoot was found to contain an uninterrupted solenostele. The relation between the development of leaf primordia and the formation of foliar gaps in the shoot stele was thus established. Furthermore, the experiment demonstrated that the fern stele is truly cauline in origin (Wardlaw, 1944). In *Primula* spp. the vascular system consists of a ring of vascular tissue interrupted by leaf-gaps. When all the very young leaf primordia of *P. polyantha* were excised over a period of time, a continuous uninterrupted ring of vascular tissue was observed. The result was thus closely comparable with that obtained in the corresponding experiment with *Dryopteris*, and comparable conclusions seem permissible.

In the ferns, as in flowering plants, lateral buds tend to be inhibited in their formation and development by growth-regulating substances proceeding from the actively growing apical meristem; but if the apical meristem is destroyed, e.g. by puncturing, or otherwise arrested, a rapid development of lateral buds ensues in both instances. In the ferns these lateral buds originate from detached meristems or bud rudiments, these being superficial areas of prism-shaped meristematic cells which at an earlier stage formed part of the apical meristem. They become isolated from the apical meristem during the formation and growth of the leaf primordia and are left behind as the apex grows on. In a number of flowering plants in which the method of bud inception has been studied in detail, the investigators have concluded that these buds also originate from detached meristematic areas.

In the experimental investigation of phyllotaxis in flowering plants, M. & R. Snow (1931–5, 1948) have shown, (*a*) that the next leaf primordium to be formed at the apical meristem arises in the first space of a certain minimal size between and above the last formed primordia, (*b*) that the position of the new primordium within this space is determined by the two adjacent older primordia only, and not by all the primordia of the top cycle, (*c*) that the position of a new primordium can be modified at will by incising the apex in a particular position, and (*d*) that the direction of the phyllotactic spiral may be reversed by appropriate experimental treatment. In comparable experiments with *Dryopteris dilatata*, all of the above findings have been borne out.

The evidence from the experiments briefly reviewed above has led to the conclusions that, with due reservations, the apical meristem both in flowering plants and ferns may be regarded as a self-determining region, capable of forming a new axis, new lateral members, and a vascular system, without the help of the older preformed organs and tissues, provided an adequate supply of nutrients (including carbohydrates and other metabolic substances) from below is maintained.

The assumption of form is the morphological expression of physiological activity. In the light of the experimental evidence given above, it is evident that however much the apices of flowering plants and ferns may differ from each other anatomically and histologically, they show a very high degree of functional similarity, or even identity. This, perhaps, is the idea that we should have foremost in our minds rather than the histological diversity which for so long has intrigued and tantalized the anatomist. Nevertheless, since this histological diversity was the starting point of our inquiry, we cannot just pretend that it does not exist. It still remains as a problem to be tackled and resolved, and it appears that this is no more possible in terms of general physiology than it is in terms of pure anatomy. We must therefore consider if any progress can be made along other lines.

The genetical approach

It may be assumed that the distinctive arrangement of cells at the shoot apex stands in some relation to the distribution of growth in that region. On analysis, this distribution is seen to be referable to genic, physiological and physical factors and to changing spatial or geometrical relationships in an enlarging system. The action of genic factors will usually be so closely linked with the action of the other factors which become incident as to be almost inseparable from them. Nevertheless, it is important that some attempt should be made to distinguish between the several kinds of factors.

In a leptosporangiate fern, the formation of a large apical cell must in some way be due to the forces present in the growing point; and so, too, the layered construction of the flowering plant apex must be due to the forces present therein. The conclusion seems unavoidable that it is primarily to genic factors that the particular cellular pattern of an apex must be attributed. Support for this view is to be found in the considerable but scattered literature which is now becoming available.

As a general statement, for which there is abundant evidence, it may be said that the factors which primarily determine anatomical characters are inherited. A direct effect of the genic complement, or of particular genes, is seen in the character of the individual meristematic cell, e.g. in its size. In a polyploid series, for example, the cells of the apical meristem are larger in the tetraploid than in the diploid. Sinnott, Houghtaling & Blakeslee (1934) have suggested that in polyploid and heteroploid types of *Datura stramonium*, the cell volume is roughly proportional to the number of chromosomes in the nucleus. The genic effect in *Datura* was seen in an increase in cell size and cell number, in different instances, these two aspects of growth being apparently controlled independently. Cross & Johnson (1941), in a comparative study of *Vinca rosea* and a colchicine-induced tetraploid, have found that the tetraploid has a more massive stem than the diploid, thicker and greener leaves, and larger flowers. While the shoot apex in the tetraploid is generally similar to that of the diploid in histology and organographic development, it is considerably broader, this being due to an evident increase (68 per cent) in the average width of the meristematic cells. The effect of doubling the genic complement, in fact, was *to modify both the shape and size of the embryonic cells at the meristem.* In a study of the shoot apex in diploid and autotetraploid maize, Randolph, Abbe & Einset (1944) have shown that the apex of the tetraploid was larger as also were its component embryonic cells. With such changes in cell size other, extrinsic, factors are liable to be introduced into the situation, e.g. physiological effects due to changed surface-volume relationships. The studies of chimaeras by Blakeslee and his colleagues have also afforded abundant evidence of how the size of embryonic cells is changed by changes in the genetical constitution.

There is evidence that the production and inactivation of growth-regulating substances, such as indoleacetic acid, are determined by genes (vide the work of van Overbeek on maize). This conception may be extended. It may well be assumed that other important metabolic substances are likewise directly or indirectly controlled, both quantitatively and qualitatively, by specific genes or groups of genes. Adenine may be such a substance. In recent investigations Skoog & Tsui (1948) have shown that adenine promotes the formation of buds and that combinations of adenine and naphthaleneacetic acid greatly stimulate cell proliferation and enlargement of all tissues, in particular, of the pith. If, therefore, in two related species or varieties there are different, gene-determined rates of production of adenine and of auxin, it becomes understandable that the histological constitution of the two apices may be rather different.

Anatomists like Foster (1949) have realized and clearly stated that the layered apex of flowering plants is by no means as definite and invariable a structure as is sometimes thought. What we particularly want to know are the factors which determine the regular structure which we normally see in a particular apex, and what causes a departure from it. It is here that experimental data may prove of great value. Ball (1948) has shown that if indoleacetic and indolepropionic acids are applied to the shoot apex of *Tropaeolum majus* L., the densely protoplasmic cells at the extreme tip of the shoot show no reaction; but lower down, at the level of the youngest leaf primordia, very definite reactions take place. Thus a one-layered tunica was formed in place of the normal two- or three-layered tunica of this species. Since substances such as auxin are gene-controlled and subject to genic changes, the bearing of these observations on the histological constitution of the apex becomes apparent.

Studies in the field of physiological genetics thus seem likely to afford at least a partial explanation of how the meristem of one strain, hybrid or species, may differ from another in respect of cell size. We can also see that genic factors may determine the distribution of growth in the apical region as a whole. What is less evident is how the cellular pattern can be directly referred to genic action. Factors of another kind, i.e. physiological and physical factors, rather than genic factors, seem likely to be at work in producing the observed pattern. Such factors, however, become incident in relation to the action of genes. This brings us to a consideration of the relation of cells to organs and organs to cells. It is unlikely that the action of a gene, or group of genes, on organ formation is ever simple and direct. The first and direct effect of a gene will be on the physiology of the individual cell in which it is active. Metabolic changes in one cell may affect the metabolism and differentiation of adjacent cells and even of cells at a distance. Furthermore, as cells grow and divide in the formation of organs, e.g. an apical growing point or a leaf primordium, a reciprocal relationship between cells and organ will be invoked; each individual cell contributes to the construction of the organ but the mass configuration of the organ in turn affects the development of the individual cells; or as Vöchting pointed out, the development and differentiation of any particular cell depends on the position which it occupies in the developing organ. Thus a cellular reaction which can be referred initially to genic action tends to become complicated and modified by the fact that the cell is a component part of a multicellular organ. In short, genes exercise their specific effects in individual cells, interactions between cells ensue, and the reciprocal relationship between cells and organs both modifies and extends the genic effect. Once a tissue mass such as an apical growing point is formed, individual cells, or groups of cells, may be shaped and arranged by the configuration of the organ as a whole, as if, in the words of D'Arcy Thompson, a field of force were somehow involved.

If we apply these general ideas to the histologically diverse apical growing points of vascular plants, it can be seen that some of the factors determining the cellular pattern are intrinsic, i.e. genetical, factors, and that others are not. On the analysis given here the genetical factors are the primary determinants, but it is evident that much more must be known both about genic action in morphogenesis and the other factors which become incident, before we can hope to give an adequate account of the histological pattern of any single apex, or bring into some general relationship the several different types of apical construction.

Favourable materials for investigating the relation of genetic constitution to morphogenesis in higher plants are likely to be those where distinct morphological and anatomical differences accompany known differences in the genetic constitution, e.g. where character differences can be related to a single pair or a small group of genes. As the scope of such investigations widens, we may look forward to the time when the many diversely constructed apices can be brought into some general relationship.

New concepts

Notwithstanding the histological differences between the vegetative shoot apices of different classes of vascular plants, their physiological and morphogenetic activities are closely comparable. Major questions relate to the possibility of bringing the several types of shoot apex within one general conception, and to the factors involved in the production of the several histological types. Even the tunica-corpus concept of apical construction is largely formal in character and will remain of restricted value until it has been explored by the methods of physiological genetics and experimental morphology. New concepts are needed; and if this tentative essay has shown anything, it is that these must be of a comprehensive nature, since many factors are evidently involved in producing the cellular pattern at the apex.

In a recent paper, Woodger (1946) has contended that there is an urgent need for a new theoretical approach to the problem of embryonic development. He contends that although the structural plan ('Bauplan') in adult organisms of different systematic groups which are being compared may appear to be widely divergent, yet, if they were studied from a different angle, they might be seen to have a Bauplan in common. It is not inappropriate to apply this idea to the embryonic region at the shoot apex; and, following Woodger's argument, it would be essential to have an adequate theory of cell organization. Some of the elements on which such a theory might be based have been touched on here. Given such a theory and some knowledge of the factors at work in morphogenesis, it should become possible to ascertain to what extent the Bauplan determining one type of apical construction is comparable with that determining another type. The situation is perhaps not unlike that which prevailed when the vascular structure of ferns first attracted the attention of plant anatomists. Eventually, as we now know, a unified conception of stelar development emerged from what, at one time, had seemed to be a mass of complex and unrelated data. That some such general considerations may be found to apply to the shoot apices of vascular plants is the view now advanced—a view of which the merits and demerits will become apparent as new facts are obtained.

A knowledge of the *distribution of growth* is of primary importance in any comprehensive study of the shoot apex. Contemplation and active investigation of this phenomenon may well lead to new and valuable concepts. If one observes the apical and subapical regions of the adult shoot in different species, or during the ontogeny of a single species, a considerable diversity is found in the relative development of cortex, pith and vascular tissue. Thus there are shoots with a narrow cortex and wide pith, with a wide cortex and narrow pith, with no pith at all, and so on. These tissue patterns are mainly referable to the distribution of growth at the apex. So far we have little information on this subject, but by exploring certain assumptions some interesting possibilities are revealed. If, for example, there is rapid relative growth in the central tissue, i.e. the pith, then the more peripheral tissues will tend to become tangentially stretched and to be arranged in a regular layered or zoned pattern with a preponderance of anticlinal divisions. If there is rapid relative growth in the peripheral tissue the pith will tend to develop under reduced pressure and this may be reflected in the eventual large size of its cells. Rapid relative growth midway between the centre and the circumference of the shoot will have effects on the development of both pith and the outermost tissues. The shoot of *Selaginella* affords an example of rapid relative growth in the inner cortex, the result being seen in the cavity or lacuna formed at the junction of stele and cortex. In different species of vascular plants there may be considerable differences in the region of the shoot apex in which the most active growth is localized, and this may have effects on the direction of cell division, on the general cellular pattern at the apex and on the shapes and sizes of differentiating cells. Moreover, since in the development of a conical or obconical structure, such as a fern shoot, every unit added to the radius requires the addition of 6·28 units at the circumference, the need for a close analysis of the distribution of growth becomes still more apparent.

The forces at work in the apex, to which reference has already been made, most probably arise in relation to the distribution of growth. To locate the region of most active growth and to explain why it should be so, are problems that might well engage the attention of the physiologist. Two sets of factors may be involved: (i) genetical factors, which will determine cell size, production of growth-regulating substances, etc., and (ii) effects proceeding from the older region of the shoot. At the level where the vascular tissue fades out some distance below the apex, how are the materials which are being upwardly translocated in the phloem and xylem distributed to the growing

region? This may be indicated as a problem on which information is urgently required and the solution of which may
well lead to new and illuminating ideas.

In view of the close similarity, in respect of growth and
morphogenesis, between the shoot apices of ferns and seed
plants, the presence of a distinctive apical meristem in the
former suggests that there may be a physiologically equivalent
region in the latter. If this is so, all the equidimensional meristematic cells at the apex of a seed plant will not be physiologically identical, those at the extreme tip of the shoot having
special properties like the apical cell in ferns. Some evidence
bearing on this aspect has already been obtained by appropriate staining techniques and experimentally by Ball (1948).
In a different connection van Fleet (1948) has shown that
apparently similar parenchymatous cells of the cortex may be
very different in their physiological properties and reactions.
The scope for new investigations of the apex, using appropriate
physiological and biochemical techniques, again becomes
apparent.

In the ferns, the apex and its constituent cells continue to
enlarge as the young sporophyte grows to adult stature, but the
apex remains singularly unchanged in its organization and
morphogenetic activities. The large apex of the adult, in fact,
is like an enlarged replica of the young sporophyte apex. In some
seed plants which have been investigated, the vegetative apex
undergoes little change from the seedling to the adult stage, and
the actual size of the meristem and of its constituent cells is
not greatly increased (the apices of the Cycadales and other
gymnosperms afford notable exceptions to this statement). In
parentheses, it may be noted that an exceedingly illuminating
experience is to compare the meristems of ferns, gymnosperms
and flowering plants all drawn or photographed at the same
magnification; the range in cell size, as also in the size of the
meristem, is quite remarkable. So far there have been few investigations showing how the simple mass of meristematic cells
at the seedling apex—the epiphysis—develops into the typical
zoned structure of the adult vegetative apex. In some species
tunica and corpus regions are said to be detectable at an early
stage in the development of the seedling; and the pith of the
hypocotyl often becomes evident at a very early stage. In some

flowering plants, as in certain eusporangiate ferns, the apex of the young plant appears to be very inactive, growth being mainly localized in the young leaf primordia. While anatomical studies of the apex in the individual development are an essential preliminary stage, and may well yield many points of interest, they must be followed by investigations of other kinds if our understanding of the apex is to be in any degree adequate.

In concluding this essay, it is emphasized that collaboration between the geneticist, the physiologist and the morphologist is essential to an understanding of the apical meristem, and that this all-important region provides a common ground for integrating the data of different branches of botanical science.

10

A commentary on Turing's
diffusion-reaction theory
of morphogenesis

In contemporary studies of morphogenesis in plants, attention is being centred more and more on growth and the genic control of metabolism. It is held that, in the inception of new organs and differentiation of tissues, a localized accumulation of gene-determined substances may be an essential prior condition. If this view is accepted as a working hypothesis, a knowledge of factors which might determine the patternized distribution of morphogenetic substances would constitute an important advance. Indeed, it is here that one of the greatest difficulties in the investigation of morphogenesis has been encountered, for biochemical concepts alone seem inadequate to account either for the inception of pattern or the progressive organization during development. Although, as Arber (1950) has pointed out, the term *form* in its full connotation deals comprehensively with the characteristic shape of an organism, or of its parts, the term *pattern* has also been used extensively in morphological studies in recent years—a use that is not recorded in the *Oxford Dictionary*. Nevertheless, the term, being virtually self-explanatory, is a convenient one, and, in addition, it carries the implication that, as in some artistic and geometric designs, a morphological or structural development may be characterized by repetitive features. The term is also useful in specifying particular aspects of the organization which becomes manifest during development.

That specific substances, e.g. auxin, are of great importance in morphogenesis is not in dispute, but thus far no biochemical theory of the inception of pattern or of organization has received general acceptance. Such knowledge as we have of the metabolism of embryonic regions does little to explain the assumption of form or the differentiation of tissues. And although there is evidence that many of these developments are, at least to some extent, gene-determined or gene-controlled, our knowledge of the actual mechanism is both slender and speculative. One working hypothesis is that this mechanism, in its most fundamental aspect, may be sought in the laws of physical chemistry as applied to the metabolic systems found in embryonic regions. In particular, it may be helpful to inquire if anything is known regarding the physical chemistry of organic reaction systems which might account for the inception of some of the characteristic patterns in plants. In a contemporary paper, Turing (1952) has advanced a theory, based on a mathematical study of diffusion-reaction systems, which seems to go a considerable way towards providing an explanation along these lines. The botanical application of this theory is considered in the present paper.

One result of the morphological studies of the post-Darwinian period, and also of the contemporary period of renewed interest in morphogenesis, has been an appreciation of the fact that similar morphological and anatomical features may be found in organisms of quite distinct taxonomic affinity. These *homoplastic* developments, which have resulted from, or are accepted as an indication of, parallel or convergent evolution, and which have been aptly described as constituting *homologies of organization*, are of general occurrence in the plant kingdom. Indeed, the main formal and structural features in plants can be referred to a comparatively small number of kinds of pattern (see p. 123). This being so, one of the main tasks in morphogenesis is to discover and investigate the factors which determine those several kinds of pattern.

In each instance where the inception of a particular pattern is being considered, it is essential to have some leading idea, or system of ideas, that will serve as a working basis for investigations. A generally accepted explanation of the occurrence of similar morphological features in related organisms, i.e. *homogenous* developments, is that there are genes, or groups of genes, which are common to the several organisms, and that these control or determine the observed developments. Actually, relevant genetical investigations indicate that the situation may be considerably more complex. (See Harland, 1936; de Beer, 1951.) But where similar features are present in unrelated organisms, they cannot so readily be attributed to common groups of genes. In attempting to explain the phenomenon of homology of organization in unrelated organisms two possibilities may be considered: (i) similar morphological features appear because essentially *the same kind of process* is operating in each of the non-related organisms; or (ii) that essentially different processes may yield comparable morphological results. On grounds of probability, the first explanation seems preferable to the second; but, because of the very great diversity of living organisms, the second cannot, and should not, be eliminated until this is justified by the data of a full investigation. Whether we are concerned with comparable developments in related species, which are considered to be more or less directly gene-controlled, or with the homologies of organization shown by unrelated species, the visible phenomena of morphogenesis have their inception in biochemical and biophysical systems. In view of the prevalence of homology of organization in the plant kingdom, it is cogent to inquire if any reaction systems are known which are likely to be of general occurrence in living organisms; or, in other words, are there reaction systems, of a kind likely to occur in plants, which might afford a basis for the inception of certain kinds of pattern?

An example of homology of organization may be considered by way of indicating the need for a renewed investigation of these problems. In all classes of vascular plants, the root stele in cross-section is seen to consist of radiating plates of xylem alternating with bays of phloem. This pattern has its inception at the root apex and is not determined by the presence of lateral members. In small roots the xylem typically consists of one to five xylem plates: in larger roots there may be six to twenty radiating xylem plates, but the tissue pattern, though considerably more complex, is essentially a repetition of that seen in small roots. No generally accepted hypothesis relating to the inception of the pattern in root steles has yet been advanced. Familiarity with root structure may perhaps engender the impression that we understand what we see, but, in fact, we have, thus far, very little knowledge of the factors which determine the characteristic differentiation of tissues in roots. This, indeed, applies equally to all the tissue systems in plants. Goebel (1922) pointed to the repetitive occurrence of pattern during development and to the relative constancy of scale of the 'units of pattern' at the time of their inception. Thoday (1939) has indicated how this conception could be used to account for the increasing structural complexity in roots of increasing size, i.e. as the stele enlarges more units of pattern can be accommodated. If this be accepted, then the fundamental problem is to discover the factors which determine the units of pattern and their specific location.

Since every morphological and anatomical development is the result of antecedent physiological processes in which several, perhaps many, steps may be involved, any adequate theory of the inception of pattern (as in the root stele given above) must take account of the nature and properties of embryonic tissue and of the physical chemistry of the metabolites which are involved in growth and differentiation. Furthermore, if we assume, as working hypotheses, that the embryonic tissue is initially homogeneous and that qualitative as well as quantitative differences are involved in the differentiation of tissues, e.g. phloem and xylem, then our task is to account for the characteristic localization or patternized distribution of the specific metabolites that determine and precede the visible pattern. Turing's diffusion-reaction theory of morphogenesis appears to provide a means of advancing our understanding of these problems, at least in their more general aspects.

General analysis of pattern in plants

Although the range and diversity in form and structure in plants are impressive, the number of major, distinctive *kinds of pattern* is small; but each may be greatly varied in the matter of detail and as a result of allometric growth. The following general categories, or kinds, of pattern may be indicated:

(1) *Axial development*, which normally follows the early establishment of polarity (with an attendant physiological and morphological distinction between base and apex), is general in all classes, from algae to flowering plants.

(2) *Concentric construction* is exemplified by the cortex and stele in shoots and roots, by secondary thickening in shoots, by the wall and archesporium in bryophyte capsules, and so on.

(3) *Radiate construction* is typically seen in root steles, in the shoot stele in *Lycopodium*, in the vascular strands in dicotyledons, etc. The branch filaments in *Chara* and other algae, and lateral members (leaves, buds, etc.) in vascular plants may also be included in this category.

(4) *Mosaic construction* is suggested as a term to include such patterns as the distribution of stomata, the distribution of the vascular strands in monocotyledons, the arrangement of tracheids and parenchyma in certain steles, etc.

Within this general category, the term *specific locations* may prove convenient to indicate such specifically localized features as the conceptacles in brown algae, the sori of ferns, etc., in which a patternized distribution can be discerned.

The separation of these several kinds of pattern is, of course, artificial. During the growth and differentiation of a root, for example, axial, concentric and radiate developments are proceeding more or less simultaneously.

Turing's diffusion-reaction theory of morphogenesis
Turing's theory (1952) is based on a consideration of the diffusibilities and reaction rates of substances which may be involved in growth and morphogenesis. Considerable mathematical knowledge is essential to follow the theory in detail; but its main features can be indicated to, and appreciated by, the non-mathematical biologist without too much difficulty. The theory introduces no new hypotheses: on the contrary, it makes use of well-known laws of physical chemistry, and, as Turing has shown, these seem likely to be sufficient to account for many of the facts of morphogenesis. The underlying point of view, in fact, is closely akin to that expressed by D'Arcy Thompson in *Growth and Form*. It will be appreciated that a theory, based essentially on laws of physical chemistry that must apply to every growing system, is of the kind that may well account for the general occurrence of certain organizational features in plants. An essential feature of the theory is that it deals with the inception of a morphogenetic *pattern as a whole*; but it is not inconsistent with epigenetic development when other organs or parts have already been formed. Not least, it is compatible with the concepts of physiological genetics.

An indication of the theory may be given by assuming that

two interacting, pattern-forming substances, or *morphogens*, X and Y, are essential metabolites in a morphogenetic process; a third substance C, which is in the nature of an evocator and catalyst, is also involved, a pattern only appearing if its concentration is sufficiently great. It is necessary to assume: (i) that both X and Y are diffusible, and at different rates; and (ii) that there is a number of reactions involving X, Y and the catalyst C: these reactions do not merely use up the substances X and Y, but also tend to produce them from other metabolic substances (which might be called 'fuel substances') which are assumed to be abundantly present in the growing region, i.e. to some extent the morphogens are autocatalytic. If a pattern is to be produced, there is a number of conditions relating the diffusibilities and marginal reaction rates which must be satisfied. (By marginal reaction rate is meant the amount by which the reaction rate changes per unit change of concentration.) If we assume that the appropriate conditions are satisfied, and that the concentration of the catalyst-evocator is initially at a low value, but is slowly increasing, the phenomena observed will be as follows:

(i) Initially there is a state of homogeneity: both X and Y are uniformly distributed (i.e. in the embryonic tissue in which a pattern will subsequently appear), apart from some slight deviations due to Brownian movement and to chance fluctuations in the number of the X and Y molecules that have reacted in the various possible ways in various regions.

(ii) The concentrations of X and Y will vary slowly as the system adjusts itself to the changing evocator concentration. This change will also result in the fluctuations of concentration smoothing themselves out more and more slowly, and eventually the point is reached where the system is unstable, i.e. the fluctuations no longer are smoothed out: they become cumulative, and even tend to become exaggerated with the passage of time.

(iii) At this stage the morphogen concentrations form a more or less irregular wave pattern. Later, however (for instance when in some places the concentration of one morphogen is practically zero), the progressive deepening of the waves is arrested. The pattern will then regularize itself, and will eventually reach an equilibrium which is almost perfectly symmetrical. The resulting pattern may be described as a *stationary wave*.

Such a stationary wave, in a biological situation, might take the form of the accumulation of one of the morphogens in several, e.g. 3, 4, 5 or more evenly distributed loci on a one-dimensional system such as the circumference of a circle: the other morphogen will tend to accumulate at intermediate loci. Put quite simply, all this is as much as to say that *in an embryonic tissue in which the metabolic substances may initially be distributed in a homogeneous manner, a regular, patternized distribution of specific metabolites may eventually result, thus affording the basis for the inception of a morphological or histological pattern.* A patternized distribution of specific metabolites can thus take place in conformity with the laws of physical chemistry as applied to diffusion-reaction systems; and this will be true whether the morphogenetic substances are held to be specifically gene-determined, or whatever mechanism is assumed to connect such genes with the morphogens. It seems not improbable that reaction systems of the kind indicated in the theory may be of general occurrence in living organisms, but, of course, evidence that this is so is essential. A provisional acceptance of the theory would certainly afford a basis for understanding both the prevalence of homologies of organization and the diversification of basic kinds of pattern under the impact of genic factors. For, as we have seen, the patternized distribution, or specific location, of metabolites depends on the diffusibility and chemical reaction of the metabolites, some, or many, of which are specifically gene-determined.

It may be difficult for some readers to understand how, from an initial homogeneous distribution of metabolites in an embryonic region, there can be a drift into instability as described in the theory. Turing has dealt with this aspect in some detail and has indicated how the instability may be 'triggered off', for example by random disturbances.

Turing has envisaged an idealized and simplified 'model of the embryo'. 'The model takes two slightly different forms. In one of them the cell theory is recognized, but the cells are idealized into geometrical points. In the other the matter of the organisms is imagined as continuous distributed. The cells are not, however, completely ignored, for various physical and physico-chemical characteristics for the matter as a whole are assumed to have values appropriate to the cellular matter.' In these statements botanical readers will see that both the classical view of Hofmeister and de Bary—that the tissue mass as a whole determines differentiation and not the individual cells—and of physiological genetics (in which the importance of gene-controlled substances, proceeding from individual cells is emphasized), are appropriately represented. The following passage may be cited as an example of Turing's approach:

With either of the models one proceeds as with a physical theory and defines an entity called 'the state of the system'. One then describes how that state is to be determined from the state at a moment very shortly before. With either model the description of the state consists of two parts, the mechanical and the chemical. The mechanical part of the state describes the positions, masses, velocities and elastic properties of the cells, and the forces between them. In the continuous form of the theory essentially the same information is given in the form of the stress, velocity, density and elasticity of the matter. The chemical part of the state is given (in the cell form of theory) as the chemical composition of each separate cell: the diffusibility of each substance between each two adjacent cells must also be given. In the continuous form of the theory the concentrations and diffusibilities of each substance have to be given at each point. In determining the changes of state one should take into account: (i) the changes of position and velocity as given by Newton's laws of motion; (ii) the stresses as given by the elasticities and motions, also taking into account the osmotic pressures as given from the chemical data; (iii) the chemical reactions; (iv) the diffusion of the chemical substances; the region in which this diffusion is possible is given from the mechanical data.

The mathematical treatment of changes in the state of even an arbitrary and greatly simplified diffusion reaction system is unavoidably complex. For many contemporary investigators of morphogenesis, however, it is the possibility of the general result indicated by the theory, rather than the intricate details of the process, that is important; for it seems improbable that the relevant detailed investigations can be adequately carried out until great refinements have been introduced into the techniques of plant physiology. Indeed, until it is proved invalid, the theory may be tentatively accepted; for it is based on the laws of physical chemistry and on a mathematical analysis of processes which, according to the present state of knowledge, may be assumed to be going on in living organisms. It is not improbable that Turing's work may stimulate the formulation of other, alternative physico-chemical theories of morphogenesis. Such a quickening of interest will be welcome to the botanist.

That diffusion-reaction systems are present in all growing regions, indeed in all living matter, is basic to studies of metabolism. What is novel in Turing's theory is his demonstration that, under suitable conditions, many different diffusion-reaction systems will eventually give rise to stationary waves; in fact, to a patternized distribution of metabolites. Thus, in the present writer's view, the theory would appear to afford an explanation of the inception of the symmetrical, radiate histological pattern that appears adjacent to the embryonic region of the root apex. Not all kinds of pattern, however, are referable to the development of stationary waves—the major feature of Turing's theory as thus far developed—but all may eventually be related to some kind of diffusion-reaction system. The inception of polarity, i.e. of axial development, in an embryo is probably due to a particular distribution of metabolites in an initially homogeneous system; this could be regarded as a very simple case of a stationary wave. The following may be tentatively indicated as examples of pattern in plants which may perhaps be explained, in whole or in part, as the theory is more fully developed and explored: phyllotactic systems; whorled branching in algae; the distribution of procambial strands in shoots; the radiate pattern in root steles and in lycopod shoots. Turing has indicated how the dappled pattern in the skins of animals and gastrulation in the developing embryo can be explained by his theory.

In the general system of ideas incorporated in the theory there are many points of interest to the student of morphogenesis. Thus, with regard to the breakdown of symmetry and homogeneity, attention is directed to the importance of small random changes in the distribution of morphogenetic substances, i.e. irregularities and statistical fluctuations in the numbers of molecules taking part in the various reactions. The determination of polarity in the fertilized ovum of *Fucus*, for example, may be due to random changes, or to factors in the environment. In the enclosed embryos of land plants, in which polarity is determined soon after fertilization, if not before, quite small gradient effects proceeding from the gametophyte tissue could be the means of initiating the breakdown of homogeneity and the establishment of polarity. Some deviations from homogeneity in a reaction system may be of great importance in the process of differentiation; for the system may reach a state of instability in which the irregularities, or certain components of them, tend to grow. If this happens, a new and stable equilibrium is usually reached, and this may show a considerable departure from the original distribution of metabolites. Thus, in contiguous cells which are initially metabolically identical, a drift from equilibrium may take place in opposite directions as a result of statistical fluctuations in the components of the reaction system, or of small changes induced by neighbouring cells. Changes of this kind could, for instance, account for the very different developments in two adjacent, equivalent embryonic cells—a histological phenomenon of much interest to the botanist.

Unless we adopt vitalistic and teleological conceptions of living organisms, or make extensive use of the plea that there are physical laws as yet undiscovered relating to the activities of organic molecules, we must envisage a living organism as a special kind of system—invariably a very complex one—to which the general laws of physics and chemistry apply. And because of the prevalence of homologies of organization, we may well suppose, as D'Arcy Thompson has done, that certain physical processes are of very general occurrence. Such are the diffusion and reaction of metabolic substances, and, as Turing has now demonstrated by a mathematical analysis of diffusion-reaction systems, physical processes can be indicated which may be directly involved in the inception of pattern. If this theory stands up to such tests as can be devised, its great value in the study of morphogenesis is apparent: indeed, it is the kind of conception for which the botanist has been looking for a very long time. The theory, of course, can only be expected to explain some aspects of morphogenesis, for in the determination of form and structure in plants many factors of different kinds are at work, and any adequate approach to the problem must essentially be a multi-aspect one (Wardlaw, 1951).

Tests of the theory

It may be that for some time to come the theory will be provisionally accepted, at least by some biologists, because it rests on a substantial mathematical and physical basis rather than because supporting data have been obtained. Tests of its validity should certainly be sought. Because of the complexity of all morphogenetic processes, relevant experimental or observational data may be difficult to obtain, but the task should not be regarded as an impossible one. An evident primary test of the theory will consist in the closeness of its applicability to a wide range of biological materials. On these grounds, as we have seen, it may be considered valid, since it cannot be denied that some diffusion-reaction systems do give rise to stationary waves, and that such reaction systems are likely to be found in living organisms. That, of course, is not proof that stationary waves do, in fact, constitute the basis of pattern in plants, but there is the probability that this is so. The general conception is certainly not incompatible with the ideas and data of contemporary plant physiology.

Some indications of the validity of the theory by the method of prediction have already been obtained by Turing, using the digital computer. In a numerical example, in which two morphogens are considered to be present in a ring of twenty cells, he has found that a three- to four-lobed pattern would result; and in other examples he has shown that a two-dimensional pattern, such as dappling, and gastrulation in a spherical body, do arise in specified diffusion-reaction systems. Here again, it should be noted that these results, based on reactions which approximate to those considered to take place in living organisms, increase the probability that the theory is valid or adequate, but do not prove that the inception of pattern is due to the assumed reaction systems.

In conclusion, it should perhaps be noted that the conception of diffusion-reaction systems as the basis for morphogenesis raises its own difficulties. Not the least of these lies in the complexity and multiplicity of the processes which have to be considered. Thus, in any embryonic region such as a shoot apex, many different reactions are evidently going on either simultaneously or in close succession, while different reactions may be taking place in contiguous tissues; and all these, together with the other factors which affect morphogenesis, result in the orderly, harmonious and characteristic development of the individual plant. The cautious investigator may well ask if situations of such complexity can indeed be unravelled and comprehended, while the professed student of morphogenesis, accustomed as he is to work with visible and tangible materials, may find that contemplation of complex diffusion-reaction systems takes him into regions of thought with which he is unfamiliar—the realm of the unpicturable. Still, as it seems, if we are to break new ground and get to the root of the matter— having scratched so long at the surface—we must bear in mind that physiological processes always precede the appearance of new organs and tissues, and that it is with these processes that we are primarily concerned. That being so, a new approach along the lines indicated in the diffusion-reaction theory should be considered on its merits.

I I
Comparative observations on the shoot apices of vascular plants

Perhaps the most concrete result of the numerous recent and contemporary studies of shoot apices in vascular plants has been to emphasize their histological diversity. Seven different 'types' have been specified, ranging from those with a single conspicuous initial cell to those with a small-celled, zoned construction (Wardlaw, 1950b). Accordingly, while such new types of apical construction as may be discovered will carry their own interest, extensive anatomical and histological data are now available for comparative study. What is now needed, as a first step, is a consideration of these materials from the genetical, physiological, physical and mathematical points of view, and a synthesis of the ideas and conclusions which emerge from these studies; for only in this way can a fuller understanding of the phenomena of morphogenesis and organization in vascular plants be obtained. While contemplation of different apices reveals their great structural diversity, it also shows that all, nevertheless, have important features in common: they all occupy a distal position, give rise to a vascularized axis with regularly disposed lateral members (leaves and buds), afford evidence of a harmonious or regulated process of development, and may persist in the embryonic condition throughout the life of the plant. Moreover, when very differently constituted apices are subjected to the same experimental treatments, closely comparable developments ensue (Wardlaw, 1950a,b). The present paper is concerned with a comparative study of apices taken from different classes of vascular plants and with the causes which may underlie the homologies of organization which they show; for in a study of this kind the aim should be not only to discover the factors which determine the organization in any one type of apex but also to account for the fact that all apices have major organizational features in common.

A note on homology of organization. Homology in biology means 'correspondence in type of structure due to common descent' (1870, *Oxford Dictionary*). Organs which are comparable in position and initial development, and to some extent in form and structure, are said to be homologous. Species with homologous members were held by comparative morphologists to have descended from a common ancestor. According to Bower (1948, p. 344), the classification of parts 'must be based upon their origin, and upon the place which they take relatively to other parts at the time when they first appear . . . *those parts of the individual, or of different individuals, species or genera, are distinguished as homologous which have the same relation to the whole plant body, whatever their function or external conditions may be.* The strictest conception of homology is that designated as *homogeny.* Lankester defined as homogenous those structures which are genetically related in so far as they have a single representative in a common ancestor.' Where closely comparable organs or structures are found in organisms of quite different systematic affinity, the phenomenon is ascribed to parallel evolution, and has been described as *homoplasy* (Lankester, 1870) or as exemplifying *homology of organization* (Goebel, 1897; Lang, 1915). As a working procedure in comparative studies, the writer suggests that comparable organs or structures should, in the first instance, be regarded as homologies of organization. The main task of the biologist is to ascertain the factors, both extrinsic and intrinsic, which bring them about. Recent cytological studies by Manton (1950) show how cautious we must be in regarding similar structures in apparently closely related genera as being truly homogenous.

The comparative examination of shoot apices
Among vascular plants apices are known with (i) a single very conspicuous, superficial apical or initial cell, as in leptosporangiate ferns; (ii) several conspicuous superficial initial cells as in eusporangiate ferns and *Selaginella*; (iii) several inconspicuous superficial initial cells as in *Lycopodium*; (iv) a weakly zoned or layered construction; (v) a fluctuating layered construction; (vi) a highly definite layered construction, (iv–vi) as

in flowering plants; (vii) a radiating construction apparently
originating from a subsurface group of mother cells, as in
cycads, *Ginkgo* and some flowering plants. A scrutiny of these
and other apices shows that there are evident differences in the
size, shape and organization of the apex as a whole, in the size
of the embryonic cells, and in the differentiation of tissues.
At a glance, the large relative size of the tissue mass of the
apices of *Microcycas* (a cycad), of *Cyathea* (a tree fern), and of
Echinopsis (a cactus) may be noted, and the small relative size
of the apices of *Elaeis* (a tree palm), of *Phaseolus* (a herba-
ceous dicotyledon) and of *Equisetum* and *Selaginella*. As to
external shape, the conical apex of *Cyathea* may be compared
with the rounded apex of *Phaseolus* and with the almost flat
apex in *Echinopsis* and *Acer*. *Iris* and *Selaginella* may be cited
as examples of apices of bilateral symmetry. The diversity
in cellular pattern and the range in cell size and gross size are
perhaps the most notable points that emerge from a comparative study of apices. In the matter of differentiation, whereas the incipient vascular tissue can be seen immediately below
the superficial apical meristem in ferns, this is not usually
evident in the apices of flowering plants. *Larix* affords an example of rows of cells of distinctive metabolism close to the
apex. Confronted with this diversity, it is perhaps not surprising
that investigators have made but little progress in explaining or
interpreting the organization of apices, one aim of which must
be to see if the several types can be brought into some general
scheme.

(*a*) *The size of meristematic cells*   If de Bary's aphorism, that
organs determine cells and not cells organs, were accepted as
absolute, the primary investigation of apices would relate to
their gross size and shape. The relation between cell and organ,
however, is a reciprocal one. From the standpoint of physiological genetics, which occupies a central place in studies of
morphogenesis, it is appropriate to begin, as in embryological
studies, by considering the individual meristem or embryonic
cells; for the initial action of any gene, or collection of genes, is
a biochemical one, and it takes place within an individual cell
and extends to other cells. Accordingly, although the physiological activity and differentiation of individual cells are

affected by their position in the tissue mass, the size, shape and metabolism of the meristem cells are important in determining the histological organization of the apex.

The hypothesis that the size and metabolism of meristem cells are primarily determined by factors in the genetic constitution is supported by a growing body of evidence. Thus Randolph, Abbé & Einstet (1944), Cross & Johnson (1941), Blakeslee, Bergner, Satina & Sinnott (1939), and Satina, Blakeslee & Avery (1940) have shown that changes in the size of the meristem cells in the shoot apex can be related to known changes in the chromosome complement; Cross & Johnson have shown that in tetraploid plants of *Vinca rosea* the cell size in one dimension was increased by approximately 50 per cent, as compared with the normal diploid. On the metabolic aspect, Overbeek (1935, 1938*a*, *b*) has shown that the amounts of auxin and of an auxin-destroying substance in races of maize are determined by a small number of genes. That the histogenic potentialities of meristem cells are different in ferns, gymnosperms and flowering plants is common botanical knowledge. The above hypothesis must, however, be qualified in various ways. In any species, there is no absolute size for meristem cells. Thus, in a fern such as *Cyathea Manniana*, the apical cell and the adjacent prism-shaped cells of the meristem undergo a progressive increase in size as the plant grows from the young sporophyte to the adult stage (Wardlaw, 1948). In *Microcycas*, the meristem cells of a young plant are somewhat smaller than those of a large one, but there is no increase in cell size as in *Cyathea*, even though the adult shoot apex becomes very large indeed (Foster, 1943). In some, possibly many, dicotyledons, the shoot apex shows little change in the size of its cells from the seedling to the adult (Ball, 1949).

A further indication of the importance of individual meristem cells is afforded by the fact that in embryogeny, as also in the development of buds, a small but characteristic apex is formed as the result of the growth and division of a quite small group of meristematic cells, i.e. a considerable tissue mass is not necessarily involved.

In this section attention has been directed to the size of meristem cells as contributing to the organization of the shoot apex. While it is held that cell size is probably determined by

genetical factors, it is recognized that the mechanism involved
presents problems of great complexity and difficulty.

(*b*) *The histological constitution of the apex*   During the past century, investigators of shoot apices have tended to give their
adherence to the apical cell theory of Naegeli, to the histogen
theory of Hanstein, or the tunica-corpus theory of Schmidt.
Each theory has its justification and its limitations. All are
descriptive rather than explanatory. As the descriptive survey
of apices, though by no means complete, is adequate, the
emphasis should now be transferred to a study of factors which
may determine the histological organization, or pattern, in
different species. In other words: How do the units of construction—the meristem cells—come to be arranged in a
particular way? A study of the histological developments of
embryos and of buds may be expected to yield some indication
of the factors which are at work. In the present state of knowledge this subject can only be considered in its more general
aspects.

D'Arcy Thompson (1917, 1942) has advanced the view that
growing systems, like all physical systems, tend towards
equilibrium; they are described as being in a state of dynamic
equilibrium. He has also suggested that when an embryonic
cell enlarges, the free energy increases, and that cell division is
the means whereby the approximate equilibrium of the system
is restored. The amount of enlargement which an embryonic
cell can undergo before division takes place will be different
for different species: it will depend on the initial cell size and
metabolism, on the metabolism of the tissue of which it is a part,
and on factors in the environment. As we have seen, the meristem cells of some species are much larger than those of others,
and we may inquire if there is any constant relation between the
distinctive cellular pattern of an apex and the size of its constituent cells. Meristems with large apical cells, e.g. *Cyathea* or
Dryopteris, do not have a layered construction, whereas a
layered construction is characteristic of many but not of all
small-celled apices. *Equisetum* has a small apex composed of
small meristem cells: it has, nevertheless, a distinctive apical
cell and it has not a zoned construction. As a general statement,
it may be said that the type of apical construction does not

depend on absolute cell size alone, but that cell size must be considered. The phenomenon is complex and involves a whole nexus of factors and relationships, including the size of the meristem cells and the distribution of growth in the apical region as a whole, i.e. cell and organ relationships must be considered concurrently. In a uniform environment, the distribution of growth, involving metabolic and other factors, is primarily determined by genetical factors. It may be assumed that the meristem cells of different genera and species have characteristic differences in their metabolism; these may be quantitative, or qualitative, or both. Now the actively growing meristem cells of the apex constitute a locus of utilization of the nutrients supplied from below and by-products of their metabolism will diffuse or otherwise move basipetally; i.e. polar or axial gradients are set up. In the apex, where the vascular tissue is still in the incipient, undifferentiated stage, the upwardly moving nutrients are no longer confined to the xylem and phloem as in the mature regions of the shoot, but will tend to diffuse upwards, over the whole cross-section, moving most rapidly to centres of growth where utilization is greatest, and where cellular resistance to such movement is least. Thus, as an example, the movement of nutrients to an actively growing leaf primordium will be more rapid than to a slowly growing shoot apex. And if it be further assumed that metabolism is not uniform in the apex but that there is a greater utilization of one metabolite in one region and of another metabolite in another, it can be seen how a complex system of gradients will be established. The diffusion-reaction theory of morphogenesis as advanced by Turing (1952) would have its place in this context. Furthermore, some cells are superficial in position, some subsurface, and some centrally placed. The supply and utilization of nutrients, the dispersal of the by-products of metabolism, and the partial pressures of carbon dioxide and oxygen, will vary from layer to layer and from cell to cell according to their position, and will affect both the growth of the individual cell and the distribution of growth in the apex as a whole. Physical factors, e.g. tensions and pressures which become incident, will also affect differentially the growth and expansion of cells in different regions. Since the genetical constitution varies from species to species, different manifestations of differential growth and different histological configurations will follow as a natural consequence.

In the present state of knowledge, the foregoing account of the histological constitution of the shoot apex is unavoidably of a very general kind, but it is clear, at least to the writer, that no 'explanation' is likely to be adequate unless it refers concurrently to the growth of the apex as a whole (i.e. as an organ) and to the gene-controlled metabolism of its constituent cells. One of the factors involved in the elongation of shoots is the supply of auxin moving downwards from the apex. If the nutrients essential for growth are not limiting, and if other conditions are satisfied, the auxin, at a certain critical concentration for the species, is implicated in the marked elongation of the shoot, i.e. there is differential growth, the polar or axial component being greater than the radial component. On the other hand, in short tuberous shoots, axial and radial growth may be about equal. Similarly, in the shoot apex, in relation to the specific metabolism, longitudinal growth may considerably exceed radial growth, in which case an elongated, conical, or dome-shaped apex will result; the converse would result in a broad, flattened apex; and many intermediate conditions can readily be envisaged. In a survey of selected monocotyledon apices, Stant (1952) has shown that they can be arranged in a series from broad flat ones to narrow pointed ones, the shape being determined by the relative magnitudes of the vertical and radial components of growth.

There is evidence that the shape of the apex and its layered construction may be partly due to the upward pressure exerted by the pith. In *Echinopsis*, the slightly convex apex consists of several outer layers of small, approximately equidimensional cells with larger cells on the flanks, a layer of incipient vascular tissue, and a conspicuous rib-meristem in a state of active division. The indications are that the pith and its rib-meristem are exerting considerable pressure on the surrounding tissues, including the cells of the tunica above; as a result the contour of the apex is in close conformity with that of the upper limits of the pith. In large apices of *Dryopteris*, and in many zoned apices, a similar relationship may be observed. If this relationship is a general one, we should expect a small non-medullated shoot to have a flattened rather than a conical or dome-shaped apex. That this is so can be seen in a number of proto-stelic ferns and in *Psilotum* and *Lycopodium*. The relationship, however, is by no means invariable, for some protostelic ferns have a conical apex, and others with a large pith show only a slight convexity.

(c) *The apical cell group*   In plants with only one or a small number of initial cells at the apex, these cells are usually relatively large and conspicuous. This phenomenon has long exercised the minds of morphologists and it cannot be said that anything approaching an adequate explanation has yet been advanced. Since there is more rapid growth on the flanks of the apex than at its summit, it might be thought that the most distal cell, or cells, would be under reduced pressure and therefore free to grow to larger size. There are, however, objections to this conception. That the size at which the apical cell in a fern divides is determined both by factors within the cell (e.g. the ratio of internal to surface energy, if we accept D'Arcy Thomp-son's view), and in the cellular environment, is shown (i) by the regular sequence in which the partition walls are laid down, and (ii) by the fact that the apical cell enlarges concomitantly with the ontogenetic enlargement of the shoot. Physical factors, then, are probably important determinants of the cellular pattern of the fern apex. It would appear, however, that there is some major factor, as yet undetected, which determines the large relative size of the apical cell. A clue to this factor is perhaps afforded by a consideration of leaf formation in ferns, where a large apical cell is also present from an early stage (Wardlaw, 1949*a*).

The leaf primordium in *Dryopteris* has its inception as a group of equivalent, superficial prism-shaped meristematic cells on the flanks of the conical apical meristem. One of these cells soon enlarges and is easily recognized as the apical cell of the primordium; all the subsequent development of the leaf is due to the growth and division of this cell and its segments. Thus, in the ferns, certain meristem cells have physiological properties not shared by the adjacent meristem cells. We may therefore inquire if there is any particular or unique circumstance associated with the inception of the large apical cell, whether it be of a leaf primordium or a shoot.

The writer (Wardlaw, 1949*b*, 1950*c*) has shown experimentally that there is considerable support for the view that the
apical cell group and the several incipient leaf primordia
round the apex, can be regarded as growth centres, each with
its surrounding physiological field, and that a new growth
centre cannot originate within any of the existing fields, but
can originate in the superficial meristem cells outside them.
Thus, in a growing apex of *Dryopteris*, when the space between
two leaf primordia becomes sufficiently large so that an area of
the superficial meristem is free from the inhibitory effects of the
two neighbouring fields, a new leaf primordium with its large
apical cell will appear. In the formation of lateral buds from
detached meristems, a not dissimilar state of affairs may be
observed. The outgrowing bud has at first a superficial meristem
of equivalent prism-shaped cells, but soon, in the centre of this
meristem, one of the cells develops into a large apical cell. In
short, in main shoots, buds and leaves, the cell which enlarges
and becomes the apical cell is situated at or near the centre of
the meristematic area, i.e. it occupies a unique position. This is
probably also true of the inception of the apical cell in
roots.

Other things being equal, the metabolism and physiological
condition of cells at the centre of a meristematic area will differ
from those of the more peripheral cells, e.g. the products of metabolism will tend to be present in higher concentrations at the
centre than at the margin. While considerations such as these
would help to explain why the central cell, or group of cells,
might be different physiologically from the surrounding cells,
they do not explain the large relative size of the fern apical
cell. That the apical cell in the fern shoot has distinctive physiological properties has been shown in experiments where that
cell alone was punctured: but formation, previously inhibited,
ensued.

In the enlargement of a quadrant of a sphere, as in the
developing zygote of a fern, D'Arcy Thompson has shown that,
in accordance with the theory of cell division by walls of
minimal area, a 'three-sided' apical cell would normally be
formed under conditions of uniform growth. He has also
demonstrated that uniform and symmetrical growth in an
organism (or an organ) by no means implies a uniform or

symmetrical growth of the individual cells. Such cellular differences, moreover, could arise without any change or variation in the surface tension in the cell walls.

The foregoing considerations have also an application to the small-celled, layered apices of dicotyledons. Although the equidimensional cells of the tunica in the dicotyledon apex appear to be identical, it is improbable that they are physiologically alike. From experimental and other evidence it is known that they are not. Thus, apices are known in which the most distal (and central) cells of the tunica are distinct in their staining reactions from adjacent cells on the flanks. Their response to applications of auxin is also different. In short, the cells at the extreme tip are held to be physiologically distinctive and to constitute the centre and focus of the morphogenetic activities of the apex.

(d) *The apex of cycads* Johnson (1951) has recognized four types of apical meristem in the gymnosperms. Of these the very large apices of the cycads are of special interest. According to Foster (1943), the apex of an adult plant of *Microcycas* shows a distal *initiation zone*, below which lies a zone of *central mother cells*; the initiation zone extends laterally into the *superficial peripheral zone*; the central mother cells give rise to a *rib meristem* below and to an *inner peripheral zone* laterally, this also arising from dividing cells like those of the rib meristem. These large cycad apices have also been described in slightly modified terms (Johnson, 1951). Foster's illustrations of the seedling apex suggest a possible line along which an analysis of the complex adult apex may be sought. The seedling has a distinctive apical cell group, below which there are enlarging cells which appear to be undergoing incipient parenchymatization. These are the cells which in the large apices are recognized as the *central mother cells*: they are described by Foster as being 'greatly enlarged and highly vacuolated'. A possible explanation of their origin is that auxin is moving basipetally in the shoot from the distal meristematic initiation zone and also centripetally and basipetally from the growing leaf primordia. This factor, in conjunction with the upward movement of nutrients, may account for the incipient parenchymatization; and these factors of growth, together with the mutual pressure of growing tissues, may be responsible for the inception of the rib and peripheral meristems at the base and lateral sides respectively of the central mother cells. As Johnson (1951) has suggested, an increase in the anticlinal divisions beneath the subapical initial zone, together with the development of the rib meristem in a smaller apex, would account for the presence of the central mother cell zone in a large apex. The underlying questions would then relate to the factors which determine this particular distribution of growth.

(e) *Some aspects of the geometry of the zoned apex*   Even if we accept the view that the zoned or layered apex is primarily due to factors in the genetical constitution, we have still to inquire into the mechanics of the process which results in the visible histological pattern.

As an approximate description, a zoned apex in transverse section consists of a system of concentric *cell layers*, each layer, especially the outermost ones which constitute the tunica, being composed of closely comparable meristematic cells; and such an apex, at the summit of an adult shoot, has attained to its characteristic size or bulk by the enlargement of the considerably smaller apex of the embryo or seedling. A consideration of some of the geometrical properties of concentric cell layers, i.e. of a system of concentric circles, reveals some interesting points that are relevant to these investigations. In the diagram (fig. 1), each concentric circle has a radius which differs from that of its neighbours by 1 linear unit. The assumption is also made that each layer is divided by radial walls into 'cells', the inner (or periclinal) wall of each being 1 unit in length. Since the circumference of any circle $= 2\pi r$ and its area $= \pi r^2$, the following relationships obtain:

(1) For each linear unit added to the radius, 6·28 linear units are added to the circumference; hence the inner wall of *zone C* is 6·28 units longer than the inner wall of *zone B* (and the same relationship holds for their respective outer walls).

(2) The areas of any two adjacent concentric zones will differ by 6·28 square units and, on the arbitrary cellular basis proposed, will differ by 6·28 'cells'. Thus *zone H* at the periphery has an area of 47·1 or 15π sq. units and is composed of 43·96 'cells', whereas 'subepidermal' *zone G* has an area of 40·82 or

13π sq. units and is composed of 37·68 'cells'. (As a simple calculation will show, the arbitrary 'cells' of *zone G* are slightly larger than those of *zone H*.)

(3) If there is a uniform rate of radial growth over the whole radius, as *zone A* enlarges to the size of *zone B*, *zone B* will have enlarged to the size of *zone C*, and so on, and there will be no disturbance of equilibrium, i.e. neither tensions nor pressures will be set up between adjacent zones. Similarly, if equal increments of area are added to *zones A, B, C*, etc. (e.g. 6·28 sq. units) the equilibrium of the system will remain unchanged.

(4) If every 'cell' in each zone grows and divides at the same rate, i.e. undergoes one division by a radial wall in the same period of time, the relationships between adjacent concentric circles will not be maintained: zones initially adjacent will become separated by interpolated zones of unit radius if the size of the arbitrary 'cells' remains unchanged. Thus, if every cell in *zone H* divides once, so that two cells with an inner

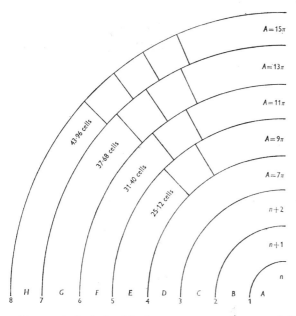

1 Geometrical relationships between concentric zones. *A, B, C*, etc., successive concentric zones; *n*, *n*+1, *n*+2, successive radii; indication of area, $H = 15\pi$ (i.e. $64\pi - 49\pi$).

(periclinal) wall of 1 linear unit are formed, the new *zone H^1*
will consist of $2 \times 43 \cdot 96$ 'cells' $= 87 \cdot 92$, and its radius will
increase from 7 linear units to 14 linear units; whereas *zone G*
after 'cell' division will consist of $2 \times 37 \cdot 68$ 'cells' $= 75 \cdot 36$
'cells' and the radius of *zone G^1* will be 12 linear units.

The several points indicated above apply equally to a dome-
shaped apex, of approximately hemispherical shape, as seen in
longitudinal section.

In the literature on shoot apices the superficial tissue or
tunica is sometimes described as a region of 'surface growth',
whereas the inner tissue or corpus is said to represent 'volume
growth'. The data set out above show how misleading are such
conceptions or forms of statement. In a growing and expanding
cylindrical organ, if the more superficial layers are not to
become disrupted, 2π units must be added at the surface for
every radial increase of one linear unit. If we assume that an
apex, as seen in transverse section, has a tunica of 3 linear units
of thickness surrounding a corpus of 4 linear units of radius (the
whole apex thus having a radius of 7 linear units) the cross-
sectional area of the tunica will be $\pi(7^2 - 4^2) = 33\pi$ sq. units,
whereas the area of the corpus will be 16π sq. units (fig. 2). If

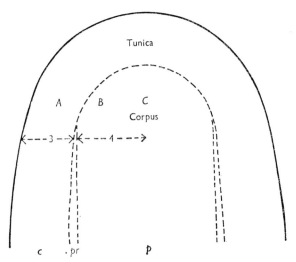

2 Geometrical relationships between tunica and corpus. *A*, tunica (cortex,
c); *B*, incipient vascular tissue, or prevascular tissue *pr*; *C*, corpus (pith, *p*).

now, both tunica and corpus expanded radially by one linear unit, then the total cross-sectional area would increase from 49π sq. units to 81π sq. units, the latter comprising a tunica of 56π sq. units and a corpus of 25 sq. units. If one radial unit were added to the corpus and if the thickness of the tunica remained unchanged, the new cross-sectional area would be 64π sq. units, comprising a tunica of 39π sq. units and a corpus of 25π sq. units.

The writer (Wardlaw, 1952) has suggested that, between the subapical region, and the region of maturation, the upwardly-moving nutrients, hitherto confined to the vascular column, will tend to diffuse upwards, outwards and inwards, i.e. the apical region will be supplied with nutrients diffusing upwards over the whole cross-sectional area of the shoot. If some such conception as this approximates to the truth, and if we were dealing with a purely physical system, the superficial and most distal cells of the apical and subapical regions would be furthest removed from the foci or sources of the diffusion gradients and would receive the smallest supplies of nutrients; whereas those nearest to the sources would receive most, and might therefore be expected to grow more rapidly. In such a system, the more slowly growing peripheral and distal cells would be subjected to an outwardly directed pressure and to tangential stress: as a result the distal region of the shoot would tend to assume a hemispherical or subhemispherical shape and its component cells a zoned or layered arrangement. They would also tend to be distended tangentially, being subject to a tangential tensile stress, and cell division would normally be by anticlinal walls. Limiting supplies of nutrients to the more peripheral tissues, and the induction of tangental tensile stress, may be among the factors that limit the radial enlargement of the primary shoot. It may also be inferred that the more central tissues of the shoot will be under pressure, i.e. under a compressive stress, and cell divisions will not be confined to one particular plane.

As a fact, we know very little about the distribution of nutrients, to and in, the apical region of the shoot. The foregoing arguments merely indicate how some of the relevant problems may be envisaged. It is, however, apparent that the more peripheral tissues in a growing shoot must be well supplied with metabolites if they are to keep pace with the expansion of the tissues within. The hypothesis that the outer-most layers on the flanks of the apex grow so rapidly that they form folds which become leaf primordia (Schuepp, 1914; Priestley & Scott, 1933) seems inherently improbable in the light of the analysis given here.

In the foregoing geometrical treatment of the zoned apex it was assumed that there was a uniform distribution of the meta-bolites on which growth depends. It is improbable that such a distribution does in fact occur. On the contrary, the distribution of the several nutrients at the apex is likely to be a differential one. A survey of the apices of vascular plants indicates that different species are characterized by different distributions of growth. Thus, whereas some shoots have a central, non-medullated vascular strand surrounded by a wide cortex, others may have a small pith, and yet others a very wide pith, these several tissue regions being differentiated at the apex. These several types of construction can be observed in both leafy and leafless shoots. Anatomical observations show that growth is not uniformly distributed over the cross-section of the shoot. In species of *Selaginella*, for example, the wide lacuna round the stele is demonstrably due to the fact that there is much more rapid growth in the parenchymatous tissue of the inner cortex than in the stele (Wardlaw, 1925). The range in possible histological patterns may be indicated by reference to fig. 2. (A particular distribution of tissues has had to be taken for granted but, in fact, a basic problem is to investigate the in-ception of such a pattern.) *A* is the tunica or incipient cortex; *B* is the region at the junction of the tunica and the corpus; it is in this region that the procambium (or prestelar tissue) is formed; *C* is the corpus. We may now suppose that, in different instances, in relation to factors in the genetical constitution, growth is most rapid in *A*, or *B*, or *C*.

(i) If the tunica, *A*, is the region of most rapid growth, the component cells will divide by both periclinal and anticlinal walls, and there will be no precise zonation. As regions *B* and *C* will develop under tensile stress, we might expect to find either a wide thin vascular cylinder, or a highly disrupted one, while the pith would expand into large, highly vacuolated parenchymatous cells.

(ii) If *B* is the region of most rapid growth, the tunica will develop under pressure from within and will accordingly have a regular, zoned construction. The vascular system may show considerable development, or it may be left suspended in a lacuna, as in *Selaginella*. The pith will be as in (i).

(iii) If C is the region of most rapid growth, the tunica will have a highly regular, layered construction, the vascular cylinder will tend to be disrupted and there will be a wide pith of relatively small parenchymatous cells.

Various combinations of these growth relationships are also possible.

Although these ideas and inferences are necessarily of a tentative character and have been treated in a static rather than a dynamic fashion, they indicate the possibility of analysing apices in terms of the gene-controlled distribution of growth, of physical factors which become incident, and of geometrical relationships. Furthermore, by appropriate investigations of apices of related species and genera, of known genetical constitution, it may be possible to gain some insight into the mechanism of genic action in histogenesis.

Discussion

A morphological study of representative members of the Psilopsida, Lycopsida, Sphenopsida and Pteropsida (comprising ferns, gymnosperms and angiosperms)—the four main groups of vascular plants according to modern ideas—shows that although very different kinds of organization are involved, all have, nevertheless, certain major features in common. In each group the plant consists of a vascularized axis on which (with the exception of some Psilopsida) lateral organs are disposed in a regular manner; the axis and its lateral members being the products of the growth of an apical meristem. The apical meristems of different genera and species may show considerable differences in their cellular pattern or organization, not only as between the four major divisions but also within these divisions. These differences are held to be due primarily to differences in genetical constitution. The marked homologies of organization shown by all shoot apices, especially in respect of the organs and tissues to which they give rise, prompt the question as to whether the leafy-shoot type of organization in different taxonomic groups is to be attributed primarily (*a*) to the

L

presence of identical or equivalent genes in the hereditary constitution of all of them, or (*b*) to parallel evolution in quite different genetical systems, or (*c*) to the action of extrinsic, or non-genetical factors. It may be that this problem will prove to be insoluble but, in view of its importance in evolutionary and morphogenetic theory, some analysis of it should be attempted. (It is only a matter of convenience to speak of separating the action of intrinsic and extrinsic factors. They are, in fact, inseparable; but in any particular situation one may be of greater importance than the other.)

While it is recognized that genetical factors are primarily involved both in the specific and more general aspects of morphogenesis, non-genetical factors of various kinds, i.e. physiological processes of a general kind, e.g. metabolic gradients, factors in the environment and other physical factors, and size and spatial relationships, are also involved. Even if the biochemical actions determined by, or associated with, particular genes could be specified, it would still be necessary to investigate their interaction with other genes and with the several extrinsic factors before an adequate account of the mechanism of morphogenesis could be given. It is conceivable that some of the main organizational features in plants may be referable to non-genetical factors, though the substratum in which these factors work is always specific and gene-determined. Polarity and axial development, the division of cells by walls of minimal area, the inception and self-determination of the apical meristem, the effect of older leaves on the inception and growth of younger ones, reciprocal relationships between organs and cells, various correlation phenomena—all important phenomena in morphogenesis and of widespread occurrence in the Plant Kingdom—are largely non-genetical in character. A view that emerges from such considerations is that some of the major homologies of organization in vascular plants may be attributable to extrinsic factors and relationships, and to the genetical and physiological equipment common to green plants, i.e. already present in the green algal ancestor(s) from which vascular plants are thought to have evolved.

The probable importance of non-genetical factors in morphogenesis is borne out by studies of the apices and somatic organization in non-vascular plants. Thus, as Lang (1915)

pointed out, the gametophyte in the mosses, which is an axial
structure with regularly disposed lateral appendages and an
incipient conducting strand, formed from a well-defined apical
growing point, affords a remarkable parallel with the axial
sporophyte of vascular plants. In some of the larger parenchymatous brown algae, e.g. the Fucales, we find some quite
remarkable parallelisms with the developments seen in vascular
plants. Thus, in *Fucus, Halidrys, Sargassum, Seirococcus* or
Ascophyllum, not only can an apical cell be distinguished at the
growing tip of the thallus, but it is seen to be the originating cell
in a clearly defined apical meristem, the process of segmentation in the meristem being in general agreement with what is
found in the ferns. Furthermore, in *Ascophyllum* as in leptosporangiate ferns, the lateral buds, or side branches, originate from
detached meristems, i.e. from superficial areas of meristematic
cells which at one time formed part of the apical meristem (for
illustrations see Fritsch, 1945*a, b*). In the brown algae, moreover, the organization is essentially axial, the lateral members
being formed in an orderly sequence; and although there is no
true conducting strand, a column of distinctive, elongated cells
extends from the apical meristem downwards in the thallus.
Lastly, in such algae as *Sargassum*, the lateral members show a
morphological development closely comparable with that of a
leaf subtending a branch in its axil. The homology of organization in brown algae and vascular plants is thus very close
indeed, and hence it may be inferred that this is due either to
the presence of parallel or equivalent groups of genes, or to the
same extrinsic factors and relationships.

The view that extrinsic factors and size and spatial relationships are important in morphogenesis is supported by experimental evidence. Thus it has been shown (i) that the position
in which a new leaf primordium will arise can be modified at
will by appropriate surgical treatment of the apical meristem
(Snow & Snow, 1931, 1933, 1935, 1947; Wardlaw, 1949*b, c*);
(ii) that the phyllotactic system may be modified, e.g. a
decussate system may be transformed into a spiral one, or the
direction of a spiral system may be reversed (Snow & Snow,
1931, 1933, 1935); (iii) that buds may be caused to arise in leaf
positions; (iv) that leaves may be caused to arise in bud
positions (Wardlaw, 1949*b*, 1950*c*); and (v) that the shape and

orientation of leaf primordia can be modified at will (Sussex, 1951).

Lastly, the reader may be referred to a recent paper by Johnson (1951) in which a point of view not unlike that expressed here is indicated: in studying apices the visible histological organization should be regarded not merely as a structural feature to be described and classified—though it may have its value in the phylogenetic seriation of related organisms —but as affording an indication of the underlying physiological processes. In the present paper an attempt has been made to indicate new avenues along which advances may be made.

12
Evidence relating to the diffusion-reaction theory of morphogenesis

In contemporary studies of morphogenesis in plants, certain basic assumptions are probably generally accepted. Among these are: (i) that the zygote, spore, regenerative cell, or meristem is a very complex organic reaction system which obeys the laws of physical chemistry; (ii) that a patternized distribution of metabolites precedes and determines each morphological or histological 'stage' of development; (iii) that the reaction system in embryonic regions such as the shoot apex becomes modified in characteristic ways as the plant increases in size and matures. How the patternized distribution of metabolic substances is brought about is one of the most fundamental and complex problems in biology. The patterns themselves may belong to several categories, e.g. those in which diffusible substances become concentrated in particular loci as a result of the working of the reaction system, and those in which centres of metabolic activity are already determined by some feature in the inherent organization of the cell or tissue.

Turing (1952) has proposed a diffusion-reaction theory of morphogenesis, based on well-known laws of physical chemistry, in which he has indicated how, in an embryonic tissue with an initially randomized or homogeneous distribution of diffusible reacting substances, a regular, stable, patternized distribution of metabolites may nevertheless be produced; the inception of this pattern being the initial phase of the ensuing morphological and histological developments. A simplified account of this theory has been given by the writer (Wardlaw, 1953a), and its application to morphogenesis in plants discussed. Turing's theory, if it can be justified, may account for such phenomena as polarity, whorled and Fibonacci phyllotaxis, the radiate differentiation of tissues in the root-stele, and so on.

As thus far developed, the theory relates to relatively simple and symmetrical patterns. Its further elaboration to more complex patterns will undoubtedly be difficult; and its mathematical presentation is likely to be beyond the non-mathematical botanical investigator. Nevertheless, once the central idea of the theory has been grasped, one can understand that a reaction system, which initially gives rise to relatively simple patterns, is likely to become more complex and to give rise to more complex ones as development proceeds. Moreover, asymmetries in the environment of the reaction system, i.e. the adjacent organs and tissues, may induce characteristic asymmetries in the system and in the patterns to which it gives rise.

Turing's theory is evidently of great importance in the investigation of morphogenesis, for it has a very general application to problems of organismal form and structure; and even if it cannot be justified, it certainly seems to point in the right direction and the work expended on it may lead to the formulation of an alternative and more appropriate theory. The most pressing need at this stage—to consider how the theory can be tested—is the central purpose of the present paper. Only a few examples which appear to offer scope for further work are considered here; but, out of their experience, readers will no doubt be able to indicate many other examples to which the theory seems to apply.

Tests of the diffusion-reaction theory
The validity of the theory may be tested in a number of ways. The general test, i.e. its applicability to a wide range of biological materials, has already been discussed (Wardlaw, 1953a).
Prediction
Indications of the probable validity of the theory by the method of prediction have been obtained by Turing (1952) using the digital computer. In one example, in which realistic values were given to the factors involved, two morphogens (morphogenetic substances) were considered to be present in a reaction system consisting of a ring of twenty cells: it was found that various

lobed patterns, e.g. a three- or four-lobed pattern, could result from the working of the system. In other examples Turing has shown that a two-dimensional pattern such as dappling, and gastrulation in a spherical body, could result from the working of specified diffusion-reaction systems.

Analytical investigations

During the ontogenetic development of the individuals of some species, the number of organs formed simultaneously at the apical meristem increases. For example, in the stellate Rubiaceae, the seedling has two cotyledons; but on proceeding upwards in the main shoot of an adult plant, or in a lateral branch, the number of leaves in the successive whorls increases till as many as nine may be present, as in goosegrass (*Galium aparine* L.). It seems probable that the number of leaf primordia formed at the apex at any particular stage of development is directly related to the size of the apical meristem. The latter may be envisaged as a reaction system in which, at intervals, a patternized distribution of metabolites takes place, this being the antecedent phase to the visible outgrowth of the leaf primordia. If now, it can be shown (*a*) that leaf primordia at the time of their inception vary in size only within narrow limits, or vary in size in a manner corresponding with variations in the apical meristem, e.g. as a result of increase in size of its constituent meristematic cells, (*b*) that there is a close relationship between the size of the meristem and the number of primordia formed by it and (*c*) that the presence of particular metabolites can be demonstrated at the loci of leaf inception and not in the tissues between them, the conception of the meristem as a diffusion-reaction system would appear to afford an explanation of the observed developments. In many plant materials it will probably be found that the ratio *number of leaves : size of meristem* only approximates to a steady value when allowance is made for changes in the size of meristematic cells and of primordia during the ontogenetic development. This is certainly true of the cellular construction of the meristem in ferns.

Many of the plant materials which might be used to test the theory by the method of analysis are of the kind indicated above. In the embryo of Equisetum three leaves are typically formed at the first node. As the young sporophyte grows, its apex enlarges and progressively larger numbers of primordia are formed at each node till, in the adult shoot of some species, a very large number indeed may be present. Here again, the essential information relates to the size of the apex when the leaf primordia are just becoming visible. Various algae in which branch filaments arise in whorls may also prove interesting and useful in this connection.

In different species of *Pinus*, and within particular species, the number of cotyledons present in the embryo may be very variable. Chamberlain (1935) has illustrated a group of embryos in which there is an evident proportionality between the diameter of the axis and the number of cotyledons. Here, again, the crucial information will relate to the size of the apical meristem at the time when the cotyledon primordia are just becoming visible. Information of great relevance to the present inquiry has been given by Buchholz (1946) on the size and morphology of the embryo in *Pinus ponderosa*. Measurements of embryos, endosperm and seeds have shown (*a*) that the embryos of large seeds, *at the time of shoot apex and cotyledon differentiation*, are much larger than those of small seeds, and (*b*) that there is a direct relationship between the size of the seed and the number of cotyledons in the contained embryo. Thus, embryos in small seeds may have 6 cotyledons, whereas those in large ones may have 15 or 16. When the seed sizes were arranged in a graded series, the embryos of the smallest grades had mean cotyledon numbers of 8·3 and 8·6 (range, 6–10 cotyledons) and the largest grade had a mean cotyledon number of 12·1 (range 9–16 cotyledons). There is thus a close correlation between embryo size and number of cotyledons. Buchholz himself remarks: 'It might be supposed that there is more room for a large number of cotyledons on the shoulders of large embryos than on small ones. This would be expected if the plumule primordium of a large embryo were larger than in a small embryo and the cotyledonary primordia did not share proportionately in the increase in size of the embryo as a whole.'

The banana inflorescence consists of a main axis bearing spirally arranged bracts in the axils of which are outgrowths supporting flattened groups of flowers (cincinni). The flower-bud primordia appear as mounds of meristematic tissue arranged in a row or in two alternating rows on the floral cushion. When the banana fruits are mature, they are sometimes referred to as 'fingers' on the 'hand', i.e. floral cushion. Some hands contain many more fingers than others, those closest to the morphological apex of the bunch containing fewest. At an early stage in their development, flower-bud primordia are of approximately equal size but the floral cushions on which they are borne may be of different sizes. And while the primordia are not formed simultaneously (Fahn, 1953), they are formed in such rapid succession along the cushion as to suggest that the latter is the locus of a diffusion-reaction system which yields a patternized distribution of growth centres.

Species in which the roots range from small to large size without secondary thickening afford materials for testing the theory in terms of the histological pattern. In the Marattiaceae, e.g. *Angiopteris*, *Marattia*, etc., small roots may have 3–5 xylem rays, whereas large ones may have 16–20 xylem rays. This differentiation can first be discerned in the incipient vascular cylinder a little below the root apex. At this cross-sectional level we may suppose that a diffusion-reaction system, of the kind that may yield a patternized distribution of metabolites, is present, the results of its activity being seen in the inception of alternating protoxylem and protophloem groups. Observation of the sizes of steles and of the numbers of their protoxylem groups at the earliest stage of differentiation, seems likely to yield information of relevance to the present inquiry. A close correlation between the radius of the stele and the number of xylem rays would lend support to Turing's theory. A general result of this kind has already been obtained for mature roots (Wardlaw, 1928), and further information relating to the size-structure relationship at the apex is now being sought.

In anomalous roots some anatomical developments of special interest in the present connection have been observed. In the normal roots of *Caltha palustris* four xylem rays are typically differentiated in the apical region. Moss (1924), however, found that in enlarged roots with flattened apices, the undifferentiated stele was also proportionately enlarged. In some of these fasciated roots two or more separate four-rayed steles had been differentiated; but in others, a single, large, ribbon-shaped stele, with a characteristic wavy band of xylem and alternating phloem had been differentiated. These developments are what might be expected if the reaction system in the enlarged, flattened stele continued to produce its characteristic patternized distribution of the precursors of xylem and phloem.

Experimental investigations

Experimental tests of the theory will consist essentially in interfering with the reaction system in a known manner, at some critical stage, the effects being seen in modified morphological and histological developments. A satisfactory test, in which some prediction of the result should be possible, would require (*a*) that some of the major factors in the system were sufficiently well understood, and (*b*) that the system could be modified under strict control. In the present state of knowledge, however, it is doubtful if any morphogenetic process is sufficiently well understood to make a crucial experiment possible. As far as the writer is aware, no experimental work has yet been undertaken in direct relation to the theory. In experiments undertaken for other reasons, however, the results obtained appear to have a bearing on the theory and suggest possibilities for further useful work. As a general indication of the writer's ideas on the experimental approach, it may be possible to modify a reaction system experimentally, with consequential effects on the morphological and/or histological pattern which it produces, by one of the following procedures: (*a*) by increasing or decreasing its spatial distribution, e.g. in embryos and apices grown to different sizes in different cultural conditions; (*b*) by physical distortion or interruption of the embryonic tissue in which the reaction system is working; (*c*) by changing the concentration of one or more of the major metabolites in the system; (*d*) by introducing new growth-regulating or morphogenetic substances into the system; (*e*) by poisoning the system, or a part of it.

Formation of rhizomorph initials. Garrett (1953) has shown that when the fungus *Armillaria mellea* is grown in agar plate culture, the formation of its rhizomorph initials is determined by nutritional factors. When a 4-mm. disk of mycelium, taken from the margin of a colony, is laid on an agar culture medium, there is an outward growth of hyphae within 24 hr. at 25° C., and after 7 days rhizomorph initials are formed in a circle on the margin of the inoculum disk. The rhizomorphs formed from these initials grow vigorously and soon extend radially to the margin of the colony, imparting a deeply lobed appearance to it. Now, Garrett has found that the formation and subsequent growth of the rhizomorphs are affected by an interaction between the carbohydrate and nitrogen in the medium: under some conditions many more rhizomorph initials are formed than subsequently develop. He also observed that no rhizomorph initials, other than those which originated round the margin of the inocvlum, were ever formed. When 1 or 2 per cent dextrose was added to the basic medium (which contained 0·5 per cent malt extract), many more rhizomorph initials were formed; but the addition of 4 per cent dextrose significantly depressed the number of initials. A significant interaction between the dextrose concentrations in the inoculum disk and in the agar-plate medium was also noted (Tables 1 and 2, both from Garrett).

Table 1. *Number of rhizomorph initials around inoculum disk*

Percentage of supplementary dextrose

In inoculum	In growth medium				Mean
	0	**1**	**2**	**4**	($\pm$ **(1·525)**)
0	1	39	27	23	23
1	5	21	23	19	17
2	7	24	27	19	19
4	10	21	26	21	19
Mean ($\pm$ 1·525)	6	26	26	20	—

Standard error for individual treatment means = 3·05.

Table 2. *Number of rhizomorph initials around inoculum disk*

Percentage of nitrogen in media

	0·02	**0·04**	**0·08**	**0·16**	**0·32**	**0·64**
1% dextrose	11	17	21	22	17	16
2% dextrose	11	17	24	22	23	24

Standard error for individual treatment means = 1·06.

Table 2 shows that by varying the carbohydrate and nitrogen contents of the medium, the number of rhizomorph initials formed can be varied. At the 1 and 2 per cent levels of dextrose, the number of rhizomorphs increased up to 0·08 per cent nitrogen (supplied as peptone), and decreased at the highest nitrogen levels with 1 per cent dextrose but not with 2 per cent. In experiments in which the normal development of colonies was mechanically disturbed in various ways, the results indicated that a certain threshold nutrient status of the medium is necessary for the inception of rhizomorphs, that these originate simultaneously, and that once a rhizomorph has begun to grow the inception of other initials in the vicinity is inhibited. Garrett's explanation of this latter phenomenon is that the uptake of nutrients by the fringing hyphae of the existing rhizomorphs, and the depletion of the nutrients in the agar beyond this fringe by diffusion, preclude the formation of new rhizomorphs.

In relation to the diffusion-reaction theory, the principal interest lies in the inception of the rhizomorph initials. In the present writer's view, the enlarging mycelial colony on an agar substratum may be regarded as an organismal reaction system. When certain nutrients (or factors) are present in suitable concentrations, the reaction system (or colony) gives rise to a number of regularly distributed growth centres, i.e. rhizomorph initials; and by varying some of the factors in the system, the number of evenly distributed rhizomorph initials can be increased or decreased at will. Thus, on the basis of the valuable indications given by Garrett, it would seem that by using *Armillaria* and other fungi, e.g. those forming sclerotia on culture media, experiments could be designed to test the diffusion-reaction theory.

Bud formation in tissue cultures. Some tissue cultures can be kept in a relatively undifferentiated condition by growing them on suitable media under certain conditions. If, however, certain substances are added to the medium, loci of active cell division are established and in due course buds may be formed from them. It would be interesting to know if these growth centres constitute a regular pattern, and if this pattern can be varied by modifying the composition of the medium.

Embryo culture. Young embryos of *Capsella bursa-pastoris* can be dissected from the ovules and grown in pure culture. In some media Rijven (1950) found that as many as six cotyledons may be formed in some embryos; i.e. the distal peripheral region of the small club-shaped embryo, which *in ovulo* usually yields two centres of growth, may be modified experimentally so that a larger number of growth centres is differentiated. Such findings suggest that the distal region of the young embryo functions as a reaction system and that the patternized distribution of metabolites to which it usually gives rise may be modified by varying some of the factors in the system. Comparable observations have been made on excised embryos of other species; Sanders (1950), for example, has observed multiple and fasciated embryos in artificial culture. More intensive experimental study of such materials may yield information on which the validity of the theory can be further assessed.

Chemical and physical treatments of apical meristems. The shoot apical meristem may be regarded as a reaction system: the orderly manner in which its lateral members are formed may be attributed to an antecedent patternized distribution of metabolites. If this view is justified, then, in favourable materials, it should be possible to modify the reaction system, in characteristic ways, by appropriate chemical and physical treatments. As we still know very little about the metabolism of the apex, it is difficult to know what treatments to apply, and still more difficult to predict the results of such treatments as may be applied. Nevertheless, a growing body of evidence shows that the normal reactions of the apical meristem can be modified by various treatments and that changes in the number, distribution, size and form of the lateral members can be brought about. Such induced changes do not, of course, prove that the apical meristem functions as a diffusion-reaction system, but an interpretation along those lines is not precluded; and it may eventually prove to be a satisfying one.

De Waard & Roodenburg (1948) and Gorter (1949, 1951) have shown that when a solution of 2,3,5-triiodobenzoic acid is applied to the vegetative apices of the tomato, some remarkable departures from the normal development ensue, e.g. the formation of leaves is inhibited. The most distal region of the apex also becomes inhibited, but the subjacent ring of meristematic tissue may continue to grow and form a ring fasciation or hollow tubular shoot. On further growth, this ring meristem begins to form leaf- or flower-bud primordia both on its inner and outer flanks. Now, in the relatively limited space on the normal shoot apex, only a small group of primordia, usually three, constitutes the top cycle of primordia. But, in a ring

meristem, considerably more space is available on the outer
and inner flanks, the reaction system is also modified in some
way, and many small, evenly distributed primordia are formed.

When 2,3,6-trichlorobenzoic acid was applied to seedling
plants of *Phaseolus multiflorus* by way of the epicotyl, the cotyledon axillary buds became curiously modified. Their growth
was relatively inextensive, they produced no leaves, but they
became greatly swollen (Wardlaw, 1953*b*). It was found that
the growth and normal morphogenetic activity of the apex had
been almost completely inhibited, but extensive root formation
had been induced in the pericycle. These out-growing roots,
some 16–17 of which could be seen in a cross-section, constituted
a radial pattern of great regularity. Here, the pericycle, in which
meristematic potentiality is retained, may be regarded as a reaction system. In normally developing shoots the pericycle remains morphogenetically quiescent; but if it is activated as a
result of applications of 2,3,6-trichlorobenzoic acid, it apparently yields a patternized distribution of those metabolites which
are directly involved in the inception of root initials, and a new
anatomical pattern results. Other instances are known in which
regular patterns of root formation have been induced by the
application of growth-regulating substances to the cut distal end
of the shoot (for a review of the literature *see* Wardlaw, 1952).
Such materials appear to be suitable for further investigations
of the theory.

The constitution of the shoot apex may be modified by sub-
jecting it to radiation by X-rays. Van Heijningen, Blaau &
Hartsema (1928) have described the results of exposing tulip
bulbs to X-rays in different ways. Some bulbs were exposed
about the time of initiation of the floral parts from the transformed vegetative apex, this being a period of marked meristematic activity; others were exposed after the embryonic
floral members had been formed. The apices of the axillary buds
in the axils of the scales were also irradiated. Treated apices
showed various inhibitions and anomalous developments. A
characteristic deviation from the normal consisted in the division of an original growth centre, which would have given rise
to a large organ, e.g. a petal, into a number of small growth
centres ('tissue centres'), the result being the formation of
lobed and warty outgrowths. All the abnormalities observed

were interpreted by these investigators as affording evidence of various degrees of inhibition.

When shoot apices are subjected to surgical treatments as in the experiments of Snow & Snow, Ball and the writer (reviewed in Wardlaw, 1952), the spatial disposition of the reaction system is modified. In these experiments it has been shown that the normal phyllotactic sequence can be modified at will in various ways, e.g. leaf primordia can be induced in other than their normal positions, buds can be induced in leaf positions, and leaves in bud positions. In one experiment Snow has shown that when the apex of the normal decussate shoot of *Epilobium hirsutum* was split by a diagonal incision, two shoots, each with spiral phyllotaxis were formed. The results of the several surgical experiments could be interpreted in terms of the disturbance of the meristem reaction system.

Mutants and polyploids. Mutants and polyploids occur in nature but they can also be induced by experimental means. Some relevant evidence based on a study of them may therefore be considered under the general heading of this section.

The principal factors in an organismal reaction system are the gene-determined metabolites. If, now, gene-mutation takes place or is induced, if the normal action of a gene is modified by a 'position effect', or if the number of chromosomes is modified by polyploidy or aneuploidy, the factorial constitution of the reaction system will also be modified, and this may have consequential effects on the metabolic patterns and morphogenetic developments to which it gives rise. The morphogenetic investigation of related plants of known genetic constitution may thus yield information of direct relevance to the diffusion reaction theory.

Marsden-Jones (1933) and Perje (1952) have shown that *Ranunculus ficaria* occurs as diploid and tetraploid races, the former being fertile, the latter largely sterile and multiplying vegetatively by bulbils in the leaf axils. *R. ficaria*, in short, apparently exists as a number of different races. In an investigation of the variation in the quantitative characters of the flowers from different localities in Sweden, Perje has shown that in individuals from different populations, which are mainly tetraploids, there are marked differences in the numbers of sepals (3–5), petals (7–16), stamens (19–38), and pistils (13–32).

When the number of stamens is large, so also is the number of pistils. These morphological developments are chiefly determined by the genetical constitution, but environmental factors are also involved. In the present connection, it would be interesting to have the facts relating to the apical meristem and early floral development in different races which have been investigated genetically. It seems probable that a large apical meristem is present in those tetraploid individuals with a large number of floral parts; and it might be possible to trace these developments to modifications of the reaction system of the parental diploid form under the impact of one or more mutant genes or gene-complexes.

Not all polyploids give rise to a larger number of floral members. Instances are known where the apices of tetraploids are larger than those of diploids, but the floral members, though of larger size, show no increase in number.

Harrison (1953) has given some facts about the morphology and anatomy of *Nuphar intermedia* which may be relevant here. This species comprises plants which fall between the generally accepted variation ranges of *N. lutea* and *N. pumila*. Moreover, there is a similarity between wild plants of this species and artificial hybrids between *N. lutea* and *N. pumila*. Hence the view that *N. intermedia* is not a pure breeding species but a collection of hybrids. In all these species the chromosome number is $2n = 34$. Harrison has compared the floral characters of the three species. The size and number of the petals and carpels, both of which show meristic variation, are of special interest here (Table 3).

Table 3. *Meristic and size data for floral characters in British* Nuphar *taxa* (all size measurements in mm.)

Taxon sample	Petal Number	Length	Carpel Number	Diameter
N. pumila agg.				
(P1 + P2)	11·42 ± 0·17	4·74 ± 0·11	9·42 ± 0·11	7·96 ± 0·14
P1	11·63 ± 0·21		9·26 ± 0·15	7·88 ± 0·19
P2	10·86 ± 0·13		9·54 ± 0·16	8·36 ± 0·15
P3 (Avinlochan)	10·56 ± 0·41	5·00 ± 0·13	10·90 ± 0·20	10·11 ± 0·23
N. intermedia	9·82 ± 0·25	8·22 ± 0·34	11·58 ± 0·18	9·20 ± 0·18
N. lutea agg.				
(L1–12)	15·11 ± 0·22	15·12 ± 0·25	15·78 ± 0·19	12·40 ± 0·22

The mean petal number for *N. intermedia* appears to be significantly less than that in the *N. pumila* aggregate; that in *N. lutea* is relatively high. Mean stamen numbers of *c.* 105, *c.* 66 and *c.* 52 are indicated for *N. lutea*, *N. intermedia* and *N. pumila* respectively. The pistil size and carpel number in *N. intermedia* are intermediate between the two other species. These materials afford evidence of a correlation between pistil size and the number of its radiating stigmata. Moreover, in *N. lutea*, the stigmatic rays may be very variable in form; in some plants, or races, the rays are narrow and well separated; in others they are broad and coalescent along most of their length, thereby constituting a virtually continuous stigmatic surface. These differences are indicative of genotypic differences. Sections of the peduncles also reveal materials for comparison: in *N. pumila*, *N. intermedia* and *N. lutea*, with diameters of 26, 32 and 36 mm. respectively, the uniformly distributed vascular strands number 8, 14 and 19 respectively; an 'introgressed' *N. pumila* with a diameter of 29 mm. had ten vascular strands. The ratios *number of vascular strands : cross-sectional area of peduncle* (taken as r^2) are as follows:

N. pumila	21	*N. intermedia*	18
N. pumila ('introgressed')	21	*N. lutea*	17

The meristic variation in all the materials considered above is determined during the growth and metabolism of the apical meristem. Comprehensive morphogenetic analyses of the floral developments, together with genetical investigations of these species, and of other comparable materials, seem likely to yield information by which the diffusion-reaction theory can be further tested.

Conclusions

In each of the examples considered here, Turing's theory appears to afford some explanation of the observed developments; at least, the theory does not appear to be inconsistent with these observations. Perhaps at this stage, the principal value of the theory is that it indicates the general direction in which satisfying explanations of morphogenetic developments are likely to be found; for whatever theory eventually proves to be acceptable, it will almost certainly be based on the physical chemistry of complex organic systems. At the present time our knowledge of metabolic processes in embryonic and meristematic tissues is very limited, while the mechanisms and rates of movement of nutrient and morphogenetic substances in such tissues, and their reactions with the 'fine structure' of cells and with the fixed centres of metabolism which may be present in cells, are still very obscure. The validation of the theory will require evidence (i) that diffusion-reaction systems are actually present in the meristems of plants, and (ii) that these systems operate in such a way as to yield a patternized distribution of metabolites, thus affording the basis for morphological and histological developments. What Turing's theory does is to indicate the possibility that morphological and histological patterns do, or could, originate as a result of the working of certain diffusion reaction systems. In this paper the writer has tried to show that there are many morphological and histological developments, of very different kinds, which could be explained by the theory and which, thus far, have not been satisfactorily explained by any other theory.

13
Experimental investigation of leaf formation, symmetry and orientation in ferns

Investigations of the inception and development of leaf primordia in *Dryopteris dilatata* have yielded the following information. (Primordia are referred to as P_1, P_2, P_3, etc., P_1 being the youngest visible primordium; I_1 indicates the site of the next primordium to be formed and I_2 the next after that. The tetrahedral cell at the tip or summit of the shoot and its recently formed segments are referred to as the 'apical cell group'):

(i) Leaf primordia are always formed on an apical meristem.

(ii) A leaf primordium typically originates near the base of the apical cone. It consists of a circular to elliptical group of the distinctive, superficial prism-shaped cells of the apical meristem, which grows out, forming a low mound of tissue; the small-celled tissue within is also in a state of active division.

(iii) A young primordium, situated between and above two older ones, occupies only a small part of the free surface of the apical cone.

(iv) On further development, an approximately central, superficial cell of the primordium now becomes much enlarged. During this process, this 'leaf apical cell' acquires its characteristic 'two-sided' shape, its orientation being such that the segments produced by its division lie in the tangential plane. These give rise to the marginal meristems from which the tissues of the primordium are formed.

(v) The alignment of the leaf apical cell and of its partition walls at cell division is thus at right-angles to the principal direction of growth in the subtending apical cone, that is, the tangential direction; and it may well be that the shape and disposition of the leaf apical cell are directly attributable to the relatively large tangential component of growth in the leaf site. Once the two-sided apical cell has been established, the foliar nature of the primordium is irreversibly determined.

(vi) In relation to the position in which a leaf primordium originates, that is, on the side of a cone, growth on its abaxial side is more rapid than on its adaxial side; and this difference soon becomes accentuated as the primordium passes into the actively growing and rapidly widening subapical region of the shoot (fig. 1). It thus appears that the orientation, dorsiventral symmetry and shape of a developing primordium may be referred to the growth activity of the region of the shoot apex in which it originates and to the disposition and activity of its own apical and marginal meristems.

(vii) While a primordium still forms part of the apical cone, its growth is relatively slow; but once it is in the subapical region, its growth and morphological development are rapid. The sooner a primordium comes to occupy a subapical position, that is, as a result of the growth of the shoot apex, the more rapid will be its enlargement and morphological elaboration.

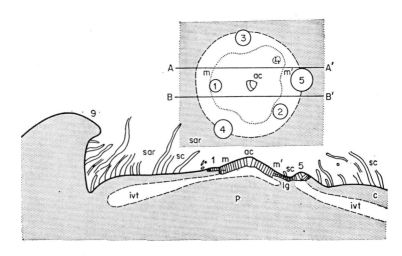

1 *Dryopteris dilatata*, the apical cone as seen in median longitudinal section, traversing the youngest visible primordium (1) and the older ones (5, 9). The apical cell, *ac*, and the distinctive prism-shaped cells of the apical meristem *m–m'* are shown. *sc*, scale; *c*, cortex; *p*, pith; *ivt*, incipient vascular tissue; *lg*, leaf-gap; *sar*, sub-apical region. (Semi-diagrammatic ×30.)

Thus, a P_3 primordium on a small apex of low phyllotaxis is larger and more elaborate than a P_3 on a large apex of high phyllotaxis.

(viii) The actual inception of leaf primordia is an obscure and complex phenomenon. In ferns, the indications are that a primordium is formed at a growth centre, that is, a locus of special metabolism. The position of this centre is apparently controlled by the apical cell group at the summit of the shoot and by the two adjacent primordia of the top whorl. The growth centre consists of a circular group of seven to nine meristem cells in which we may suppose there is an accumulation of the substances that determine the outgrowth of a lateral organ.

On the basis of experimental observations, I have already attempted[1] to relate leaf symmetry and orientation in ferns to the position in which the primordium originates, and to possible inhibitive effects proceeding basipetally from the apical cell group. Thus it was shown that if the I_1 position is isolated from the upper part of the apical meristem by a deep tangential incision, a bud, and not a leaf, is formed. In general terms it was thought that, as a result of this operation, inhibitive effects from the apical cell group would be precluded, and a levelling up of the growth-rates on the adaxial and abaxial sides of the primordium might take place. If so, the development of a radially symmetrical organ, that is, a shoot bud, might be expected. Cutter[2] has since shown that if P_1 and P_2 are similarly isolated before the characteristic two-sided initial cell has been formed, they, too, can be transformed into buds; and Wardlaw and Cutter[3] have shown that the prevascular tissue, which immediately underlies the prism-shaped cells of the meristem, is important in these several morphogenetic developments.

In some recent experiments an attempt has been made to ascertain whether the dorsiventral symmetry and orientation of leaf primordia can be attributed: (a) to a specific localized effect, on the adaxial side of the primordium, of an inhibiting substance proceeding from the apical cell group; or (b) to the regulated growth which is characteristic of the intact shoot apex as a whole. In relation to (b), it may be noted that there is a steady acceleration in the rate of growth from the apex to the base of the apical cone, and a rapid acceleration of growth in the region of transition from the base of the cone to the sub-apical

region (fig. 1). Primordia originate near the base of the cone
and are soon affected by this active region of transition. Wide
and deep tangential incisions above P_1 and I_1, whereby the
formation of buds can be induced, would evidently preclude
(*a*) and cause serious disturbances to (*b*). Here is it relevant to
note that when bud formation is induced in a leaf site, for
example, at I_1, or by the transformation of a leaf primordium,
for example, P_1, the bud not only quickly develops a large
apical meristem but also shows strong growth on its adaxial
side.

Of a considerable and varied series of new surgical experiments which have now been carried out, the two most directly
relevant to the problems of leaf and bud formation are illustrated diagrammatically in fig. 2. In one series, small deep
tangential incisions, severing the incipient vascular tissue, were
made immediately above I_1, I_2, P_1 and P_2. If dorsiventrality in
a primordium is due to a limitation of the growth-rate on its
adaxial side, that is, as a result of the basipetal movement of a
growth-regulating substance from the apical cell group, some
bud formation might be expected in this experiment. In the

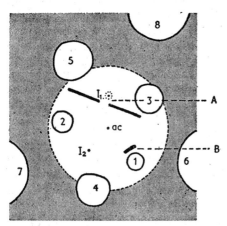

2 *Dryopteris dilatata*. Apex as seen from above, showing leaf primordia 1–8
and the positions of I_1, the next primordium to be formed, I_2, the next after
I_1, and the apical cell, *ac*. (A) Two wide and deep tangential incisions have
been made above I_1 so as to leave a 'bridge' of intact tissue between the
primordium site and the apical cell. (B) A small deep incision has been made
on the adaxial side of P_1.

other series, two deep and extensive tangential incisions were made just above a primordium (or primordium site), so as to leave a 'bridge' of intact tissue between it and the apical cell group. If dorsiventrality is mainly due to a localized (adaxial) inhibitional effect proceeding from the apical cell group, then leaves should still be formed; but if it is a result of the regulated growth of the apex as a whole (or of the sector in which the primordium is situated), then, in relation to the extensive restriction of both the acropetal and basipetal movement of metabolic substances by the tangential incisions, and the consequential radical disturbance of the normal distribution of growth in the apical cone, some bud formation might well be expected.

The results of these two experiments are set out in Tables 1 and 2, the records relating to specimens in which the apical cell group remained intact and quite undamaged. (These experiments are of some delicacy and not easy to perform successfully.)

The evidence from the first set of experiments (Table 1) is

Table 1 *Effect of Small Adaxial Incisions*

Primordium	Leaf formed	Bud formed	No development
I_2	2		
I_1	6		2
P_1	9		1
P_2	3		
Total	**20**	**0**	**3**

Note. In other specimens, in which the shoot apex had accidentally been slightly damaged, 13 leaves and 6 buds were obtained.

Table 2 *Tangential Incisions with Intact Bridge*

Primordium	Leaf formed	Bud formed	No development
I_2	3		
I_1	7	5	3
P_1	9	2	1
P_2	1	1	
Total	**20**	**8**	**4**

Note. Where the 'bridge' had been eliminated, that is, by the two incisions joining up, or where the shoot apex had been damaged, even very slightly, buds were usually formed. Where the 'bridge' was wide, leaves were usually obtained, though occasionally a bud was formed.

conclusive: small adaxial incisions have in no instance resulted in bud formation. It thus appears that the dorsiventrality of leaves cannot be attributed to the localized action of a growth-regulating substance on the developing primordium.

The evidence from the second series, though less conclusive, is none the less important. In at least some apices (actually 25 per cent of all primordia and 40 per cent of I_1's), bud formation can be induced when the apical cell is undamaged and when intact tissue extends between it and the primordium (fig. 3). In those apices in which buds were formed, the experimental region was marked by very conspicuous growth activity and the buds were of large size (fig. 3). Late-season apices, which were relatively inactive, typically yielded leaves.

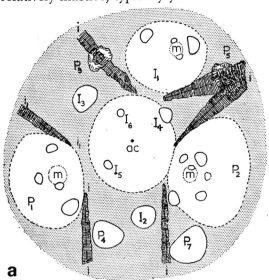

a

3a *Dryopteris dilatata.* Apex, as seen from above, showing the large buds which have been induced in the I_1, P_1 and P_2 positions when these positions were partially isolated from the shoot apical cell by two wide tangential incisions. In each instance a 'bridge' of intact tissue was present on the adaxial side of the primordium. *ac*, position of the apical cell of the main apex. P_1–P_7, primordia present at the beginning of the experiment, P_1 and P_2 having been transformed into large buds. The locus of I_1 has also developed into a large bud, ii–i_1i_1, tangential incisions. I_2–I_6, new leaf primordia formed during the course of the experiment. *m*, apical meristem of induced buds. The broken lines indicate approximately the bases of the main apex and of the buds. (Diagrammatic camera lucida tracing, × 15.)

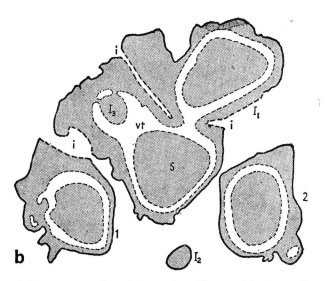

3b Transverse section of the specimen illustrated in fig. 3a. Cortex and pith, stippled; *s*, axis and stele of main shoot; *i, i* incisions; *vt*, vascular tissue. Figure 3b shows that the large solenostelic bud which developed in the I_1 position is in vascular continuity with the main axis, *s*. Similarly, in sections lower down, the solenostelic buds formed from P_1 and P_2 are seen to be in vascular continuity. At that level, the pith of bud I_1 is confluent with that of the main axis. (Semi-diagrammatic, ×25.)

Where buds were obtained, as in fig. 3, microtome sections showed that the bud was in vascular continuity with the parent shoot in the region of the 'bridge'.

In the light of these experimental results, to what factors and relationships may the organization of the fern apex and its orderly morphogenetic activity be attributed?

There is considerable support for the view that the apical cell group is a region of special metabolism, as is also each growth centre which may become a leaf, or, if subjected to experimental treatment, a bud. Products of the metabolism of the apical cell group, moving basipetally by way of the prevascular tissue, may excerise a regulative effect on the growth of the apical and sub-apical regions. Thus, in the upper part of the apex, where the concentration of the growth-regulating substance will be high, the rate of growth of the meristem cells may be kept low; but lower down, and in the region of transition to the sub-apical region, where the concentration of the growth-regulating substance will be much lower, growth may be promoted. This hypothetical substance, which might be an auxin or hormone, or a precursor of one or other of these, is likely to be essential in some way to protein synthesis, as Skoog[4] has recently suggested. An important feature of this conception is that the substance will produce its organizing and morphogenetic effects by the way it affects the growth of individual cells, each of which will react in a characteristic way according to its physiological state, rather than by its action on primordia as such.

An alternative view might take the form that the apical cell group is not a locus of special metabolism, the organization of the apex being due to competition among its constituent cells for the nutrients involved in growth. As these nutrients move acropetally from the older regions below, with differential utilization of the several substances at different levels *en route*, the supply will diminish progressively and change qualitatively as the shoot tip is approached.

The most satisfying explanation, however, probably lies in a combination of the two views. Thus destruction or isolation of the shoot tip would simultaneously remove both its regulative and competitive activities. In the 'bridge' experiment described above, the two wide tangential cuts go a long way towards isolating the shoot tip, and, as we have seen, marked growth activity may be induced on the adaxial side of primordia and buds may be formed. On the other hand, as even a cursory inspection will show, a small incision above a primordium will do little to modify the normal movement of metabolites. As rapid wound-healing, with formation of parenchyma and an outgrowth of scales, takes place at the injured tip and in proximity to incisions, it is evident that both the quantity and quality of the nutrients being drawn upwards are different from those normally utilized in the growth of meristem cells. Scale and parenchyma formation, indeed, are characteristic of the normal growth developments on the abaxial side of actively developing primordia. Furthermore, in apices with a damaged tip, young primordia, already determined as leaves, may be observed facing and curving outwards instead of inwards, in relation to the high rate of growth on the adaxial side.

In conclusion, it may be noted that in the ferns the apical cell is the focal point in the distal meristem and is essential for the continuing growth of the shoot. It may also be regarded as the ultimate determiner of the orientation and symmetry of leaf primordia. But such effects are apparently not produced by its direct action but are mediated through the organization and physiological activity of the apex as a whole.

14
The chemical concept
of organization
in plants

A considerable volume of recent and contemporary work, including descriptive, analytical and experimental studies, has been directed to the problems of morphogenesis in ferns and other vascular plants, the general aim being to advance knowledge of factors and relationships which may determine the characteristic form and structure of representative species. Individual investigations have been concerned with such topics as the inception of leaves and buds, their characteristic symmetry and mode of growth, the positions which they occupy on the shoot, the inception and further differentiation of the vascular system, and so on. Botanists are also becoming increasingly interested not only in the several phases and aspects of the individual development, but in the individual in its entirety, i.e. in its *organization*. In the Plant Kingdom as a whole, and in the development of any individual plant from a zygote, spore or other generative cell, it is now recognized that there is not only the overall organization of the adult: there is evidence that organization is manifested at the several different levels of development.

To explore and elaborate the concept of organization may be indicated as one of the most important tasks of our time; and this will be true whether the primary interest is in the phylogenetic or causal investigation of plants. In that a satisfying concept is likely to require the synthesis of many different kinds of information, it is unavoidably one that is difficult to grasp and to define. For the present, the term may be taken to connote the distinctive form, structure and physiological activity, which develop with great regularity and fidelity in the individual plant of a species in its usual environment, the starting-point being the organized structure of the specific protoplasm of the zygote, spore, or other generative cell or cell group. This, of course, is a very general definition, propounded somewhat from the morphological standpoint, and it may well become more or less radically modified as our understanding of the problem increases. Each level of organization will require definition in appropriate terms. At the cellular level, for example in a newly fertilized ovum, a scrutiny of the immediately ensuing development suggests that the cytoplasm has a characteristic organization; and this would require definition in terms quite different from those which would be appropriate to the histological organization of an apical growing-point. Again, an undifferentiated tissue culture, i.e. a callus, is unorganized at the morphological level, but inherent in the individual cells there are the distinctive attributes or properties of the species and, given appropriate stimuli and conditions of growth, cell groups will, in some cases, form characteristic buds and plantlings.

In attempting to define more clearly what constitutes organization in plants—an organism being an organization—it should be recognized that a phase of speculation and theorizing must necessarily precede the more tangible phase of solid achievement. In the present paper, accordingly, the writer has tried to indicate some assumptions which seem likely to be essential, certainly useful, if we are to advance. It is recognized that in biological science speculations and hypotheses are only likely to lead to new advances if (i) it can be shown that they have a wide applicability to the phenomena under consideration, and (ii) they can be tested experimentally. In general, our understanding of organization will be elaborated and rendered more precise by extending investigations of morphogenesis (including embryogenesis) on the widest basis, experimental work being introduced wherever possible. For a review of theories of organization in plants, the reader is referred to Wardlaw (1952).

Basic assumptions relating to organization
At this point it seems worth setting forth a number of conceptions, some of which have already become included in the

general corpus of knowledge, which seem likely to lead to new work on, and a fuller comprehension of, organization.

(i) Gene-controlled metabolism, and the physical properties of certain of its products, determine the characteristic development of the individuals of each species, the zygote, spore, generative cell, or meristem, being envisaged as a complex organic (or organismal) reaction system which obeys the laws of physical chemistry.

(ii) To an extent that cannot yet be defined, the 'fine structure' of the protoplasm, and the presence of fixed centres of special metabolism in the cell, may constitute part of the specific organization and affect its further manifestation as development proceeds.

(iii) A patternized distribution of metabolites, which is characteristic of each 'stage' of the essentially continuous ontogenetic process, precedes and determines the visible histological and morphological developments.

(iv) Under the impact of genetical and environmental factors, and of organismal relationships already established (such as polarity, correlations, nutrient-supplying capacity, etc.), the reaction system of an embryo or shoot apex becomes modified in characteristic ways during the development of the individual —a conception which could account for the harmonious and epigenetic development of plants.

(v) The nutritional status of, and the supply of nutrients (including morphogenetic substances) to, embryonic or meristematic regions are important determinants of the eventual morphological and histological developments.

(vi) Since (a) organization is manifested at the cell, tissue, organ, and organism levels, (b) the genetical constitution of even the simplest organism is not simple and not all genes are in action at the same time, and (c) many different factors and relationships are involved in growth and the assumption of form, it seems improbable that any simple or micro-theory of organization will prove to be satisfying.

The interpretation of embryos as reaction systems
The principal morphological and histological facts of the embryonic development in all classes of plants have now been ascertained with some degree of precision (see Wardlaw, 1955).

With these facts before us, an attempt should be made to explain the embryonic development in terms of growth; and this task is made the more attractive by the fact that, as a comparative study shows, embryos from all the major taxonomic groups have many features in common. It is now suggested that these homologies of organization can be explained, at least in part, by the basic assumptions set out in the preceding section, i.e. by envisaging the developing zygote as a reaction system.

In the zygotic development, in species selected from algae to angiosperms, the following phenomena are of general occurrence:

(i) In the mature or newly fertilized ovum, the distribution of metabolites may be relatively homogeneous but it quickly becomes heterogeneous: and with or without an attendant elongation of the zygote, an accumulation of different metabolites takes place at diametrically opposite points, the polarity of the new organism being thereby established. Bonner (1952), however, has stated that polarity may have different physical bases in different instances. Where the ovum is enclosed, the physiological activity of the surrounding tissue is probably important in determining its polarity. In the free-floating zygotes of algae, factors in the environment may induce the reactions which lead to the establishment of polarity.

(ii) In the polarized zygote, the apical or distal pole becomes the principal locus of protein synthesis, growth and morphogenesis, whereas the basal or proximal pole is characterized by the accumulation of osmotically active substances, its cells becoming vacuolated and distended.

(iii) The first division of the zygote is typically by a wall at right angles to the axis, cell division being possibly stimulated by the increase in size (though this is not invariable in embryos) and the instability associated with the drift to cytoplasmic or metabolic heterogeneity. As cell division tends to restore equilibrium in the system, as D'Arcy Thompson has suggested, the position of the partition wall will be such that the forces present in the two daughter cells will be balanced. According to the nature and distribution of the forces involved, the zygote may be more or less equally divided, or it may be divided into a small, densely protoplasmic distal cell and a larger basal cell.

The nature of this division is thus ultimately determined by the specific metabolism of the zygote. (The possibility that changes in the nucleus itself are active determinants in cell division must also be envisaged.)

(iv) During the further growth of the embryo, the positions of the successive partition walls are in general conformity with Errera's law of cell division by walls of minimal area.

(v) As the embryonic development proceeds, factors in the genetical constitution and in the environment, which hitherto had played a minor part, become active; growth is specifically allometric or differential; and the embryo begins to assume a distinctive form. An immense diversity of form is thus possible, but, with some exceptions, e.g. colonial algae, axial development is a general concomitant of the establishment of polarity.

(vi) While the embryo is still small, it shows an acropetal gradient of decreasing cell size. With the exception of those algae which grow by means of an intercalary meristem, the distal region of the axis, which may remain perennially embryonic, becomes organized histologically as an apical growing-point.

(vii) Nutrients are taken up from the environment by the more basal tissues of the embryo and translocated to the apex, though absorption may take place over the whole surface up to a certain stage of development. Primary growth is in the nature of an accretionary process, the older tissues becoming firm and rigid and showing various characteristic concentric and radiate differentiation patterns.

The factors which determine the embryonic development must be much more fully explored before embryological data can be used with safety in taxonomy and phylogeny. What seems to be most needed at the present time is a general theory of embryogenesis and a statement of the principles of embryonic development (see Woodger, 1945, 1948). One basic assumption which seems to be necessary is that the zygote of any species is a highly complex and specific organic reaction system.

The initial distribution of metabolites in the ovum or zygote may be relatively homogeneous but a drift towards heterogeneity soon begins. (Thus far, evidence relating to the chemical and physical constitution of the ovum or zygote is very scanty.) The polarized distribution of metabolites, which underlies the filamentous or axial development of the zygote, is of profound importance in the ensuing embryonic development. The polarity of the young embryo, and all the morphological and histological developments by which its development is characterized, are primarily referable to an antecedent patternized distribution of metabolites. The inception of a biochemical pattern in the zygote, and the successive orderly changes in it during the embryonic development, culminating in the visible, specific organization, are related in some way to the components of the reaction system, to the impact of physical and environmental factors, and to various organismal relationships which develop during growth. A basic problem in embryogenesis and morphogenesis is thus to discover how these patternized distributions of metabolites and, in particular, of morphogenetic substances, are brought about. Turing (1952) has given one indication of how an initially homogeneous diffusion reaction system, of the kind that may be present in a developing zygote, may become heterogeneous and give rise to a regular patternized distribution of metabolites, thus affording a basis for a morphological or histological pattern (see also Wardlaw, 1953); and other physico-chemical explanations may yet be advanced. Each species, as it seems, is a unique physical system and must be so investigated.

Any biochemical theory of embryogenesis will unavoidably be a vast over-simplification of the processes actually involved in it. To know how different metabolites may react and become distributed so as to constitute a particular pattern is to describe only a part of the morphogenetic process: the energy relations within the system, and the physical properties of the reacting substances and of the final products must also be known (Needham 1942).

Application of basic assumptions to the shoot apex
The structure and morphogenetic activity of the shoot apex in vascular plants have been closely investigated during recent years. In studying morphogenetic processes in ferns such as *Dryopteris* or *Adiantum*, it is necessary to have exact knowledge of the histological constitution of the shoot apex. This, however, is no more than the first stage in a much more comprehensive investigation: the meristem is a region of growth and therefore the essential tasks are to discover the facts concerning its

N

metabolic activity and to relate them to the observed morphological and histological developments. The hypothesis now
proposed is that these visible manifestations of meristematic
activity are determined by a patternized distribution of
metabolites.

The meristem reaction system must be envisaged as being of
very considerable complexity. Experimental studies and histological observations justify the view, already advanced by
Schoute (1936) and now elaborated, that the shoot apex in
ferns and flowering plants consists of several zones, or regions,
of distinctive metabolism, each zone merging and interacting
with the adjacent zone or zones. The physiology of these zones,
however, awaits investigation. These regions, e.g. in *Dryopteris*,
may be indicated as follows: (*a*) The distal region comprising
the apical cell and its adjacent segments. (*b*) The subdistal
region, consisting of distinctive, prism-shaped meristematic cells,
formed from segments of the apical cell; in this region the
patternized distribution of the metabolites associated with the
inception of leaf primordia is beginning to take place though
no distinctive histological or morphological changes can yet
be detected. (*c*) The organogenic region, subjacent to (*b*), in
which the outgrowth of leaf primordia can be observed, i.e.
the results of the antecedent patternized distribution of metabolites become visually evident; in this region the development
of bud rudiments is typically inhibited; the inception of tissue
differentiation is in regions (*b*) and (*c*). (*d*) The subapical region,
subjacent to (*c*), in which there is a rapid increase in the
diameter of the shoot and leaf primordia begin to grow actively;
also, the further differentiation of the tissue systems becomes
conspicuous.

A somewhat different example of the way in which the distribution of metabolites may determine the growth behaviour
of the shoot apex is illustrated by the investigations of Bennet-
Clark & Ball (1951) and by Ball (1953) of the rhizome of
Aegopodium podograria. Their experimental studies suggest that
the diageotropism of the rhizome may be attributed to the
action of two hormones working in opposition, both being
affected by gravity. If a rhizome is inverted, or is immersed in
a solution of B-indoleacetic acid, the distribution of hormones
would appear to be altered and the rhizome apex responds by

characteristic growth changes. In dorsiventral ferns such as *Polypodium vulgare, Pteridium aquilinum, Drynaria quercifolia, Marsilea drummondii*, etc., it is necessary to account not only for the dorsiventrality of the rhizome but also for the formation of leaves on the dorsal side only and roots on the ventral side only.

In the ferns and flowering plants, the need for the several basic assumptions may be indicated by some further examples. When we study the shoot apices of ferns such as *Dryopteris, Matteuccia* or *Onoclea*, we think in terms of an exposed conical or dome-shaped terminal meristem to which the metabolites needed for growth are moving acropetally and from which products of metabolism are moving basipetally and laterally; while metabolic substances associated with the growing leaf primordia may move acropetally, laterally and inwards. But in some other ferns, e.g. *Ophioglossum vulgatum* (Wardlaw, 1953) and in a number of flowering plants, buds can be induced endogenously. In *O. vulgatum* the writer has illustrated and described the induction of endogenous buds in the middle cortex of the root and in the medulla of decapitated shoots. In both situations certain parenchymatous cells become centres of growth; they divide rapidly and form an ellipsoidal mass of densely protoplasmic cells. Within this tissue mass, a shoot meristem and the first leaf become differentiated more or less simultaneously, and a little later the initials of the first root can be distinguished. It is difficult to see how these developments are to be explained unless we assume that, prior to the visible manifestation of the several organs, there has been a characteristic distribution of the specific metabolic determinants of shoot, leaf and root; but how this distribution is actually brought about is a very difficult problem. If, however, we assume that the initial meristematic cell mass is a reaction system in a particular environment, it is possible to perceive, at least in a general way, that the metabolic substances associated with the inception of particular organs might come to be located in particular positions. It is recognized that this statement lacks precision and is a vast over-simplification of the situation. Nevertheless, some attempt must be made to explain the observed developments. If we do not make use of the assumptions indicated here, in the hope that they may be confirmed by further work, or replaced by better ones, we are left with the experimental data as interesting but entirely unexplained records.

Investigations of endogenous buds in *O. vulgatum* have yielded other points of interest relating to organization. The details of the embryogeny are usually accepted as exemplifying in part the organization of the species. Within limits this view is no doubt justified; but it is important to bear in mind that an embryo is the product of a zygote (or genotypic cell) in a particular environment, and its morphology may therefore be more or less affected by the nutrient status of the environment. In *O. vulgatum*, important differences have been noted between the developments in induced buds and in young embryos. The embryo, which is slowly and feebly nourished by an underground mycorrhizic gametophyte—though, from the heavy starch deposits, carbohydrates would not appear to be limiting —first forms a large and conspicuous root and only later its shoot apex and first leaf. But induced buds, which are abundantly nourished by the adjacent sporophyte tissue, grow rapidly and first form the shoot apex and a leaf, and only later the first root. Other interesting relationships between nutrition and organogenesis have been established by Wetmore (1950) and Allsopp (1953).

Discussion and outlook
The concept of organization, both in its general and special aspects, is likely to prove of increasing interest to botanists in the coming years. A particular merit of the theme is its essentially integrative character, the data of physiological-genetics, physiology, morphology and histology all being essential to its elaboration. The need for this synthesis of data from many different sources makes the concept a difficult one to grasp and to state in a satisfactory manner. Some botanists, indeed, may consider that we are not yet ready to attempt such a synthesis. That may be true. But, in the writer's view, a beginning should not be unduly delayed. If the concept does no more than provide a focus for assimilating the rapidly accumulating data of different kinds, it will have served a useful purpose. Moreover, it seems probable that many botanists, each making his contribution in a particular field, would welcome a general theory of organization. Essential features of such a theory would be

that it afforded, or seemed likely to afford, a satisfying explanation of the morphological, histological and cytological phenomena observed during the development of a selected plant and that it could assimilate the data of physiological genetics, experimental morphology and other branches that contribute to our knowledge of morphogenesis. The theory must also account for the harmonious transition from one embryonic or development stage to the next.

A comprehensive and adequate theory of organization is of paramount interest to the morphologist. But such a theory is probably of not less interest and importance to the physiologist, for it may well be that, in the future, the integration of botanical science—if indeed it becomes integrated—may take place round the concepts of growth and organization. Hersch (1941) has drawn an interesting contrast between what he calls the 'substance-minded' and the 'relation-minded' man. The morphologist traditionally, but not invariably, exemplifies the former type of observer: he tends to think of the developmental pattern in terms of the visible structural features seen at successive 'stages' during development; his records are illustrations and descriptions of these 'stages'. But if the morphologist is a student of morphogenesis, his outlook is rather different: he regards the pattern observed at any particular stage in development as the expression of the factors which produced it, and hence he thinks in terms of processes and relationships of various kinds. As Hersch has said, 'the relation-minded morphologist tends to think in terms of the non-picturable. If the problem of the developmental pattern is similar to the problems of the more exact sciences, then no doubt, in time, a system of equations will be developed to facilitate our thinking about it.' That may well be true. Nevertheless, for a long time to come, and until a very different generation of botanists occupies the scene, an acceptable theory of organization must in some measure satisfy both the substance-minded and the relation-minded observer. This can perhaps be achieved if some of the basic conceptions outlined here are accepted.

In the pattern-forming, reaction system envisaged here, different genes may have greater or less importance at different phases. It may be assumed that the reaction system of a particular species normally comprises certain ingredients, e.g.

particular enzymes, proteins or protein precursors, growth-regulating substances, carbohydrates, and so on. The embryonic pattern or organization at any particular stage in the onto-genetic development will be determined by the substances which have entered into the reaction system up to that stage and by environmental and other factors. As the reaction system changes during the development of the individual organism, so also will the stable pattern to which it gives rise. The progressive elabor-ation of form and structure, or the 'reduction' of a more com-plex form, the harmonious transition from one developmental stage to the next, the epigenetic nature of plant development, and the manifestation of organization or integrated wholeness, are each and all the expression of a specific reaction system which changes continuously as growth proceeds.

A chemical theory of organization advanced on the basis of present knowledge will almost certainly be an over-simplified statement of the processes actually involved. To know that certain metabolites may become distributed during a series of reactions so as to constitute a pattern is to know only a part of the processes of the assumption of form: the biophysical properties of the various substances must also be known. The evidence is consistent with the view that the actual determin-ation of form and the differentiation of structure are due to the action of physical factors, and to the physical properties of biological substances (D'Arcy Thompson, 1917, 1942). The segmentation pattern found in the embryos in different classes of plants, for example, is usually in close agreement with Errera's law of cell division by walls of minimal area, the posi-tion and orientation of the new walls being determined by the forces which develop in the enlarging organism. Needham (1942) has suggested that in biology we are primarily con-cerned with energy and organization. If so, it would appear that a reaction theory, comprehensively developed along the lines indicated here, may account for the inception of form and structure in the individual species and for its regulated, integrated and harmonious development, this being what we mean when we speak of organization.

In a recent book, Bonner (1952) regards morphogenesis, or the emergence of an organismal pattern, as being referable essentially to the progressive, constructive processes of growth, morphogenetic movement and differentiation; and to the limiting of these processes in various ways by intrinsic and extrinsic factors which check, guide and canalize them. He also considers that a sustaining aim in the study of morphogenesis should be a search for a micro-theory (or micro-structure theory) which would explain in a satisfying manner the general phenomena of development; for such a theory is more likely to afford an insight into the inner mechanism of the relevant phenomena than any other kind of theory. Such a micro-theory does not necessarily imply that all the phenomena of biology are to be brought down to physics and chemistry: some kind of intermediate micro-theory may prove quite satisfactory. Bonner does not actually advance any micro-theory, but he indicates that the kind that would be satisfactory would be one in which, as in crystal structure, the organism could be envisaged as being essentially constructed of small repetitive units. In the present writer's view, it seems improbable that there can be any simple theory of morphogenesis. A micro-theory, as proposed by Bonner, would be valuable as far as it goes, but it would have its place as one component of a complex system. In the general conception indicated here, each individual development is based on a specific reaction system and each species is a unique physical system. But this reaction system, like the epigenetic morphological entity to which it gives rise, undergoes a series of changes during development, each successive state of the system being partly determined by the preceding ones, and in turn partly determining those that follow. It is true that, in some species of vascular plants, e.g. a fern, once the juvenile stage is passed and a regular phyllotactic system established, the further development of the shoot takes place with great regularity and could be regarded as being essentially an accretion of repetitive units. In such instances the conception of repetitive units is a valuable one (see Goebel, 1922). The inception of repetitive morphological or structural units is accounted for in Turing's theory.

The theory of organization outlined here may contribute towards a better understanding of the very general phenomenon of parallel development in plants. It may well be the means of 'bridging' at least some of the 'gaps' that seem to occur in all phylogenetic systems. Thus, in a reaction system, it is possible to see how a mutant gene, or the substance which it determines, may produce small changes, or even large ones, in the system, with concomitant morphological effects. It is known from genetical studies that comparatively small changes in the genetical constitution may sometimes be attended by quite extensive morphological changes. A fuller exploration of this aspect may show that some morphological features, which are regarded by the taxonomist and phylogenist as being of great importance, may be due to quite small changes in the underlying reaction system and may be considerably less important than has been thought.

Some such considerations as these should help to clear our minds for the next phase in the investigation of embryogenesis and morphogenesis. Growth and organization are essentially dynamic phenomena. When we study them we are essentially concerned with processes, of which the visible morphological and histological features are the result. There is still much important work for the experimental morphologist and taxonomist to do, but perhaps more than ever before, morphology and physiology should be regarded as aspects of one discipline. Certainly morphologists will do well to bear constantly in mind what the fundamental processes are, and to relate their observations to them wherever possible. Meanwhile, the concepts of growth and organization taken together afford a useful focal point for effecting some integration of the many and diverse contributions to our knowledge of morphogenesis in plants.

15
Generalizations on the apical meristem

Although the vegetative shoot apical meristems of vascular plants show very evident differences in histological organization, they all give rise to a vascularized leafy axis. That such apices have some organogenic property in common thus presents itself as an inescapable fact.[1] There is therefore a need for concepts that will have a general application to apices, irrespective of their histological patterns.

During recent years a group of French botanists, under the leadership of Plantefol[2] and Buvat,[3,4] have been active in histological investigations of vegetative (and floral) meristems of vascular plants, this work having been stimulated by a new theory of phyllotaxis by Plantefol. Among their main conclusions, the following are of special interest: (i) Leaf primordia originate in a superficial ring of actively meristematic tissue, described as the *anneau initial* (initial or initiating ring), which is situated around the base of the inert, or relatively inert, distal apical zone. This *meristematic ring* is regarded as a self-perpetuating embryonic tissue. It is held to be the true vegetative meristem, which would, therefore, be sub-terminal and lateral, and not distal as has so long been thought. The most distal zone of the apex is thus considered to be unimportant in histogenesis and to have no effective part in organogenesis during the vegetative phase. (ii) The inception of leaf primordia in the meristematic ring is part of the theory of phyllotaxis proposed by Plantefol.[2] This states that the phyllotaxis of most plants can be referred to a small number of foliar helices. At the extremity of each of these, in the meristematic ring, there is a generative centre which gives rise to all the leaf primordia

Nature, Vol. 178, p. 1427, 29 Dec. 1956.

on that helix. The activities of the several generative centres are harmonized by an organizer which is present in the apex. A generative centre is held to be a physiological, rather than a recognizable, morphological unit. While some investigators[5,6] claim that the theory has been verified by experimental treatments of the apex, others[7,9] have expressed or implied disagreement with it.

In a contemporary monograph from the French school, Camefort[10] has taken up my point about the need for general concepts relating to the shoot apex,[1] and has suggested that a meristematic ring (*anneau initial*) may perhaps be present in the apices of all vascular plants. He has, moreover, suggested that, within this concept, it may be possible to interpret the structural differences in the apices of Spermatophyta. He points out that the difference in activity between the allegedly inert apical zone and the meristematic ring may vary according to the species. Where this difference is profound, as in some gymnosperms, for example, in *Picea*, the distal region tends to be homogeneous and a tunica is absent; where it is less pronounced, as in the angiosperms, a tunica is present. In his view, the absence or presence of a tunica is thus associated with a more or less pronounced difference in activity between the apical region and the meristematic ring. Where the summit of the apex is histologically heterogeneous, as in *Taxus*, *Ginkgo*, and *Cedrus*, the superficial cells are cytologically less inert than the underlying 'central' cells. The histological patterns in such apices could be regarded as being intermediate in character between angiosperms (with a tunica), and those gymnosperms which have a homogeneous distal region. In all this diversity of apical construction, says Camefort, the presence of a meristematic ring is the permanent, fundamental fact; the homogeneous meristem, the heterogeneous meristem and the stratified meristem are indicative of successively smaller differences respectively between the activity of the meristematic ring and the apical zone.

Camefort has undoubtedly put forward an interesting generalization which deserves to be more fully investigated. In 1953, I[11] had already indicated that, in attempting to understand the characteristic histological organization of any apex, the size and metabolic properties of the embryonic cells, and the

distribution of growth in the apex as a whole, must be considered. Also, although the French workers have stressed the
activity of the meristematic ring and its importance as the site
of leaf inception, this idea is not in itself new. Already, in 1936,
Schoute[12] had advanced the idea that the apex consists of
several physiological zones, or levels, which in my[13] terminology and conception would include, in basipetal sequence: (i)
the most distal region, which, in ferns, would include the apical-
cell group; (ii) the subdistal region in which the patternized
distribution of metabolites associated with the inception of
growth-centres takes place; (iii) the organogenic region in
which the formation of primordia from the growth-centres can
be observed; (iv) the sub-apical region in which conspicuous
enlargement takes place; (v) the region of maturation. The
meristematic ring of the French school would comprise regions
(ii) and (iii). The inertness attributed to the most distal region
is, it need scarcely be said, a very controversial matter and
has already been the subject of criticism.[14-16]

The morphological phenomenon of greatest generality is that
the shoot apices of almost all vascular plants give rise to regularly spaced leaf primordia. This, of course, has long been recognized. The observation, however, acquires a new importance
if it is restated with reference to the antecedent and underlying physiological processes, namely, that in the peripheral tissue
of the sub-distal region of all apices, in which the embryonic
cells are seemingly homogeneous, a patternized distribution of
metabolites apparently takes place. This results in the accumulation, concentration, or formation, in more or less well-defined
and characteristically spaced loci—often described as growth-
centres—of substances actively involved in growth and cell
division. Turing[17] advanced a physicochemical theory to explain how a pattern of metabolites might be brought about in
an embryonic tissue in which the substances were initially more
or less homogeneously distributed, and I[13,18,19] have suggested
how this, or some similar physicochemical theory, might apply
to very diverse organogenic and histogenic developments in
plants. Turing, moreover, postulated that quite differently
constituted reaction systems may nevertheless give rise to closely
comparable morphogenetic patterns. If this concept can be
shown to be valid, it will evidently go a long way towards

explaining the many parallelisms of development (or homologies of organization) to be observed in plants. In particular,
it will have an important application to the general phenomenon of apical activity now under consideration. The growth-
centres, however we may explain their origin, usually grow out
as localized mounds on the surface of the apex and develop as
dorsiventral leaf primordia. It is a very remarkable fact, and
again one of very wide generality, that, in microphyllous and
megaphyllous species alike, the primordium at its inception is
multicellular and comprises both surface and sub-surface cells.
Even the very small, non-vascularized, scale-like microphylls of
Psilotum are covered by this general statement. In the angiosperms also, the different organs of the flower all have their
inception in similar multi-cellular growth centres.

It is relevant to inquire to what extent the generalization
that all shoot apices have the inherent property of giving rise
to foliar primordia is borne out by the fossil evidence. Here we
come upon a very remarkable state of affairs, remarkable not
because the evidence has been neglected by botanists, but rather
because concentration on, and perhaps over-emphasis of, the
inferences that can be drawn from it from one point of view,
seem to have obscured other not less valid inferences. Thus, the
ancient Psilophytales, in particular such organisms as the
leafless species of *Rhynia*, have been accepted by many botanists
both as affording an evident basis for the phylogenetic treatment of the vascular plants, and as providing essential evidence
relating to the evolution of the leafy shoot. The discovery of
Palæozoic plants, with a bifurcating, leafless and rootless axis,
such as *Rhynia*, has enabled morphologists to envisage a 'prototype' or 'archetype' from which the more advanced classes of
vascular plants, all characterized by leafy shoots, might have
originated. But, on the evidence, *entirely leafless* vascular plants,
whether on the 'evolutionary upgrade', or as the result of 'reduction' from some leaf-bearing ancestor, appear to be very
rare. The only completely leafless vascular plants known to us
belong to the Psilophytales, for example, species of *Rhynia*,
Horneophyton, *Zosterophyllum*, etc. Other species considered to be
of this affinity, however, had an axis covered with evident
microphylls, for example, *Asteroxylon mackiei* and others had
small scale-like lateral organs. The very ancient fossil, *Barag-*

wanathia longifolia, of Silurian age, had an axis with abundant microphylls. Many vascular plants, contemporary with the Psilophytales, but belonging to other classes, are known, and these all had leaves. It is therefore an open question whether the entirely leafless genera of Psilophytales were not exceptional in their organization, rather than archetypic of all other classes of vascular plants as has so often been assumed or implied. Indeed, the possibility that they may have been reduced forms in relation to special environmental conditions cannot be excluded.[20, 21] The existence of other classes of vascular plants of at least the same antiquity as the Psilophytales, all characterized by foliar development, widens the scope of the generalization under consideration. Lastly, without entering into details, it may also be noted that in the gametophytes of mosses, and in some of the larger brown algae, organized apical meristems are present which give rise to regularly spaced lateral organs.

As we have seen, the importance attributed to the meristematic ring in organogenesis derives from Plantefol's phyllotaxis theory of multiple helices. According to the theory, each leaf-generating centre moves acropetally (by some mechanism not explicitly specified) along a helical path in the apex. Its actual location is in the meristematic ring where, from time to time during the growth of the apex, it gives rise to leaf primordia.[22] The majority of dicotyledons are considered to have two foliar helices, these having their beginning at the inception of the two cotyledons; and if there are three cotyledons, then there are three helix-generating centres. As an apex enlarges during the ontogenetic development, the number of helices may increase, this being attributed by the French workers either to the formation of two centres from an original centre or to the intercalation of a new centre between two existing ones.

Thus, as I understand the theory, in order to account for *the inception of the several helices*, it is necessary to assume that a pattern of growth-centres can originate spontaneously in the apex, beginning with that of the embryo. As I have attempted to show, and have indeed emphasized, this property, or functional activity, of apices is a phenomenon of the widest generality in plants. It is particularly evident in the embryos of species of *Pinus*, for example, *P. ponderosa*, where, in direct relation to the size of the apical meristem, 6–16 evenly spaced cotyledon primordia originate simultaneously in a ring around the meristem, there being no pre-existing lateral organs to exercise regulative effects of any kind. Again, in Rubiaceae such as *Galium aparine*, and in species of *Equisetum*, etc., the ontogenetic development is characterized by an increase in the number of foliar members at successively higher nodes. These and many other instances afford evidence that characteristically spaced growth-centres are determined by factors inherent in the apex, this, in my view, being one of the basic and essential features of the organogenic development of the leafy shoot. Where leaf primordia are already present, the morphogenetic situation in the apex is much more complex. Observation and experiment indicate that the positions of growth-centres (and of the primordia to which they give rise) are affected by the young primordia which surround the apical meristem, a new primordium typically originating above and between two adjacent primordia.[23] It is also relevant to note that primordia always originate in the apical meristem at some characteristic distance below the extreme tip (or centre). This distance varies with changes in the size of the apex during the onto-genetic development, and it can be modified in characteristic ways by various experimental treatments. The overall organi-zation of the apex, and the integrated physiological activities of its component regions, must, therefore, also be given due con-sideration in the analysis of organogenesis and phyllotaxis.

It will be noted that while ideas relating to growth-centres, or to the helix-leaf-generating centres of the French school, may be stated in different terms according to the individual in-vestigator's approach, there appears to be a considerable measure of agreement on several points, notably, that growth-centres have a real existence as loci of special physiological activity, that they are formed at some characteristic distance below the tip of the meristem, and that they are characteristic-ally spaced around the apex. In my view, however, a growth-centre is a discrete physiological unit occupying a particular position in the meristem, so that there are as many growth-centres as there are leaves, whereas in Plantefol's theory, each leaf-generating centre retains its pristine physiological properties and moves acropetally around the apex, giving rise at intervals to leaf primordia. The latter collectively constitute a foliar helix.

Snow[7,8,24] has shown that the theory of multiple helices is open to criticism on several grounds. We may also note that Plantefol's theory brings its own difficulties when he attempts to explain floral ontogenesis;[25] for whereas the sepals are regarded by him as foliar members arising on the vegetative helices, the petals, which in many flowers also have the form and structure of leaves, are considered not to be homologous with leaves but organs *sui generis*. Camefort[10] has not attempted to answer Snow's criticisms, and, indeed, states (p. 102) that the only causal theory of phyllotaxis that attributes a direct role to the vegetative apex in the positioning of leaves is that of Plantefol. In this, he appears to overlook the results of experimental investigations in which the self-determining nature of the apex in organogenesis has been demonstrated and duly emphasized,[26-28] as well as views based on Turing's theory.[13, 17-19]

When considered in terms of the underlying physiological mechanism, the theory of multiple helices raises new problems of great complexity. This, of course, is not in itself an argument against the theory, and, indeed, we may recognize that the investigation of apical organization and organogenesis not only bristles with difficulties, but also that these may well increase rather than decrease with new accessions of knowledge. As I have indicated, the postulated acropetally moving leaf-generating centres of the French school must have their beginning in a pattern of growth-centres. If, therefore, the concepts (*a*) of the inception of discrete growth-centres in the sub-distal region, as determined by genetical and other factors and relationships, (*b*) of the regulative effects of existing primordia on new ones, and (*c*) of the functioning of the shoot apex as a regulated and integrated whole, are taken together, the theory of phyllotaxis advanced by Plantefol appears to be unnecessary, while the new facts ascertained by the French workers can be assimilated to other views on apical growth and organization.

16
The inception of
leaf primordia

In earlier chapters general aspects of growth and the processes associated with the differentiation of tissues have been considered. This chapter is concerned with the *inception* of leaf primordia in living species. When we consider how much attention botanists have given to this and related problems, the task seems simple enough: a closer scrutiny, however, shows that the processes involved are complex and that they are still far from being adequately understood. In that vascular plants consist essentially of a rooted axis bearing leaves, the formation of these lateral organs in an orderly sequence may be recognized as a major expression of the apical organization and activity. This subject has also interesting evolutionary implications, since some of the earliest and most primitive vascular plants were leafless. The further growth and morphological development of leaf primordia are considered in later chapters.

The leaf a product of the apical meristem

A leaf primordium is always a product of the apical meristem. We have no evidence of leaf primordia not arising from cells of the meristem; and this is true whether we are concerned with the microphylls of some pteridophyte species or with the gametophytic leaves of bryophytes. The first leaves of embryos—the cotyledons—also arise from distal embryonic tissue and they, too, in their symmetry and orientation, stand in a definite relationship to the apex, whether it is actually visible or merely nascent.

The shoot apical meristem, which consists essentially of cells in the embryonic state, is a region of complex metabolism, and is supplied from below with the materials utilized in its growth

The Growth of Leaves, proceedings of the Third Easter School in Agriculture Science, University of Nottingham, 1956.

and formative activities. Apices, which may be conical, paraboloidal or flat, typically give rise on their flanks, and at some characteristic distance from the tip or centre, to regularly spaced, discrete outgrowths which become leaves. It thus appears that leaf primordia arise in localized sites in which there has been either (*a*) a localized accumulation of certain metabolic materials or (*b*) a release from an earlier state of inhibition; or both processes may be involved. As the visible leaf primordium is evidently a locus of active growth, it is reasonable to attribute its inception to a restricted area of special metabolism on the meristem, i.e. to a growth centre. In a fern such as *Dryopteris*, with a large conical apex, the leaf growth centres are relatively small and occur in situations relatively remote from the shoot apical cell, i.e. at the base of the cone and close to the region of transition to the broad sub-apical region (fig. 1). In some flowering plants, on the other hand, it is evident that the growth centres must lie very close to the tip or centre of the meristem, e.g. in *Acer* and other species in which the youngest primordia are of large size relative to the meristem (figs. 4, 6, 7). In others, e.g. *Rosa, Ranunculus, Nuphar, Solanum* or *Lupinus* they may be less close, and in yet others, e.g. *Hippuris, Elodea,* the Gramineæ, *Lycopodium* and *Equisetum*, they occupy positions low down on the flanks of the meristem (fig. 2, 3, 5). As growth centres may differ in number, position, size and shape in species from different taxonomic groups, a considerable diversity in primordium morphology is to be expected.

Since the inception of growth centres is a manifestation of apical organization and reactivity, they must be determined by the specific genetical constitution. Apical organization, both in its physiological and morphological aspects, may, however, change in characteristic ways during the development of the individual plant, i.e. from the embryonic stage onwards; it may also be modified by environmental factors and, in particular, by nutritional factors. Hence leaf primordia, and the mature organs into which they develop, may show characteristic ontogenetic or experimentally induced morphological changes.

The inception of primordia in ferns
That loci of special metabolic activity are present before the leaf primordia in ferns are formed can scarcely be doubted, though,

so far, the specific protoplasmic features of the cells in these loci, leaf sites or growth centres, have not been demonstrated microscopically. As soon as a leaf site begins to grow actively, e.g. in *Dryopteris*, there is an outgrowth of tissue and a primordium begins to be visible under a binocular microscope as a low mound of approximately circular outline, rising from the otherwise smooth flat surface near the base of the apical cone. This is the result of the enlargement of a group of some 8–10 of the superficial prism-shaped cells of the apical meristem, together with the growth and division of a group of underlying, somewhat smaller cells. As development proceeds, the primordium occupies positions successively further away from the shoot apical cell and, in general conformity with the growth of the sector of the apex on which it is borne, it widens radially and tangentially, becomes strongly curved on its adaxial face. While the *Dryopteris* primordium* is still on the apical cone, as P_1 or early P_2, i.e. during its first or early second plastochrone, it has probably not yet been determined as a leaf (as distinct from a bud rudiment); but, as it passes on to the broad sub-apical region, one of its more centrally-placed superficial cells begins to enlarge and, by a characteristic segmentation, becomes the conspicuous lens-shaped apical cell, typical of the fern leaf primordium. The orientation of this leaf apical cell is such that its main axis is approximately radial and its lateral segments lie in the tangential plane of the shoot. These segments retain their embryonic nature and constitute the marginal meristems which contribute largely to the further development of the primordium. From this stage onwards the primordium is seen to be an organ of dorsiventral symmetry. The tangential extension of the primordium at the base of the cone, and especially lower down in the region of transition to the sub-apex, may be one of the factors determining the characteristic orientation of its apical cell. From an early stage there are, moreover, evident differences in the rates of growth on the adaxial and abaxial sides or faces of the primordium which contribute to its characteristic dorsiventrality on further enlargement. Incipient or prevascular tissue can be demonstrated

---

* The usual phyllotaxis symbols are used: P_1 is the youngest visible primordium; P_2 is next older, and so on; I_1 is the site of the next primordium to be formed, or the primordium that subsequently arises in that site.

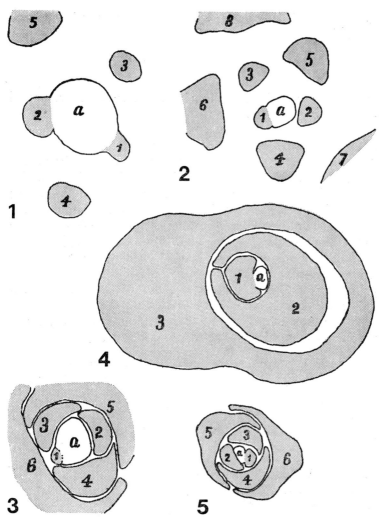

1 *Dryopteris dilatata* (*D. aristata*). Transverse section of a thinnish rhizome at about the level of the axils of the two youngest leaf primordia, 1 and 2 (diagrammatic). The youngest primordium originates low down on the flanks of the conical apex and the transverse cross-sectional area of the latter is large relative to that of the youngest primordium. *a*, apical meristem; 1, 2, 3, 4, 5, leaf primordia in the order of their increasing age (referred to as P_1, P_2, etc., in the text). **2** *Nuphar lutea*. Transverse section as in *Dryopteris*, the young primordia are quite widely separated on the apex. **3** *Ranunculus acris*. Transverse section as in fig. 1. As in *Dryopteris*, the apical meristem is of large size relative to the youngest primordium, which has its inception low down on the flank of the meristem. **4** *Magnolia lennei*. Transverse section almost through the summit of the apex, showing the very large relative size of the last-formed primordium which originates high up on the apex. In sections slightly lower down, it can be seen that the primordium completely envelopes the apex. **5** *Rosa multiflora*. Transverse section as in fig. 1. The apices of many dicotyledons show relationships of this general kind. (All reproduced at same magnification, ×50.)

in the young leaf primordium and is seen to be conjoined with that of the shoot.

Leaf inception in flowering plants

Although the histological details may be very different, the inception of the leaf primordium in dicotyledons is not unlike that in ferns in its main features. In the dicotyledons the primordium first becomes visible as an outgrowth on the flank of the apical meristem. In some species it may originate so close to the centre of the meristem and be of such large relative size that the meristem undergoes marked changes in its shape and size as each new primordium, or, in decussate systems, pair of primordia, is formed (figs. 4, 6). But in other dicotyledon apices the relationship of the nascent primordium to the meristem may be not unlike that in ferns, and many intermediate relationships are known (figs. 2, 3, 5).

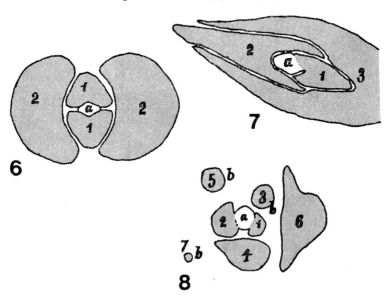

6 *Phoradendron flavescens.* Transverse section, as in fig. 1, showing the very large relative size of the youngest primordia in a decussate system (from a section by Dr. E. G. Cutter). **7** *Iris pseudacorus.* T.S. apex as in previous diagrams. **8** *Nymphaea micranthra.* Transverse section at the level of the axil of P_1. Primordia 3 and 5, which are very young flower buds, are of radial symmetry, but occupy leaf sites. (All drawn to the same scale, $\times 50$.)

The young primordium, which may be tangentially elliptical or already dorsiventral when it first becomes visible under the binocular microscope, is sometimes described as a foliar buttress (*soubassement foliare*: leaf foundation) (figs. 3, 4). As development proceeds, the flanking tissues of this organ merge with the cortical tissue of the axis, while a more central group of cells, which may soon be extended in the tangential plane, shows vigorous upward growth and forms the leaf axis. The latter, according to the species, may give rise to the petiole and lamina. Thus, in both the ferns and the dicotyledons, the further development of the primordium is due to the activity of centrally-placed superficial cells. This central locus of active development may perhaps be equated with the original growth centre. In some species the leaf base undergoes further development, giving rise to stipules. Since, from the very outset, leaf formation in dicotyledons is a continuous and harmonious process, it may be doubted if the concept of foliar buttress has any value other than that of a handy descriptive term in some instances. (For a good account of leaf inception in flowering plants, etc., see Esau, 1953.)

The histological details of leaf inception vary from species to species, but, in many of them, the initial outgrowth is the result of localized periclinal divisions in one or more of the sub-surface histogenic layers and of anticlinal divisions in the surface layer. The particular layer in which a growth centre has its inception, and the number of tunica layers present, will evidently determine the extent to which tunica and corpus contribute to the new primordium. As the relative proportions of tunica and corpus are fairly constant for the species and, basically, are genetically determined, the manner of inception of the primordium can be referred to, and affords some information on, the specific organization. In this connection, periclinal cytological chimaeras have proved of great use in analysing the contribution of different histogenic layers to leaf inception and development.

In many dicotyledons it can be seen that the abaxial side, even of the very young primordium, is already abutting on that region of the apex where tissue differentiation and enlargement are taking place, i.e. the growth relationships and rate of differentiation of the adaxial and abaxial sides are different virtually from the outset. From an early stage, but varying according to the species, a strand of procambial tissue is differentiated in the primordium. This becomes conjoined with the vascular tissue of the axis.

In some monocotyledons, e.g. in *Elodea* and in grasses, in which the tunica may consist of one or two histogenic layers, the primordia arise at a considerable distance below the tip of the elongated apical meristem. Leaf formation originates in periclinal divisions in the outer layers of cells, suggesting that the growth centres occupy superficial positions. A feature of the developing primordium (buttress) in grasses and other mono-cotyledons (fig. 7), as also in some dicotyledons, e.g. in *Magnolia* (fig. 4) and the Umbelliferae, and in *Ophioglossum*, is that it extends laterally, eventually more or less completely circling the shoot.

Experiments on undetermined primordia

If we do not take a leaf primordium, with its characteristic symmetry and orientation, for granted—as something 'given', in the sense that leaves were at one time regarded as organs belonging to a fundamental and unchangeable category of parts—we are apt to find ourselves asking such questions as: How does a growth centre originate? Why does it usually give rise to an organ of dorsiventral symmetry, i.e. to a leaf and not a bud? And why (with some notable exceptions) is the growth of a leaf definite whereas that of a vegetative shoot is indefinite? And so on. While none of these questions can yet be fully answered, a number of experiments have been undertaken to explore the nature and potentiality of leaf growth centres and primordia, and some interesting observations have been made. In this work, ferns such as *Dryopteris*, with large apical meristems and small, widely-separated leaf primordia, have proved particularly useful, the more so as the very young primordia remain undetermined as foliar organs for some time. The following authenticated observations* relate to *D. dilatata* (*D. aristata*), the shoot apical cell being intact and uninjured, unless otherwise stated:

(i) If a deep and wide *tangential* incision is made on the

* See Wardlaw (1952) and more recent papers in *Ann. Bot.* by Wardlaw and Cutter.—ED.

adaxial side of an I_1 site, or an undetermined P_1 primordium (and sometimes P_2 or even P_3), a bud and not a leaf is usually formed. After a similar deep cut, or after undercutting, on the abaxial side, only leaves have been obtained.

(ii) If small, localized deep incisions are made above undetermined leaf primordia, or sites, thereby interrupting direct tissue continuity with the shoot apical cell group, leaf primordia are typically formed.

(iii) If experiment (i) is carried out in such a way that a small bridge of intact tissue is left above the primordium or site, buds may sometimes be obtained in actively growing apices.

(iv) If an I_1 site, or a young primordium, e.g. P_1, P_2 or P_3, is isolated from the adjacent older primordia by deep *radial* incisions, the isolated site or primordium develops as a leaf: it shows an increased rate of growth and is soon of abnormally large relative size.

(v) If experiments (i) and (iv) are repeated with *actively growing* apices, but using very shallow incisions (i.e. which do not sever the incipient vascular tissue of the apical cone), normally developing leaf primordia are usually obtained.

(vi) If undetermined primordia, e.g. P_1 or early P_2, on *inactive* apices, are isolated on rectangular panels by shallow incisions, the primordium may disappear completely, its site becoming occupied by scales; and an I_1 site similarly isolated may fail to form a primordium. On more active apices, leaf primordia are typically obtained as in experiment (v).

(vii) If the shoot apical cell group only is destroyed, e.g. by light puncturing, leaf primordia may continue to be formed in the normal phyllotactic sequence for some time and no buds or perhaps only one, may arise on the meristem. If the distal region of the apical meristem is more extensively destroyed, two or three buds may soon be formed near the base of the cone; some leaf primordia may be formed higher up, usually in the normal phyllotactic positions.

(viii) In apices in which the distal region has been fairly extensively damaged, some of the leaf primordia formed subsequently may show the normal acropetal orientation but others may (*a*) become orientated towards one of the rapidly-growing induced buds or (*b*) have their orientation reversed, i.e. point away from the apex.

The foregoing experimental evidence is compatible with the view that (*a*) the inception of a leaf primordium in a particular position on the apical meristem, (*b*) its characteristic symmetry, (*c*) its acropetal orientation, and (*d*) its characteristic rate of growth and development, are determined and regulated by the organization and physiological activity of the apex as a whole (including the primordia on and near the meristem), the intact shoot apical cell being an essential element in the system.

Some of these surgical experiments have also been carried out on the apices of dicotyledons but, so far, there is no record of a leaf site, or of a very young leaf primordium, being caused to develop as a lateral bud. Some notable modifications in leaf development have, however, been reported, including the formation of centric or radial leaves (for a recent discussion, see Sussex, 1955). These are awl-like structures, usually of limited growth. Transverse sections show them to be approximately circular in outline, with radially disposed vascular tissues. Their characteristic development has been related to the isolation of the leaf site, by a tangential incision, from the regulative effects of the meristem, or to the fact that they may have arisen from a very small area of the meristem. When similar organs have been seen in experimental fern materials, it has usually been in circumstances and in situations making for attenuated growth. Some of the centric fern leaves appear to owe their symmetry to the suppression of dorsiventrality. Although, in dicotyledons, bud induction in leaf sites has not so far been achieved, we may note that in genera such as *Nuphar* and *Nymphaea*, lateral organs of radial symmetry, i.e. flower buds, characteristically originate in leaf sites and constitute part of the normal phyllotactic sequence (fig. 8).

Metabolic relationships in primordium inception

Since very little is known about the physiology of the shoot apex, a speculative approach to the relevant problems seems admissible at this stage. The orderly appearance of primordia in time and space indicates that there is a considerable measure of regulation in the underlying physiological processes. As metabolic materials move in to the apical meristem from the older regions of the shoot, they must pass successively through the maturing and differentiating tissues of the sub-apical region. If, as seems likely, they are utilized differentially at different levels, they will change both quantitatively and qualitatively in characteristic ways during their movement towards the extreme apex. The apical (or distal) cell group is thus likely to be a locus of special metabolism. Its by-products, which may include growth-regulating substances, will tend to move basipetally along concentration gradients, producing different effects in cells at successively lower levels. According to their position on the meristem, therefore, individual cells, and also growth centres, will be differently constituted; they will have different competences to react to morphogenetic stimuli and to utilize nutrients. Each leaf-growth centre may likewise determine characteristic centripetal and centrifugal movements of metabolic materials, its physiological field being thereby defined; and, according to its position on the meristem, it will be affected by the apical cell group and by the proximity of other growth centres. It thus becomes understandable why, in *Dryopteris*, deep and wide tangential incisions of the apical meristem above the I_1 or P_1 positions, or deep radial incisions on either side of a primordium, are attended by far-reaching changes in the normal morphological development. It has also been possible, by the direct application of various physiologically-active substances to the fern shoot apex, to induce the formation of tangentially extended double leaf primordia, of two leaf primordia in a single leaf site, of centric leaves, and of buds from young leaf primordia; the arrest of growth in meristems has also been effected. In the young embryos of some dicotyledons three or more cotyledons have been induced as a result of chemical treatment. In relation to the general metabolic scheme outlined above, it may be noted that leaf primordia are formed on the meristem just above the actively growing sub-apical region. One consequence of this is that the abaxial side of a primordium enters into a region of active growth while its adaxial face is still within the meristem, a region of less active growth. This may be one of the main relationships affecting the dorsiventral development of primordia.

In *Dryopteris*, each new leaf primordium originates in its expected position with a very high degree of regularity. A considerable body of experimental evidence indicates that when the last formed primordium (P_1) is in early plastochrone, the sites or growth centres of the next three primordia to be formed (i.e. I_1, I_2 and I_3) are already determined, the first leaf sites that it has been possible to displace by experimental means being those of I_4 and I_5. The dicotyledons also yield evidence of the presence of growth centres in characteristic positions some time prior to the outgrowth of primordia. As to the inception of growth centres in an orderly sequence, thus constituting a pattern, attention may be directed to the views that have been expressed by Bünning (1952), Turing (1952) and recently by M. and R. Snow (1955).

The foregoing observations and inferences indicate the need for close study of the constitution and reactivity of embryonic cells in different regions of the meristem. The view that the apices of ferns and flowering plants comprise several zones or regions of distinctive metabolism, each merging and interacting with the adjacent zone or zones, also deserves fuller investigation.

Primordia and the dynamic geometry of the apex

In position, organogenic activity and embryonic character, the shoot apices of all vascular plants have much in common though they may show evident histological differences. In different species, the meristem may occupy a relatively small or a relatively large part of the distal region of the shoot; it may be conical, rounded or flat; and it may be borne on a broadly extended sub-apical region or on a relatively thin cylindrical one. Leaf primordia may be of small or large size relative to the apical meristem; they may occupy all the space 'available' to them, or only a part of it; they may be of different shapes, from localized circular mounds to crescentic out-growths more or less encircling the axis; and they may arise very close to the distal region or in positions relatively remote from it. The actual arrangement may be accepted as being ultimately determined by the specific genetical constitution, together with other factors which are at work during development. The growing apex is essentially a dynamic geometrical system giving rise to lateral members, and therefore, according to the constitution and state of the system, there is scope for great diversity in the form of the leaf primordium at its inception and during its further development.

In a detailed mathematical treatment of the problems of phyllotaxis, Richards (1951) has shown that it is possible to have apices with the same divergence angle, the same plastochrone ratio, and the same proportion of new (or incipient) primordium area to apical area, but yet to have very different phyllotaxis systems. *The essential difference is in the shapes of the leaf primordia.* For a given divergence angle, the contact parastichies depend on the plastochrone ratio (from which the radial relative growth rate of the apex can be assessed) and the primordium shape, both of which are dependent on the specific organization and reactivity of the apical meristem. Richards concludes that, to be valid, a causal theory of phyllotaxis must explain the absolute distancing of growth centres from one another, the absence of primordia from the central region of the apical meristem, and the factors determining its transverse size at the time of leaf inception. These problems are equally the preoccupation of the experimental investigator of the apical meristem and its organogenic activity.

In general terms, the larger the leaf growth centre and the closer it is to the centre of the meristem, the larger will be the leaf primordium relative to the apex and the lower the phyllotaxis; and conversely, the smaller the leaf growth centre and the further it is from the centre, the smaller will be the primordium and the higher the phyllotaxis. *Acer*, and other species with decussate leaves, are examples of the first instance; *Hippuris*, *Equisetum*, *Lycopodium* and *Dryopteris* are examples of the second; and between these extremes many intermediate conditions are to be found. The grasses, on the other hand, afford examples of growth centres remote from the tip of the axis, yet with low phyllotaxis; but in them, although the primordium begins as a localized growth centre, it soon extends laterally right round the shoot. In short, consideration of the dynamic geometrical aspects again emphasizes the need for a fuller knowledge of specific apical organization and reactivity; for the inception of a primordium of characteristic form is an expression of that organization and is determined by the physiological processes which maintain the apical meristem in its active embryonic state.

17

On the organization and reactivity of the shoot apex in vascular plants

In all vascular plants, the plant in the vegetative phase consists essentially of a shoot with lateral members (usually leaves). The growth of this axis, the formation of leaf primordia in a regular sequence, the developmental harmony which becomes manifest during their further growth, and the differentiation of characteristic tissue systems, depend primarily on the activity of a distal embryonic region—the *apex* (here to be understood as comprising the essentially embryonic *apical meristem* and the subjacent region of expansion). What has long proved so puzzling to botanists may be stated quite simply: it is that, whereas certain major functional activities are common to the apices in all classes of vascular plants, e.g. their regulated organogenic activities, their maintenance of the embryonic state, etc., apices may nevertheless be demonstrably very different in such matters as their size, shape, histological constitution and pattern of tissue differentiation. Apices may be conical, paraboloidal or almost flat; their leaf primordia may be of small or large size relative to the apical meristem, and they may arise close to the apical cell or group of apical initials, i.e. to the summit or centre of the meristem, or in positions on the flanks more or less remote from it. Probably not least important for many contemporary investigators, as judged by the volume of published work, is the fact that apices show great diversity in their histological constitution, i.e. in the number and relationships of their histogenic layers, some seven or more 'apical types' having now been described (for recent reviews and literature, see Gifford, 1954; Buvat, 1955; Bersillon, 1955). Yet, notwithstanding these differences of detail, it is hard to avoid the conclusion that in

American Journal of Botany, Vol. 44, No. 2, p. 176, Feb. 1957.

their overall organization, all shoot apices are essentially alike. It should, therefore, be possible to formulate some general system of ideas which would explain, in part at least, why they show this remarkable *general homology of organization and morphogenetic activity*. Accordingly, it is pertinent to consider what conclusions emerge regarding the constitution and activity of shoot apices when due account is taken of the facts known from anatomical, experimental, genetical and other investigations.

Apical organization and reactivity as determined by genetical factors

It seems desirable at the outset to state explicitly those aspects of apical organization and reactivity which appear to be closely and primarily determined by the genetical constitution, i.e. for any particular species, and making due allowance for the effects of environmental factors and for the stage in the ontogeny, they are the anatomical, histological and cytological characters that must be accepted as 'given'. They include the following:

(1) *The size and protoplasmic constitution of the cells of the meristem.* The view that the size of meristem cells is determined by genetical factors is supported by a growing body of evidence (summarized in Wardlaw, 1952, 1953b). Since, in the ontogenetic development of any species, there is no absolute size for meristem or embryonic cells, this statement must evidently be qualified in various ways; but it serves to convey a general truth, namely, that when the individual plant has reached a certain size, under the usual conditions of growth, the cells of its apical meristem tend to be of a certain characteristic size.

(2) *The relative rates of growth in the vertical and transverse planes.* This relationship is a very important one, because it determines the shape of the apical and subapical regions. In conjunction with the factors of cell size and constitution, the specific distribution of growth may largely determine the characteristic cellular constitution of the apex, i.e., the disposition of embryonic cells in a characteristic number of histogenic layers, or whatever the cellular pattern may be. Qualifications of the kind indicated in (1) will also apply here.

(3) *The absolute size of the apex.* In all embryos the apex is initially of very small size; but whereas in some species it undergoes a relatively small ontogenetic increase in size, in other

species, e.g. in ferns and cycads, it undergoes a progressive ontogenetic enlargement, and is eventually of large relative size. Genetical factors as in (1) and (2) are closely involved in these developments, but environmental and other factors, which may affect the supply of nutrients to the apex, must be given due weight.

(4) *The positions of growth centres*. The concept of *growth centres* is considered later. Briefly, the idea is that a leaf primordium typically develops on the meristem at a locus of special metabolism, i.e., at a growth centre. The positions of such loci are specific. In different species, growth centres may be high up or low down on the meristem, and may comprise cells pertaining to relatively superficial or relatively deepseated histogenic layers. These positional relationships will affect the size and shape of primordia and their phyllotaxis.

(5) *The nature of the apical reaction system*. The apex is the seat of many complex but inter-related chemical reactions, all the necessary metabolic substances being contained in the embryonic cells or supplied from below. As the components of the *apical reaction system* are directly or indirectly gene-determined, each system is specific in character; but since many of the major metabolic processes are common to all green plants, certain components are likely to be common to different reaction systems. Accordingly, both general (or common) and specific effects may be expected to result from the working of an apical reaction system.

At this stage in the investigation of shoot apices, the foregoing statements are unavoidably no more than general indications of what may, at a later stage, become the basis of valid criteria of comparison. For the time being, they show that the diverse apical configurations described by plant anatomists can be brought into at least some relationship with the underlying physiological and genetical factors.

So far, the number of papers in which the histological constitution of the apical meristem has been related to the genetical constitution is small. Nevertheless, they support the views given here and leave little doubt that further work along these lines is likely to be rewarding (see Cross and Johnson, 1941; Randolph *et al.*, 1944; Satina *et al.*, 1940, 1941, 1943; Bain and Dermen, 1944; Dermen, 1945, 1947, 1951, 1953; Abbe and Stein, 1954;

Stein, 1955; Weber and Stein, 1954; Stein and Kroman, 1954;
Wardlaw, 1948, 1952, 1956*a*).

The apex a dynamic geometrical system

In different species, the apical meristem may occupy a relatively
small or a relatively large part of the distal region of the shoot;
it may be borne on a broad subapical region or on a relatively
thin cylindrical one; and it may be flat or almost flat, paraboloidal or conical (fig. 1). The shape of the apex and the
positions occupied by its lateral members are determined by the
distribution of growth. On proceeding from the subapical
region to the tip or centre of the shoot, the rate of vertical and
transverse growth falls off in a regular manner: in a large,
regular, conical apex, like that of *Dryopteris*, the rate of growth
falls off steadily; in various paraboloidal apices it falls off considerably more rapidly; and in flat apical meristems (other
things being equal), the growth rate at the centre of the system
may be very low indeed. These considerations are important
in relation to recent views advanced by Plantefol (1947, 1951)
and his co-workers.

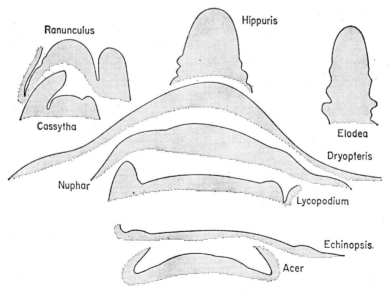

1 Longitudinal median sections of different vegetative shoot apices. (All
× 150.)

In a comprehensive treatment of phyllotaxis, Richards (1951) has shown that, for a given divergence angle, the contact parastichies depend on the plastochrone ratio (from which the radial relative growth rate of the apex can be assessed) and the shape of the nascent leaf primordium. In short, the factors which determine apical organization and leaf inception have a fundamental bearing on phyllotaxis. In his view, a valid causal theory of phyllotaxis must explain the absolute distancing of growth centres from one another, the absence of primordia from the upper (or central) region of the meristem, and the factors determining its transverse size at the time of leaf inception. As a general statement, we may say that the larger the leaf growth centre and the closer it is to the tip of the axis, the larger will be the primordium relative to the apical meristem and the lower will be the phyllotaxis; and, conversely, the smaller the leaf growth centre and the further it is from the tip, the smaller will be the primordium and the higher the phyllotaxis. In fact, the growing apex, in which we perceive biological organization, is a dynamic geometrical system.

Leaf growth centres

A leaf primordium becomes visible and recognizable on a flank or sector of the apical meristem as a result of active growth in a localized site of characteristic size and shape. It is, accordingly, reasonable to attribute its inception to a restricted region of special metabolism on, or in, the meristem, i.e. to a *growth centre*.* In a fern such as *Dryopteris*, with a large conical apex, the leaf growth centres are small and occur in situations relatively remote from the shoot apical cell. In some flowering plants, somewhat similar relationships may be observed; but in others, the growth centres originate very close to the tip of the meristem. Because growth centres differ in number, position, size and shape, in species from different taxonomic

---

* *Growth centre.* The writer should make it clear that, in his view, a *growth centre* which becomes a primordium is not necessarily a 'centre' in any strict mathematical sense. Rather it is a *locus* or *site* which may comprise a group of cells in which, as compared with the surrounding cells, there has been an accumulation of those metabolic materials (enzymes, substrates, growth-regulating substances, etc.) which, in appropriate circumstances, will make for rapid growth.

groups, a considerable diversity in primordium morphology and phyllotaxis follows as a natural consequence.

The nascent leaf primordium in ferns is the result of the enlargement of a group of some 8–10 superficial prism-shaped cells of the apical meristem, together with the active growth and division of a group of underlying, somewhat smaller cells. Although the histological details may be rather different or, indeed, very different, the inception of leaf primordia in flowering plants is generally comparable with that in ferns in its main features, the growth centres typically comprising both surface and sub-surface cells of the apical meristem. Moreover, as the writer has shown (Wardlaw, 1957a), the microphylls of *Psilotum* and *Lycopodium* also originate from comparable multi-cellular growth centres. This histological account of leaf inception could be greatly extended, but perhaps enough has been said to bring home the main point, namely, that, in all vascular plants, the inception of the lateral members which we recognize and describe as leaf primordia exemplifies a remarkable homology of organization; and we can see that, within a general system of relationships, such as that of the nascent primordium to the apical meristem, there is nevertheless great scope for diversity in the subsequent morphological development.

Since a primordium originates on the flank of a conical or paraboloidal apex, with differentiating tissue below and embryonic cells above, or, to state the matter in more general terms, in a growing system in which there is an acropetal falling-off in the growth rate, it follows that the growth relationships and rates of differentiation on the adaxial and abaxial sides of the primordium are different virtually from the outset. These relationships, in conjunction with factors which exercise regulative effects, may be held accountable for leaf symmetry and orientation.

Both in pteridophytes and seed plants, the vertical growth of the young primordium is primarily due to the activity of a centrally-placed, superficial initial cell or cells: a conspicuous leaf apical cell is present in ferns, for example, whereas there is a group of superficial initial cells in the flowering plants. Here again, within a general morphogenetic pattern, there is scope for great diversification in the further development of the foliar

o

member. And, lastly, a feature of the growth of the young primordium is the inception in it of an uninterrupted, initially acropetally developing vascular tissue whereby it is brought into a characteristic relationship with the vascular system of the parent axis. In different instances, according to the relative growth activities of the leaf primordia and the shoot apex, the vascular system of the leafy shoot may range from being predominantly foliar to predominantly axial in origin; many flowering plants illustrate the former, whereas microphyllous pteridophytes exemplify the latter state of affairs.

A leaf primordium becomes visible, in a well-defined position on the apical meristem, as a localized, discrete outgrowth, of characteristic shape and size for the species and for the stage reached by the plant in its ontogenetic development. The experimental evidence is consistent with the view that the growth centres, which are considered to precede the visible outgrowth of primordia, have a real existence and that they may be established some considerable time before any morphological development can be observed. In large apices of *Dryopteris*, for example, three growth centres (those of I_1, I_2 and I_3)* are already present when the last formed primordium (P_1)* is still in early plastochrone. It will be apparent that these growth centres must occupy positions on the meristem somewhat above the level of P_1, and that, in a growing apex, each growth centre will be left behind and appear further away from the shoot apical cell. It is now necessary to consider how the inception of growth centres, and of the leaf primordia to which they give rise, may be brought about. The factors which determine and regulate the morphogenetic activity of the apical meristem are still but little understood and, accordingly, various views, some of them very much opposed, have been put forward in attempts to explain the observed developments. But whatever view may eventually be considered most valid, it is virtually impossible to advance without making one basic assumption, namely, that the whole shoot apex, or some particular region of it, is a com-

* Common practice sanctions the use of the symbols P_1, P_2, P_3, etc., for the youngest leaf primordia on the apex of a plant, and I_1, I_2, I_3, etc. for the incipient primordia, the spaces at which they will have their origin being already recognizable. From oldest to youngest, in the order of their occurrence, the series would be P_3, P_2, P_1, I_1, I_2, I_3, etc.

plex physico-chemical reaction system, the working of which is such that it gives rise to growth centres in an orderly sequence.

The apex as a system of inter-related zones

The most comprehensive and adequate conception of the shoot apex, in the present writer's view, is that it is a system of integrated and inter-related zones. On the basis of recent experimental work, together with the views of Schoute (1936), based on anatomical studies, the following regions or zones, in basipetal sequence, may be indicated (fig. 2):

(1) *The distal region* (i.e. the summit or centre of the meristem), comprising the apical cell group in ferns, or a group of embryonic initial cells, in one or more layers, in seed plants: this is the centre or focal point of the meristem on which the integrity and sustained development of the primary axis depend.

(2) *The sub-distal region* of the apical meristem consisting of a superficial layer or layers of meristematic cells, and, according to the species, of embryonic or differentiating cells within. *This is the region in which the inception of growth centres takes place.*

(3) *The organogenic region*, subjacent to (2), is that in which the outgrowth of leaf primordia takes place and tissue differentiation has its inception or becomes more conspicuous. In the ferns, this region occupies the base of the apical cone; in the flowering plants, in different instances, it may be high up, or low down, on the apex.

(4) *The subapical region*, subjacent to (3), is usually characterized by (*a*) a considerable widening and elongation of the shoot, (*b*) a conspicuous enlargement of the primordia, and (*c*) active formation of a parenchymatous cortex and pith and further differentiation of the vascular tissue.

(5) *The region of maturation.*

In the foregoing account of the apex, the *subdistal region* is seen to be of special importance since, in it, the growth centres have their inception. In 1913, Schoute advanced the view that a growth centre is first determined and a leaf primordium subsequently organized round it. He also held that a specific substance is produced by these centres which inhibits the formation of others in the immediate vicinity; hence new primordia only arise between older ones when the space has become sufficiently large. The inhibition of new growth centres by the apical cell

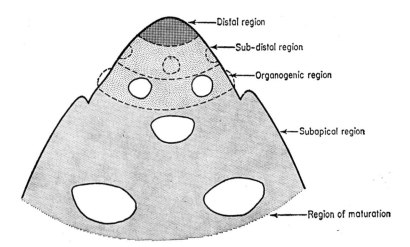

2 Diagrammatic representation of an apex, with whorled phyllotaxis, as a system of inter-related zones.

(or group of apical initial cells) was also postulated: hence the position of a new primordium on the flanks at some characteristic distance from the apical cell group. Support for the concept of growth centres has been given by Richards (1948, 1951) and Bünning (1948, 1952), while the writer (Wardlaw, 1949) and Sussex (1955), in numerous surgical experiments, have shown that there is substantial evidence for the views that growth centres do exist in the meristem and that the young primordia exercise mutually regulative effects.

Bünning (1948, 1952) has ascribed the patternized distribution of growth centres to the mutual incompatibility of regions of vigorous protoplasmic growth. In his view, competition for nutrients cannot be the decisive factor underlying or determining the distribution of growth centres; rather it is that 'a certain type of embryonic growth will not allow the same type of growth to take place nearby' but a different type of embryonic growth may proceed unimpeded. The processes in question are probably enzymatical, i.e. in the inception and development of a growth centre, a particular enzyme may become quantitatively predominant and this will lead to a diminution of the corresponding substrate in the surrounding field (Bünning). While these ideas may be applicable to a shoot apex on which organs have already been formed, it may be questioned if they explain how a patternized distribution of metabolites could be brought about in a new embryonic region in which no organs or tissue systems have yet been formed or differentiated. This, indeed, may be indicated as one of the major problems in morphogenesis. In this connection Turing's (1952) diffusion reaction theory of morphogenesis seems to be of special interest and importance, for he has indicated that an initially homogeneous physico-chemical reaction system may become heterogeneous and give rise to a patternized distribution of metabolites, and thus provide the basis for a regular and characteristic morphological or histological development. The writer (Wardlaw, 1953a, 1955a,b) has pointed out that this theory, or some similar theory resting on a physico-chemical basis, is of the kind that could have a wide and general application to the phenomena of morphogenesis. It may, for example, afford an explanation of the orderly inception of growth centres in the sub-distal zone of the apex. This theory would afford a satisfying explanation of whorled phyllotaxis, e.g. in species of *Galium*, in which 4, 5 or more, evenly-spaced primordia have their inception simultaneously on the flanks of the meristem. In such species, the members of consecutive whorls usually alternate, and hence we may infer that the eventual equilibrium of the reaction system in the sub-distal zone is affected by the positions of existing primordia immediately below. The number of new growth centres produced at any particular stage in the development of the shoot is, however, determined by the state of the reaction system in the sub-distal zone. During the ontogenetic increase in size of the apex in some species with whorled phyllotaxis, there is an increase in the number of primordia in successive whorls (Wardlaw, 1955a). Any whorl with more, or fewer, primordia than the whorl below cannot absolutely alternate with it, but it seems probable that the disposition of primordia will be such as to make for equilibrium in the apex as a whole.

If Turing's theory applies to whorled phyllotaxis, it may also be expected to apply to spiral and other systems. In this connection the writer now suggests, as a working hypothesis, that the inception of primordia in characteristic and usually predictable positions is due, partly to the activity of the reaction system located in the sub-distal region of the apex, and partly to effects exercised by the adjacent older primordia. On this basis, one can understand that an apical reaction system which gives rise to a simple pattern at an early stage in the ontogeny is likely to yield more complex variants of the same pattern during its subsequent enlargement, and that asymmetries in proximity to the reaction system, e.g. the presence of adjacent developing organs and tissues, may induce characteristic asymmetries in the system and in the patterns to which it gives rise. In large apices of *Dryopteris*, the experimental evidence indicates that the sites of growth centres of the next three primordia to be formed, i.e. I_1, I_2 and I_3, are already determined, but that the positions in which I_4 and I_5 eventually arise can be modified by appropriate surgical treatments. In some dicotyledons, as the work of M. and R. Snow has shown, it appears that only I_1 is so determined. Critical changes in some of the components of the reaction system may affect the patterns to which it gives rise. Lastly, variations in the histological patterns that have been observed in the vegetative apex itself, and the transition from the 'vegetative' to the 'floral' meristem, must sooner or later be attributed to quantitative and/or qualitative changes in the substances that are being produced in other parts of the plant and are being 'fed into' the apical reaction system.

Like any active physico-chemical system, the apex, comprising the several zones indicated above, will constantly tend towards a state of equilibrium, and this will contribute to the harmonious development of the leafy axis to which it gives rise. The apex, thus envisaged, is essentially holistic in nature.

Apical reactivity as ascertained by surgical techniques
From the evidence of surgical experiments, it has now become possible to attempt at least a general account of the reactivities of different regions of the apex with particular reference to the formation of leaves and buds. Sustained growth of the apex depends on supplies of metabolic substances from below. Hence, in fig. 3 the thickness of the arrows suggests diagrammatically, on the basis of rate of growth and utilization, the diminution of supplies as the apical cell group is approached. It may reasonably be assumed that qualitative differences are also involved. If, now, the arrows were directed downwards and their thick-

nesses reversed, the diagram would afford a representation of regulative effects, some of them inhibitive, exercised by the apical cell group. In short, the normal reactivity of the intact apex involves both acropetal and basipetal effects; and it can be seen that, as long as the apical organization persists, critical physiological differences will exist between the adaxial and abaxial sides of a primordium, or of a selected area of the apical meristem (or even of a subapical region). As the apical cell and its adjacent segments (in ferns), or the group of apical initials (in seed plants), are very slow-growing, they will exercise only a

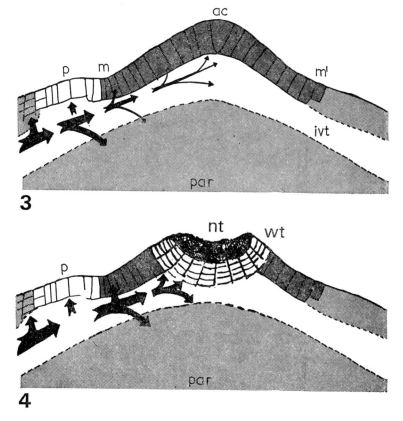

3 Diagrammatic representation of the movement of metabolic materials in a normal and **4** punctured fern apex. *ac*, apical cell; *m–m¹*, apical meristem; *p*, very young leaf primordium; *ivt*; prestelar tissue; *par*, parenchyma of pith; *nt*, necrosed tissue; *wt*, wound tissue.

small 'pull' on metabolic materials from below; but they are not completely inert and inactive as Buvat (1952, 1955) and his co-workers have asserted.

When the apical cell of *Dryopteris* is punctured without an attendant collapse of the adjacent cells, wound healing takes place very slowly and only a small amount of parenchyma is formed. In these circumstances, the organization of the apical meristem remains relatively unchanged for a considerable time, and leaf primordia continue to be formed until all the meristem is used up. The integrity of the axis is, however, destroyed, any further axial development being effected by a lateral bud induced on the meristem or on the subapical region. If a larger puncture is made at the tip, wound healing, with attendant formation of callus and parenchyma, is extensive, and it is evident that important changes in the distribution of metabolic materials must have taken place, as suggested in fig. 4; i.e., there is now much greater utilization in the distal region. Moreover, the regulative effects of the distal embryonic cells are curtailed, or may be more or less completely eliminated. The critical physiological difference between the adaxial and abaxial sides of primordia will now be diminished and may even be reversed. In these circumstances, some buds are usually induced on the meristem in leaf or bud sites and the orientation of some of the last-formed leaf primordia may be reversed.

When a wide and deep tangential incision is made above P_1 or I_1, the regulative effect of the apical cell group is eliminated in the sector below the cut and wound healing takes place in a relatively active region of the apical meristem: the critical difference between the adaxial and abaxial sides of a primordium, or site, disappears and a bud is formed. But with a *shallow* tangential incision, which does not sever the prestelar tissue, the regulative effect of the apical cell group is not removed and a leaf is formed if the apex is in active growth. When a young primordium, or site, is isolated from the adjacent, lateral, older primordia by deep radial incisions, it is released both from their hormonal regulative effects and from their greater power to utilize and therefore attract nutrients. The critical difference between the adaxial and abaxial sides still persists, however, and the primordium develops as a foliar organ of large relative size.

In reviewing these experimental data, the possibility presents itself that the regulative effects of the apical cell group or, more generally, of the distal region of the meristem, may be due to its uptake capacity as well as to its basipetal hormonal effects. A shallow tangential incision, which does not sever the prestelar tissue, would not affect this uptake: a deep one would, and would make for an accumulation of materials below the incision. Whatever views may finally prove acceptable, the experimental evidence points to the importance of the physiological activity of the cells in the most distal region of the meristem in contributing to apical organization and orderly morphogenetic activity. And while theories which stress the seeming inertness or inactivity of this region are important in that they have led to an intensification of observation and been the means of discovering new facts, they may forfeit some of their value by over-emphasis (see Wardlaw, 1957b,c). The experimental data and the facts (a) that an incipient leaf primordium is of a characteristic shape and size, and (b) that it only arises at a certain distance below the apical cell group, are strongly indicative that the most distal region of the meristem possesses special physiological properties. Any treatment, e.g. 'starvation' or mechanical injury, which diminishes or modifies the normal activity of the distal region, permits leaf primordia to be formed closer to it. In the writer's view, the formation of leaf or of bud primordia must be referred to the organization of the apex as a whole, in which nutritional and hormonal factors play essential and inseparable parts—a concept which will also apply generally to situations in which new morphogenetic effects are induced by experimental treatments.

Cytology and apical organization and activity
Plantefol (1947, 1951), Buvat (1952, 1955), Bersillon (1955) and others have shown that, as judged by the number of dividing nuclei, some regions of the apex are meristematically much more active than others; for example, leaf sites and the basal (or organogenic) region of the apical cone (or dome) are demonstrably more active than the cells near the tip or centre of the meristem. Indeed, these workers assert that, during the vegetative phase, the tip of the meristem is virtually inert, that it largely owes its origin to the subjacent region and is simply

carried upwards by its growth. Some of the views elaborated in the foregoing sections, together with what follows, may perhaps enable us to evaluate the cytological contribution.

If an apical meristem is of large size relative to its leaf primordia, its conformation will be but little affected by primordium formation, and a comparatively small amount of growth, with concomitant cell division, in the most distal cells will suffice to keep pace with more active cell division lower down and to maintain the paraboloidal or conical apex at its characteristic size. But if, on the other hand, the primordia are of large relative size and are formed close to the tip or centre of the apical meristem, the meristem will undergo very considerable changes in size and shape at the formation of each primordium. In such cases, active growth and cell division can be observed at, and close to, the tip of the meristem. Recently, Newman (1956) has shown, by the direct study of living apices of *Tropaeolum* and *Coleus*, that divisions do take place in the apical initial cells. Elsewhere the writer (Wardlaw, 1957*b*) has discussed the cytological contribution to our knowledge of apical organization in some detail against the general background of information obtained from anatomical and experimental investigations. His conclusion is that cytological investigations are likely to make for new and important accessions of fact but that the emphatic statements made by the French school are open to substantial criticism. Gifford (1954) has also advanced some cogent criticisms on this theme.

Metabolism in the apex
Comparatively little is yet known about either the general physiology of the apex or, more particularly, about its metabolism. Its basic components include carbohydrates, the precursors of the cytoplasmic and nuclear proteins and related substances, enzymes and hormonal substances. Important discoveries are now being made in the chemical study of apices by the method of partition chromatography. Many amino acids and related amides have been shown to be present either in the free state or after hydrolysis of the protein fraction. Various growth-hormonal and related substances have also been detected and separated. Some of the substances that participate in the reactions in the meristem are thus known and, in some

cases, it has been possible to indicate the probable paths of biosynthesis. In these analytical studies, the apex, because of its small size, has usually been treated as a whole. We do not yet know which substances are of special importance in maintaining the characteristic organization of the meristem or in the inception and development of growth centres. Skoog (1954) has indicated the probable importance of the nucleic acids in the regulation of growth and has shown that, in some tissue cultures at least, bud inception is closely dependent on the presence of nucleic acid constituents.

In angiosperm embryo-culture, it has been found that whereas older embryos (from the 'heart-shaped' stage onwards) can readily be grown to full size in simple media, very young excised embryos can only grow normally when the basic medium has been supplemented by various substances such as coconut milk. In the pure culture of excised fern apices in simple media of known composition, e.g. with a nitrate as the nitrogen source, Wetmore (1954) has shown that normal growth can be sustained; but it may be greatly promoted by the addition of the substances present in yeast extract. His evidence shows that the embryonic cells of the fern apex have the capacity to synthesize elaborate organic substances from simple sources, and calls attention to the importance of enzymes in the apical reaction system. The apices of various flowering plants, on the other hand, have grown but little in media in which the apices of vascular cryptogams thrived. Additions of auto-claved coconut milk and of casein hydrolysate to media enabled the apices of *Syringa* to grow, become rooted and to undergo normal development. This discovery has led to further interesting studies of the substances present in the shoot apex. (See Wetmore, 1954, for a full discussion of this important topic and for references.)

As metabolic materials move upward to the apical meristem from the older regions of the shoot, passing successively through the maturing and differentiating tissues of the subapical region, they will tend to be utilized differentially at different levels. As a result, they will change both quantitatively and qualitatively as the most distal region of the meristem is approached. If the apical (or distal) cell group is a locus of special metabolism, as the histological observations suggest, its by-products, which may include growth-regulating substances, will tend to move basipetally along concentration gradients, producing different effects in cells at successively lower levels. At different levels on the meristem, therefore, the individual cells, and the groups of cells constituting growth centres, will be somewhat differently constituted, and this may determine their competence to react to morphogenetic stimuli and their ability to compete for nutrients. Similarly, each leaf growth centre may determine characteristic centripetal and centrifugal movements of meta-bolic materials, its individual physiological field being thereby defined. From such considerations it becomes understandable why, in *Dryopteris*, deep and wide tangential incisions of the apical meristem above the I_1 or P_1 positions are attended by far-reaching modifications of the normal morphological develop-ment, i.e. the induction of a bud in a leaf site, or the transforma-tion of an undetermined leaf primordium into a bud (Wardlaw, 1949, 1955c; Cutter, 1956). Lastly, it has been possible, by the direct application of various physiologically-active substances to the fern shoot apex, to induce a premature development of scales and parenchyma, the formation of tangentially-extended double leaf primordia, of two leaf primordia in a single leaf site, of centric leaves, and of buds from young leaf primordia. Leaf inception and growth may be suppressed by chemical treatments (Wardlaw, 1955d, 1957d). In somewhat related embryological studies, Haccius (1955) has shown that the young embryos of some dicotyledons can be induced by chemical treatment to form three or more cotyledons.

Conclusion

In plants, as in animals, aspects of organization are manifested at all levels from the sub-microscopic to the cellular and throughout development from the fertilized egg to the adult state. The constitution and reactivity of the shoot apex are of special interest to botanists because they underlie and deter-mine the morphological and histological developments in the most highly elaborated plant species. The comprehensive study of the growth and form of a living organism may, indeed, be equated with an attempt to analyse its organization in its manifold aspects; and this, as may be inferred from the present paper, requires a confluence and integration of facts and concepts from many branches of botanical science and also from mathematics and the physical sciences. The study of organization is not only of great inherent interest, but it has a special importance in that it seems likely to afford the most satisfying means of reconstituting Botany—the science of plants as distinct from the science of particular biological processes—from its several specialized branches.

18

The floral meristem
as a reaction system

The morphology of flowers has long been a major preoccupation of botanists, in particular because it affords the main (though not the only) basis for the taxonomic and evolutionary treatment of the Angiosperms—the largest, most varied and most advanced sub-division of the Plant Kingdom. Interest in floral ontogenesis is also of long standing, this study having had many adherents since the middle of the nineteenth century: as a result, relevant information is now available for many species, together with much speculation as to the nature and relationships of the several floral organs. The rediscovery of Mendel's work at the beginning of the century led to numerous investigations on the effects of intercrossing on the specific form, size and colour of flowers and, in more recent years, to hypotheses on physiological-genetical aspects of floral morphogenesis. An important development during recent decades has been the demonstration that, in many species, the onset of flowering is due to photoperiodic induction, i.e. to processes set in motion by the exposure of mature green leaves to characteristic periods of light and darkness. There have thus been very considerable accessions of knowledge in this important field. When, however, we consider the diversity of flowers, and the fidelity with which each specific floral ontogenesis takes place—this constituting the highest manifestation of organization in the Plant Kingdom—and when we further consider how very little is known of the underlying causation, we can hardly fail to realize that floral morphogenesis in its many aspects not only presents a major challenge, but one which affords virtually inexhaustible opportunities for constructive thought and work, especially at the present time. The purpose of the present paper is to consider a theory which, in the writer's view, may have a general application to the phenomena of floral ontogenesis.

The characteristic developments in any species are primarily due to its genetical constitution, to factors in the environment, and to certain other factors and relationships. It is now known that characters such as the size, shape, colour and number of floral parts are determined or controlled by particular genes or groups of genes, and it is generally assumed that, in some way, the orderly floral ontogenesis is fundamentally gene-controlled. But can we indicate the physiological mechanism that brings about the onset of flowering in a vegetative shoot apex and the orderly formation and disposition of the several distinctive organs—sepals, petals, stamens and carpels? That these are questions of major importance needs no emphasis. It is perhaps timely, therefore, to consider some of them in the light of contemporary ideas.

Morphological concepts of the flower
The morphological nature of the flower has long been a subject for discussion. In the classical view, which we can trace back to early observers such as Grew, Wolff and Goethe, and which is still staunchly supported by distinguished contemporary workers, the flower is regarded as a determinate shoot, or axis, bearing appendages which are homologous with the leaves produced during the antecedent vegetative phase. In some accounts the floral organs are held to be the result of a progressive meta-morphosis of foliage leaves, with an attendant reduction or 'telescoping', of the internodes. The classical view has been vigorously opposed, by some observers on the interpretation of fossil evidence, by others on the basis of ontogenetic studies. Thus Thompson (1937), Grégoire (1938) and Plantefol (1948) explicitly deny that any homology exists between the flower and the vegetative shoot, even in those primitive families in which the homologous relationship seems most apparent and surely established (Eames (1931) and Bailey (1954)). In this contrary view the essential floral parts are considered to be organs *sui generis*. (For recent discussions and reviews, *see* Engard (1944), Douglas (1944), Esau (1953), Tepfer (1953), Bailey (1954) and Buvat (1952, 1955).)

Floral ontogeny

In the observations that follow, it will be assumed that we are concerned with species in which the vegetative shoot apex itself becomes the floral axis, i.e. as in any cymose inflorescence. The vegetative apical meristems of different species may differ in size, shape, allometric growth, histological organization and in the spatial arrangement of their leaf primordia; but they are all alike in that they give rise to a vascularized axis, potentially capable of indefinite elongation, and to characteristic and regularly-spaced leaf primordia. With the onset of flowering this growth pattern is modified in various ways, including a broadening of the distal region, as a result of more active growth and division of the distal and subjacent inner cells, and a marked diminution and eventual complete cessation of the vertical elongation of the axis. The meristem, thus modified, may now be recognized as the floral axis, meristem, or recep-tacle. On its basal margin, leaf-like perianth members are borne in one or more condensed helices or whorls, the transition from the vegetative phase being usually attended by the formation of small bracts and bracteoles. On its further development, the floral meristem, as in the hypogynous flowers of the Ranales, may undergo slight elongation, and it gives rise successively to alternating whorls, or greatly condensed spirals or helices, of stamens and then of carpels. If the most distal region of the apex is not completely used up in the formation of the carpels, it develops into parenchyma, with complete cessation of its meristematic and organogenic activities. Thus, anomalous de-velopments apart, the process of floral ontogenesis is char-acterized by an irreversible sequence of changes, new organs, including the all important stamens and carpels, being formed in an orderly sequence. The result is a fully-formed flower of distinctive organization and possessing functional unity. With suitable modifications of detail, the same general account will apply to more specialized floral types.

Basic assumptions regarding organogenesis

In attempting to explore the phenomena of floral ontogenesis, certain basic assumptions are necessary at this stage: (*a*) parti-cular metabolic substances are more or less directly involved in specific organogenic activities, e.g. in the formation of a stamen as compared with a sepal or a carpel; and (*b*) in an embryonic region or tissue, such as the peripheral tissues in an apical meristem, in which the various physiological substances may initially be more or less homogeneously distributed, a patternized distribution of particular metabolites can, and does, take place, their concentration in particular loci preceding and determining (in conjunction with other factors and relationships) the visible histological and morphological developments (Wardlaw, 1955).

The zoned apex and the inception of leaf primordia
The vegetative apex may be regarded as consisting of a number of contiguous and inter-related zones, or regions, of differing physiological activity. These include, in basipetal sequence: (*a*) *the distal region*; (*b*) *the sub-distal region*, in the more superficial tissues of which there is an accumulation of particular metabolic substances in regularly-spaced loci or *growth centres*; (*c*) *the organogenic region*, in which the leaf primordia, formed as a result of active cell growth and division in the growth centres, first become visible;* (*d*) *the sub-apical region*, in which the base of the conical or dome-shaped apical meristem broadens out conspicuously and in which there is rapid growth and differentiation of the primordia; and (*e*) the region of maturation (Schoute, 1936; Wardlaw, 1955, 1957). Since the leaf primordia which become visible in *region* (*c*) are of characteristic shape and size for the species and for the stage reached in the ontogeny, it is a reasonable inference that the *growth centres*, i.e. centres of special metabolism, from which they originate are also quite well defined. Various experiments indicate that growth centres occupy positions above and between the youngest primordia that surround the apical meristem; on the histological evidence they are located in the peripheral tissue, so that a very young primordium, on first becoming visible, appears as a low mound on the surface of the meristem and comprises cells of the tunica and outer corpus. The formation of regularly-spaced leaf primordia and, by inference, of the growth centres from which they originate, may be indicated as a phenomenon of the widest generality: it is characteristic of the apical

* Regions (*a*), (*b*) and (*c*) constitute what the present writer would define as the *apical meristem*.

P

meristem in virtually all classes of vascular plants, including those with microphylls, 'reduced' leaves, cataphylls, etc.

A characteristic activity of the sub-distal region

The *characteristic spacing* of growth centres is due to factors at work in the sub-distal region, though the *actual positions* which they occupy are, on the experimental evidence, partly determined by the last-formed primordia. The spontaneous inception of evenly-spaced primordia is seen in the two initial leaves, or cotyledons, of Dicotyledons; in some instances three and even four evenly-spaced cotyledons may be formed. In *Pinus ponderosa*, some six to sixteen cotyledon primordia may be formed simultaneously round the apical meristem according to its size. Again, in dicotyledon species with whorled phyllotaxis, such as *Galium aparine*, the seedling has two cotyledons, but, during the ontogenetic enlargement of the shoot, the number of leaves at successive nodes increases so that whorls of seven, eight and nine can be observed. We may suppose that the same general physiological mechanism is at work in species with spiral phyllotaxis.

On the basis of a comprehensive mathematical study, Turing (1952) has indicated the possibility that, in a *physico-chemical reaction system*, of the kind that may conceivably be present in an initially homogeneous embryonic region or tissue, certain of the reacting substances or their products could periodically become concentrated in uniformly-spaced loci. It will be apparent that a theory along these lines would afford a satisfying explanation of the simultaneous formation of a variable number of growth centres (and cotyledons) in *Pinus*, depending on the size of the meristem, and, more generally, of the morphogenetic activities of the sub-distal and organogenic regions of shoot and floral apices. An important feature of Turing's theory is his indication that quite differently constituted reaction systems may give rise to comparable patterns. Other investigators, e.g. Goebel (1922) and Bünning (1948, 1953) have also suggested explanations of the inception of morphogenic and histogenic patterns in seemingly homogeneous embryonic tissues. It will here be assumed that a reaction system, of the kind envisaged in Turing's theory, functions in the peripheral tissues (comprising cells of the tunica and outer corpus) of the sub-distal region. The components of the system will include metabolic substances (carbohydrates, aminoacids, enzymes, etc.) which occur in all green plants, together with other more specifically gene-determined substances which may be 'fed into' the reaction system at particular times. Quantitative and qualitative changes in the supply of metabolites to the apical reaction system may be expected during the ontogenetic development of the plant, especially at such critical periods as the onset of flowering, with consequential effects on its organogenic activity.

The reaction system in the floral meristem
Numerous investigations indicate that when flowering plants are exposed to certain periods of light and darkness, gradual or quite rapid modifications in the supply of particular metabolites to the shoot apex probably take place. Whatever the nature of this change may eventually be shown to be, it has a profound effect on the activity of the vegetative shoot apex: in fact, it brings about its transformation into a floral apex. That the reaction system in the sub-distal region is extensively, but not fundamentally, modified is indicated by the fact that it continues to give rise to a regular pattern of growth centres throughout the whole course of floral ontogenesis.

Some botanists have attributed the onset of flowering to the action of a specific substance, still unknown, sometimes referred to as 'florigen', which behaves like a catalyst in the apex and eventually induces changes of a far-reaching kind. In the sub-distal reaction system, we may envisage the 'flower-inducing substance' (or substances), not merely as effecting an initial change in the working of the system, but as initiating an orderly sequence of changes in its activity, partly by evoking successively the action of specific genes (or of the substances which they determine), and partly as a result of physiological correlation.

The ontogenesis of a hypogynous, apocarpous flower may now be considered in greater detail. When the florigenic substance enters the vegetative apex, changes in the allometric growth pattern take place. The reaction system continues to form growth centres, usually in a condensed helix, i.e. separated by very small vertical intervals, or simultaneously in a whorl. The primordia which develop from these growth centres do not, however, grow into full-sized leaves. They appear to have a restricted or limited growth potentiality and typically give rise to the small, leaf-like, outer floral organs which we recognize as sepals. In many species leaves of reduced size—bracts and bracteoles—mark the transition from the vegetative to the floral phase.

In relation to the changes that have taken place during the 'sepal phase', new, specific, gene-determined substances become active components of the reaction system. The latter, which is now located in the meristem above the level of the calyx, gives rise to a group of growth centres, alternating with the sepals. These growth centres are physiologically different from the previous group. They typically give rise to the leaf-like organs, often distinctively coloured, which we know as petals. The changes effected in the reaction system during the 'petal phase' lead to further specific genic action and to the inception of the stamens; and so on, until flower formation terminates with the formation of carpels, by which time the residual, distal region of the meristem has been more or less completely used up. It is probably to the action of particular genes (or groups of genes), as components of the reaction system, that the clear delimitation of the several phases of organ formation should be attributed.

In an apocarpous, hypogynous flower, *all* the organs first appear on the floral meristem as low mounds, each consisting of a localized group of cells, these being elements of the tunica and outer corpus (Tepfer, 1953). In position of origin and histology, these primordia are in all respects similar to leaf primordia and are, therefore, homologous with them; but that the growth centres and primordia formed during the several phases of floral ontogenesis differ in some of their physiological properties seems to be an unavoidable conclusion. According to the genetical constitution of the species, one or other of the organogenic phases may be prolonged. Thus, ten to sixteen whorls (or groups) of stamens may be formed in some species of *Ranunculus*; and organs of an intermediate, or intergrading character may be formed, e.g. the staminodes in *Aquilegia*, and the leaf-like outer perianth in some primitive Dicotyledons. Reference to a wide range of materials indicates that however different and distinctive the several floral organs may be, they all have morphological features in common. This, indeed, is what we should expect if they were all products of a reaction system which can be modified in various ways by the introduction of new factors, but only within the functional limits of the system.

Some of the ideas used here, namely, that a particular gene only becomes active when the physiological situation has become appropriate to its activity, and that the changes induced by it in the reaction system (or in the protoplasmic substratum) prepare the way for the action of certain other genes, have a prominent place in physiological-genetical theory as applied to morphogenesis (Mather, 1948).

Some advantages of the theory

In our present very limited state of knowledge of floral ontogenesis, any comprehensive theory is unavoidably highly speculative and correspondingly open to criticism. It may, however, be claimed that the conception of the floral meristem as a reaction system has a number of advantages. It attempts to bring the facts of floral ontogenesis, as already ascertained by morphologists, into a working relationship with the physiology of growth, physiological-genetics, physical chemistry and mathematics. By so doing, it indicates a way out of the tangled controversial issues regarding the nature and origin of the floral organs, these issues, as it now seems, having largely arisen because the protagonists tended to use purely morphological concepts. Some of the controversial matters, which figure so largely in the morphological literature, either become unimportant, or can be harmonized, if the concepts of organogenesis are transferred from the morphological to the physiological plane. Thus, such problems as the homology of the several floral organs and the vegetative leaves, whether one floral organ could be 'transformed' into another, and whether some organs should not properly be regarded as organs *sui generis*, disappear if we bear in mind that all the appendages of the axis are the products of the same reaction system. As we have seen, this reaction system, located in a particular region of the apical meristem, retains its property of giving rise to regularly-spaced growth centres throughout the whole development of the plant—from cotyledons to carpels. These, however, may vary in their chemical composition and potentiality for growth. Accordingly, whereas the primordia to which they give rise are homologous in respect of their position of origin, histology, symmetry and orientation *vis-à-vis* the apex, some of the mature organs into which they develop may be so different morphologically and structurally as to be considered organs *sui generis*. Again, in the normal development (i.e. teratological and experimental materials apart), it is not necessary to consider how one distinctive floral organ might have been formed as a result of the 'transformation' of another organ: for it is to differences in the metabolism of growth centres and young primordia that the visible morphological differences must be referred.

However much the androecium and gynoecium in the more advanced floral types may appear to consist of organs *sui generis*, they are nevertheless products of the continuing activity of the reaction system that has already given rise to the vegetative leaves and perianth members. Where the carpels are formed separately on the sides of a still elongating conical or paraboloidal apical meristem, it is not surprising that they resemble foliar organs and have been so described. In fact, in some primitive flowering plants, the carpels are unmistakably foliar structures (Bailey, 1954). But where, as in some epigynous flowers, the floral development is characterized by a great condensation of the vertical axis and by the upgrowth of a ring of tissue round the distal region of the apex, and where only a small number of carpels can be formed from the residual distal region, the homology of the carpels with the perianth organs may well prove difficult to establish. Nevertheless, one can see that such flowers merely present special cases of the general ontogenetic development considered in earlier sections. In short, it is held that the theory now advanced affords both a new approach and a new freedom in our attempts to explain developmental relationships in the more complex phenomena of floral ontogenesis.

It is evident from a survey of the literature that some recent and contemporary investigators share the writer's view, at least in part, that any satisfying explanation of the morphological data of floral ontogenesis must be in terms of the underlying physiological processes (*see* Schüepp, 1929; Thompson, 1937; Engard, 1944; Tepfer, 1953, etc.).

The search for validating evidence

It is now pertinent to consider if there is supporting evidence for the theory outlined in the previous sections and if it can be tested and shown to be valid.

General Applicability—Like any other comprehensive theory, there is the initial question of its general applicability to a wide range of related phenomena. A survey of the morphological data of floral development indicates that the theory could indeed have a very general application, and that explanations of the kind that have been indicated are not only essential but seem unavoidable.

Supporting Evidence—In species in which the vegetative phase is characterized by the formation of leaves of different sizes and shapes, i.e. by heteroblastic development, it has been shown experimentally that physiological differences in the shoot apex are involved, nutritional factors being important in some instances. The same general conception could also apply to the apex in the flowering phase. For many years physiologists have been trying to detect and isolate the substance or substances responsible for the induction of the flowering phase. The effects of the naturally-occurring auxins and of a wide range of synthetic activating substances on floral induction and development are now the subject of a vast literature; but, so far, the all-important 'florigenic' substance has not been definitively specified or isolated. That some particular substance(s), formed in adult leaves under the appropriate photoperiodic exposure for the species, is translocated to the apical meristem and becomes an active component of the reaction system, was strongly indicated by experiments of Wetmore, c. 1956.* The shoot apices of vegetative plants of *Xanthium*, and of several other species of Dicotyledons, which had been given the photoperiodic treatment known to induce flowering, were excised and transferred to tubes containing a standard tissue-culture medium. These apices grew and gave rise to flower buds. By contrast, the excised apices of similar plants which had been given less than the required photoperiodic treatment for flowering remained in the vegetative condition and developed as leafy

* Cited by J. P. Nitsch: *Plant Tissue and Organ Culture* (A Symposium), Delhi, 1963.

shoots. Even after the minimal effective photoperiodic exposure, the metabolic changes induced in the apical meristem were such as to set in motion the whole chain of reactions involved in floral ontogenesis.

In a recent investigation of an 'ever-flowering' strain of the cultivated strawberry, Sironval (1956) has shown that in plants which have been given the appropriate photoperiodic exposure for flower induction, as compared with those in the purely vegetative phase, qualitative differences can be detected in the chemical fraction containing the pigments. He has isolated two substances (vitamin E and a sterol) which are present in this fraction in plants entering on the flowering phase, and has shown that when they are applied to the leaves of plants in the vegetative phase, they promote the onset of flowering.

The probable importance of correlative developments in floral ontogenesis has been briefly noted. In this connection, reference may be made to experimental studies of the spider flower (*Cleome spinosa*) by Murneek (1927). In this species a phase of pistil production is normally followed by a phase of stamen production. Murneek has shown, however, that if the young pistils are removed as they are formed, the production of pistils may be continued for an abnormally long time.

Although floral development usually takes place with a high degree of regularity, it is readily conceivable that this orderly process could be disturbed, or modified, within the viable limits of the reaction system, by the impact of new factors, with effects on the ensuing morphological development. Horticultural species, which are subject to extensive intercrossing, inbreeding, polyploidization, mutation, etc., to unbalanced and sometimes excessive nutrition, and to the pervasive effects of viruses, have yielded much information that could be interpreted in terms of the theory. Such phenomena as the occurrence of petaloid stamens, the multiplication of the perianth and androecial whorls, various other meristic variations, proliferation of the gynoecium or its abortion, the reversion of the floral meristem to the vegetative state, etc., are all indicative of modifications of the normal reaction system. The classical works on teratology (cf. Worsdell, 1916) thus acquire a new interest when re-examined in terms of the theory. Cutter (1955) has described an unusual case of the sustained acropetal formation of free carpels in a tomato plant. This may have been due either to a virus or to a genetical change. In this instance, the 'gynoecial phase' of the floral development had evidently been prolonged and the residual apex modified, presumably because of some anomaly in the metabolism of the meristem. Departures from the 'normal' floral ontogenesis also occur in wild species. Some of these have been described in standard works on taxonomy and others are becoming known as a result of the contemporary interest in genetical-ecology. Marsden-Jones (1933) and Perje (1952) have shown that *Ranunculus ficaria* exists as a member of diploid and tetraploid races, affording a range of genotypes with a considerable variation in floral construction, e.g. sepals 3–5, petals 7–16, stamens 19–38, carpels 13–32. They noted that flowers with numerous stamens typically have a large number of carpels. Similarly, Harrison (1953) has given interesting information on meristic variation in some species of *Nuphar*.

According to the theory, there are differences in the metabolism, and therefore differences in the cytoplasm, of the cells of growth centres at different levels on the floral apex. In this connection, Mather (1944) has shown that in a distylic species of *Primula*, where the stamens in the 'pin' and 'thrum' flowers *are formed at different levels on the floral meristem*, the incompatibility properties of the pollen cells cannot be understood by reference to their genotypes but to their cytoplasms. The tristylic species *Lythrum salicaria* affords another example of the same kind.

Experimental Investigations—Experimental tests of the theory will consist essentially in interfering with the floral reaction system at critical stages in a known manner, followed by observation of such morphological modifications as may be induced. The main impediment to crucial experiments lies in the fact that so very little is known of the factors that are at work in the reaction system.

Since the principal factors in the apical reaction system are likely to be gene-determined metabolites, the close investigation of floral ontogenesis in related plants of known genetical constitution, e.g. the mutants, polyploids, aneuploids, etc., of a 'normal' diploid, may yield information of direct relevance to the theory. Plants in which heritable changes in floral ontogenesis have been induced by exposure to ionizing radiations may also yield useful information (Gunckel and Sparrow, 1954; Gunckel, 1956).

A considerable literature indicates that significant changes in floral morphology can be induced by treating the leaves or the soil with natural or synthetic hormonal substances. Detailed studies of the effects of such substances on the individual organogenic processes may yield information of special interest in the present context. Indeed, since floral organogenesis is determined by metabolism, experiments involving chemical treatments, together with studies of metabolism in the meristem, are essential if we are to advance. In attempts to validate the theory, the aim will be to introduce new components into the floral reaction system during one or other of its successive phases, and to observe their effects on the ensuing organogenic developments. That extensive changes can be induced in the fern shoot apex by direct chemical treatments, has already been demonstrated by the writer (Wardlaw, 1955, 1957).

The reaction system in the floral meristem can also be interfered with physically, in particular by surgical treatments. Here the main aim, as in the numerous successful experiments with the vegetative shoot apex, will be to modify normal developmental relationships by the partial or complete isolation of one part of the floral meristem from another. Reference has been made to an experiment by Murneek and Cusick (1956) has shown that if the floral meristem of *Primula* is incised in certain ways at particular stages in its development, characteristic and predictable organogenic effects are produced. The results so far obtained are compatible with the concept of the floral meristem as a reaction system which passes through an irreversible sequence of phases.

Discussion and conclusions

If the phenomena of floral ontogenesis have not received the attention they deserve, one may well suspect that the major impediment, or deterrent, has been nothing more than the sheer diversity and super-abundance of material: it has, indeed, been difficult to know how and where to begin. It is the writer's hope that the present paper may perhaps give direction and encourage new work in this important field.

Attempts to investigate the proximate causes of floral ontogenesis carry their own interest and may contribute to a better understanding of phylogenetic and evolutionary relationships, since, fundamentally, the evolution of floral forms is the visible expression of, and is determined by, the evolution of the under-lying organismal reaction systems.

The formation of dorsiventral lateral organs on the shoot apical meristem is a phenomenon of the widest generality in vascular plants. Individual leaves, however, even on the same plant, may exhibit very considerable differences in form and structure. The size, shape, position and development of leaves, and of the growth centres from which they originate, depend on the specific genetical constitution, on the stage reached in the ontogenetic development of the plant, and on environmental factors. These observations also apply to the floral organs. Although the apical reaction system is profoundly modified in some ways at the onset of flowering, it remains essentially un-changed in what appears to be its primary or basic morpho-genetic activity. In the more primitive dicotyledons it can be seen that carpels and stamens, as well as sepals and petals, have the appearance and histological organization of foliar organs. Some of these general relationships are also apparent in more highly evolved floral types, though they tend to be obscured by the great condensation of the axis and by changes in the dis-tribution of growth in the meristem.

As Engard (1944) has clearly pointed out, there are no 'transitions' at the morphological level, but there are 'physio-logical transitions', i.e. metabolic changes in the axis. Foster (1929, 1935) and Schuepp (1929) have shown that there are also important differences in the distribution of growth in different primordia, e.g. in cataphylls as compared with foliage leaves. The task, as it now presents itself, is to explain how an apical reaction system is constituted and how it becomes modi-fied, phase by phase, in such a way that it both maintains a general pattern of morphogenetic activity and yet gives rise to successive groups of distinctive floral organs. The heteroblastic development already referred to probably affords an important clue. In relation to the changing metabolism of the floral apex, the growth centres formed after the perianth phase give rise to organs which, though still showing some foliar characters, are of very distinctive morphology. Even the most advanced types of flower, i.e. those with a sunken gynoecium and a small number of carpels, can be interpreted along the lines indicated above.

In considering floral ontogenesis in the more advanced members of the Dicotyledons and Monocotyledons, it is worth bearing in mind that relatively small changes in genotype may sometimes lead to quite extensive differences in phenotype. This can be seen, for example, in the peloric (or actinomorphic) as compared with the zygomorphic flowers in *Antirrhinum*, in the pin and thrum flowers of *Primula*, and so on. Moreover, if a particular mutant gene (or block of genes) affects the reaction system in some important way at an early stage in floral ontogenesis, the later phases of development may well be characterized by morphological features which are conspicuously different from those of the original species. The evidence indicates that some kinds of genetical change, with consequential modifications in the reaction system and attendant correlative developments, are more likely to occur than certain other kinds. In this connection Stebbins (1950) has suggested that the several kinds of floral specialization resulting from mutation and selection, e.g. hypogyny to epigyny, polypetaly to gamopetaly, etc., have probably occurred independently in differing groups.

19
Reflections on the unity of the embryonic tissues in ferns*

Our knowledge of the long-term process of evolution is based on comparative morphological studies of living species and their fossil correlatives. In his classical phylogenetic investigations of the ferns, Bower referred to twelve *criteria of comparison*, most of these relating to morphological and anatomical features of the sporophyte, viz. its embryology, initial histological constitution, the external morphology and vascular anatomy of the shoot, the architecture and venation of the leaf, the dermal appendages, the position and structure of the sorus, sporangial and spore characters and the output of spores. Using these criteria, he was able to present a picture of the evolution of this important group (1923–8). Although inevitably subject to criticism as new facts have come to light, e.g. from the cytological investigations of Manton (1950 *et seq.*), and the systematic studies of Copeland (1947), Holttum (1954) and others, some features of his phylogenetic scheme seem likely to remain of permanent value, and it is certain that some of his more general reflections, e.g. his emphasis on the prevalence of evolutionary parallelism, i.e. of homologies of organization, in ferns, will continue to interest and inspire workers in this field.

It is evident, as indeed it was to Bower, that the several criteria can be studied from a different point of view, namely, that of the underlying causation. In each instance, where some characteristic morphological or anatomical feature is being considered, it may be asked: How does this specific form, or structure, come into being? What factors are primarily involved in

Phytomorphology, Vol. 8, Nos. 3, 4, Dec. 1958, p. 323.

* Being the substance of a paper given before Sect. K, British Association Advancement of Science, Glasgow, 29 August 1958.

its development? To answer these questions there must be investigations of the genetical constitution and of the histological organization and potentialities for growth, differentiation, and development of embryonic cells, tissues and regions. Now, it is a fact that all the major sporophytic organs included in Bower's criteria, and the meristematic tissues from which they originate, can be referred to the organization and morphogenetic activity of the shoot apical meristem. Comprehensive morphological and experimental studies of this region, and of the organs and tissues formed from it, are therefore essential to more fundamental causal concepts and investigations. Among other things, they are likely to extend and deepen our knowledge of the characters used in comparative studies. The development of the zygote presents its own special problems; but at a very early stage in embryogenesis, especially in leptosporangiate ferns, a nascent shoot apex is formed and continues to function throughout the life of the plant as the primary morphogenetic region.

During the phyletic period, with the earlier inspired observations of Naegeli and others as a guide, botanists recognized the importance of the shoot apex and were profoundly interested in it. The emphasis placed on particular aspects, however, differed considerably from that of contemporary investigators. They were preoccupied with the nature and phylogenetic interpretation of the 'apical initials', i.e. whether there was a single '2-sided', '3-sided', or '4-sided' apical cell, or whether several apical initials were present; and they paid close attention to the mode of segmentation of these initials and to the cell lineages derived from them, especially in relation to the inception of the different tissue systems. There can be no doubt as to the excellence of much of this observational work. In contemporary morphogenesis it is recognized that the inception, nature, and histogenic activity of the apical cell (or cells) do indeed present very interesting and important genetical, physical and physiological problems—still largely unsolved—but, as experimental studies have progressed, it has become increasingly clear that a more adequate understanding of the functional activities of the apex requires more broadly-based concepts. In its capacity for self-maintenance as the primary embryonic and morphogenetic region of the plant, *the shoot apex, comprising the distal apical meristem and the sub-apical region, is now recognized as an integrated whole, and as a region of very considerable complexity which functions in an orderly and characteristic manner because of its specific organization* (Wardlaw, 1952, 1957). With these general observations before us, the unity of the embryonic tissues throughout the plant may now be considered.

Leaf and bud primordia

The conspicuous histological features of a fern apex, e.g. that of *Dryopteris* sp., are the large apical cell and the superficial prism-shaped cells formed by the division of its segments. Together these constitute the *apical meristem*—the primary and persistent embryonic region of the axis from which originate leaves, buds, scales and the tissue systems.

Leaf primordia always originate from the apical meristem. In *Dryopteris*, a group of some 7–10 of the prism-shaped cells, situated near the base of the apical cone, together with the underlying cells derived from them, begins to grow more rapidly than the adjacent regions. The result is a small, oval outgrowth. As this primordial mound grows, the most centrally-placed superficial cell enlarges conspicuously and becomes the leaf apical cell, its lens shape and orientation being such that its segments are tangential to the shoot apex. The whole of the further *primary* development of the primordium is due to the growth and division of this cell and the *marginal meristems* formed from its segments.

Buds, which on further growth give rise to lateral branches, may be induced at the apical meristem of leptosporangiate ferns, but they are not normally formed there. As can be readily demonstrated in ferns, such as *Onoclea* and *Matteuccia*, buds typically appear, in interfoliar positions, some distance below the apical meristem and in older regions of the axis. They originate from superficial areas of prism-shaped meristem cells which, at an earlier stage, formed part of the apical meristem. Incipient leaf and bud primordia are histologically identical, and, in fact, buds can be induced in leaf sites and *vice versa*. These observations are important from the causal standpoint and in relation to views on the origin and evolution of leaves and branches.

The further development of leaf primordia

By its formation of lateral segments, the '2-sided' leaf apical cell gives rise to distinctive marginal meristems, *these having a histological organization generally comparable with that of the shoot apex*; i.e. the marginal meristem not only comprises the marginal apical cells or initials, but also the subjacent organized meristematic tissue derived from them. In the small oval, semi-circular or dichotomizing laminae of young sporophytes, the distal meristem loses its physiological dominance, lamina formation being largely due to the activity of the marginal meristems. In the elongating leaves of the adult plant, on the other hand, the distal apical meristem maintains its dominance until the primary phase of leaf formation ceases. In species with pinnate leaves, the marginal meristems become interrupted in a regular manner by a parenchymatous development, the persisting groups of meristematic cells becoming the apices of the pinnae. Since the differentiation of the veins is also directly associated with the distal and marginal meristems, these meristems are primarily responsible for what Bower described as the 'architecture and venation of the leaf'.

The origin of the sorus

The position and structure of the sorus and related developments are among the most important of Bower's criteria. In his phylogenetic reconstruction, two major lines of descent, referred to as the *Marginales* and the *Superficiales*, are indicated. The former are held to be the more primitive, the sori originating in, and occupying at maturity, marginal or intramarginal positions on the lamina; the latter, with sori disposed superficially on the under surface of the leaf, are regarded as more advanced. Not all investigators agree with this conception: contemporary trends are to include many of the advanced members of both the *Marginales* and *Superficiales* in the Polypodiaceae, or otherwise to disregard Bower's use of this criterion. Interesting light is shed on this matter when sori are examined from the causal standpoint, i.e. how and where they originate and how their inception affects the further growth of the fertile leaf.

A sorus is usually described as having its origin in a *placenta* or *receptacle*. Neither term is particularly apt. If we inquire what a receptacle is in relation to its origin, the answer is quite simple. *It is a special kind of meristem*, in fact, a *fertile or sporogenous meristem*, capable of forming characteristic organs, and *originating from the organized marginal meristem as defined above* (Wardlaw, 1959). In ferns in which both young and mature sori occupy conspicuously marginal positions, each overlying a vein-ending, these statements are evidently true: the discrete sorus results from a direct transformation of a group of the marginal initials and adjacent meristem cells. But even where the mature sori are conspicuously superficial, the incipient soral meristems can be traced to the marginal meristems. All soral meristems, in brief, originate there, the diversity in their eventual positions in different species being referable to the organization of the marginal meristem and to growth relationships in it during and after the onset of the sporogenous phase. This is particularly evident in those ferns in which the fertile and vegetative leaves, or pinnae, are markedly different. The three main growth relationships may be briefly indicated as follows:

(i) The marginal initials and closely associated meristem cells, at an early or at a later stage in the formation of the lamina, are transformed directly into a soral meristem, the marginal initial cell becoming the apical cell of the sorus, e.g. as in *gradate* species. *The mature sorus is marginal.*

(ii) A group of the sub-marginal meristem cells becomes the soral meristem. According to the extent to which the further marginal growth of the leaf is correlatively inhibited by soral development, the adult sorus may occupy a position close to the margin or more remote from it. *The mature sorus is intra-marginal or superficial.*

(iii) The soral meristem originates as in (ii) above but with no attendant, marked restriction of marginal growth. The development of the soral meristem may also be relatively delayed. In both circumstances *the result is a superficial sorus*. The relevant histogenic details can be followed without undue difficulty in Dryopteroid and Blechnoid ferns.

Discussion

Beginning with the concept that, in the evolution of the ferns, the marginal position of the sorus was probably the primitive condition, Bower attributed the superficial position to a 'phyletic slide' of the sorus from the margin to the undersurface of the leaf. His evidence indicated that this process had probably taken place independently in different lines of descent. These conclusions, based entirely on comparative studies, still seem valid; for, as indicated above, *all* sori are marginal in origin and many genetical changes, as we now know, involve changes in the relative rates of growth in contiguous organs and tissues. From the standpoint of contemporary morphogenesis, an important investigation would be to ascertain, in appropriate materials, how differences in genetical constitution affect metabolism and growth relationships in the marginal meristem and in the eventual position of the sorus. Some exceedingly interesting observations on changes in the relative position of the sorus, associated with changes in genetical constitution due to mutation and hybridization, have been reported by Andersson-Kottö (1929, 1938) and other workers: the differences recorded are evidently referable to changes in the allometric, or differential, growth pattern. The eventual position of the sorus depends on the nature of the correlative developments in the leaf margin at the onset of the sporogenous phase. Other important regulated developments can also be referred directly or indirectly to the soral meristem, e.g. the formation and orientation of the indusium and the differentiation of localized, receptacular vascular tissue. These several developments, in fact, exemplify relationships of the same general kind that are found at the shoot apical meristem; and in this connection the unity of origin of the several embryonic tissues may be recalled.

While contemporary and future investigations of the embryonic regions of ferns will undoubtedly advance our knowledge of the process of development, it remains to be seen to what extent they will modify phylogenetic views based on comparative morphological investigations. The details of Bower's phylogenetic scheme, especially his treatment of the more recent, advanced groups, have already been criticized as a result of new cytological and taxonomic studies; e.g. genera, which Bower considered to be closely related, have been shown by Manton (1950 *et seq.*) to have completely different basic chromosome numbers. But what of his more general conclusions, in particular those in which he emphasized the prevalence of parallel evolutionary development? The cytological evidence cannot be gainsaid: it may be frankly recognized that some of Bower's suggested phylogenetic relationships are not valid. But the similarities of form and structure on which he based his conclusions are real and call for explanation, perhaps even the more so because of the known differences in chromosome complements. As we have seen, most of the characters used in the comparative study of ferns can be referred to the organization, growth, and morphogenetic activity of primary embryonic regions, these being the initial, tangible expression of the genetical constitution. Using the contemporary idiom, we might say that gene-controlled biochemical factors are the primary determinants of Bower's criteria. But is there anything really new in this? The same *general* idea is implicit in comparative morphology—Hofmeister stressed the over-riding importance of inherent factors in all morphogenesis—and it is worth recalling that genes are only known from the biochemical or structural effects which they produce. The comparative and causal approaches, in fact, are by no means as far apart as is sometimes thought; and it remains to be seen how far Bower's more general ideas on fern evolution will be modified by new investigations.

There is a further important point. The form and structure of an organism cannot be attributed to the genetical constitution alone: other, extrinsic factors, such as those in the environment, and various physical relationships of a general kind which become incident during metabolism and growth, may be important proximate causes of formal and structural developments. Bower himself realized that causal inquiries, especially those dealing with physical factors or mathematical relationships, might have disquieting effects on the value of his criteria, and therefore on phylogenetic theory. He showed, for example, that stelar elaboration in ferns—which in passing, we can now relate to the nutrition, size and activity of the apical meristem—goes hand in hand with an actual increase in size; and he noted, aphoristically, that to the extent that morphological and structural features can be attributed to extrinsic, i.e. non-genetical factors, so do they lose their value in phylesis, i.e. as evidence of change during descent. This, however, is a thesis on which a great deal more has yet to be done. There are indications that it may, indeed, become one of the most important topics in contemporary biology, a view that is emphasized when one reflects on the numerous instances of close parallelisms of development, or homologies of organization, in unrelated as well as in related organisms. Causal factors in morphogenesis are not all of a kind: in addition to the genes, which may be described as causal factors which act with biochemical specificity, there are the many extrinsic factors and relationships of the kind that D'Arcy Thompson so elegantly expounded. The difficulty is that since organisms develop as integrated wholes, and since the zygote or spore from which development proceeds possesses specific organization, intrinsic and extrinsic factors are well nigh inseparable in practice; and the difficulty still remains even when we say, as indeed we must, that genetical factors are primary. Nevertheless, an attempt must be made to ascertain the distinctive effects of the several kinds of morphogenetic factors; for only by so doing can the essential feature of evolution, i.e. the evolution of genetical systems, or of organismal reaction systems, be properly assessed.

The general views, which have emerged from studies of the organization and activities of the meristems of ferns, seem likely to have a not less important application to seed plants.

20

Some reflections on Errera's law

Sans doute, les savants, comme tous les hommes, sont attachés à leurs idées—et comme tous les hommes aussi, ils sont le plus attachés aux concepts qu'ils peuvent le moins démontrer.
L. Errera: *Les Savants*

Pour réussir dans la science, il faut douter; pour réussir dans la vie, il faut être sûr.
L. Errera: *Philosophie*

To have been invited to participate in commemorating the birth of Léo Errera is indeed a great honour. To speak of his work is a delight, but this is not without its difficulties. For here was a scholar and investigator of the highest originality and distinction at a time when there were many great scholars. Nor does the matter rest there. In the life and work of Léo Errera we perceive not merely one great talent, but many: an investigator of theoretical, practical and applied aspects of botany, a philosopher in biology, a scholar of wide interests, a percipient commentator of aphoristic terseness, and a poet. For those who pay tribute to Léo Errera there is, indeed, a virtual *embarras de choix*. In the circumstances, I propose to do what I think Errera himself might well have done, namely, to select a single penetrating theme, attempt to indicate its wide application, and follow it to its logical conclusion. Of the many topics in Errera's writings on which I might have spoken, I have selected his fundamental contribution to our knowledge of the shapes of cells and of the characteristic pattern of wall formation in

Commémoration Léo Errera Université Libre de Bruxelles, 10, 11 et 12 Septembre 1958.

dividing cells—in short, what is now generally known as Errera's Law—the principle that cells tend to divide by walls of minimal area.

Errera's law: historical aspect

It is appropriate, at the outset, to indicate the historical setting. Hofmeister, Sachs and other botanists had already been concerned with the positions in which new walls were formed from the biological point of view, but the possible relationship of these phenomena to physical laws was in no sense closely envisaged by them. However, there were contemporary physicists and mathematicians who were examining kindred phenomena in their more strictly physical aspect. They had indicated that a homogeneous, weightless liquid film, like that of a soap bubble, cannot persist unless it constitutes a surface of mean constant curvature, i.e. a surface of minimal area for the volume enclosed. In 1886 Berthold, in his *Protoplasmamechanik*, adopted the *principle of minimal areas* and compared the forms of many cells, and the disposition of their partition walls, with those assumed by a system of weightless films under the influence of surface tension. But he did not actually ascribe the biological phenomena to this mechanical cause. It was Errera who, in the same year, and apparently quite independently, definitely ascribed to the cell wall the properties of a semi-liquid film; he deduced that it must be subject to ordinary physical laws and must assume configurations in conformity with the principle of minimal areas. These momentous conclusions were set out in a short paper in the Comptes Rendues, 1886, entitled: 'Sur une condition fondamentale d'équilibre de cellules vivantes'. His actual statement, now referred to as *Errera's Law*, is simple and direct: A cell wall, at the moment of its formation, tends to assume the form which a weightless, liquid film would assume under the same conditions. As we have seen, surfaces of minimal area, for the volumes enclosed, would consequently be involved. To Errera we therefore owe the further fundamental rule that 'the incipient partition wall of a dividing cell tends to be such that its area is the least possible by which the given space-content can be enclosed' (D'Arcy Thompson, 1917; 1942).

During the next two decades Errera's views were more or less vigorously opposed by a number of biologists, and a reserved attitude persisted even into the later editions of Strasburger's well-known textbook. Errera's views, however, have steadily gained ground and, with the passage of time, have received strong support. Some embryologists have affirmed that some of the segmentations during the embryological development are purely physical in nature, i.e. that the histological pattern of plants and animals is essentially referable to molecular physics —a view very different from that entertained by other, comparative embryologists who regarded the position of each new wall as being of phylogenetic significance. As Errera himself pointed out, the principle could have great importance in that it enables us to understand many organic forms and to relate 'cellular architecture' to molecular physics. And, indeed, we may ask: If we do not accept Errera's explanation, what other satisfying explanation is in sight?

I have always regarded it as a curious fact that Errera's Law, which sheds so much light on the phenomena of cell shape, division and pattern, has received so little attention in standard textbooks. It would be a simple, though perhaps invidious, exercise to indicate the number of well-known books on plant anatomy and histology in which there is no reference whatsoever, or, at best, only a passing or indirect reference, to Errera's momentous pronouncement. As a student, when I was profoundly interested in developing tissue systems, I do not recall any mention of Errera's Law. But I recall that one day Professor Bower placed in my hands a copy of the first edition of D'Arcy Thompson's fascinating book *On Growth and Form* (1917) and there, to my delight, I came upon the writings of Léo Errera whom we honour today.

The law and plant embryogenesis
Illustrations of Errera's Law are to be encountered everywhere in embryonic and meristematic tissues. In particular, the early embryogenesis in all classes of plants bears witness to the generality and value of his concept. The development of the nearly spherical leptosporangiate fern embryo up to, and beyond, the octant stage, approximates closely to the idealized division of a sphere into cells of equal volume by walls of minimal area. Critics of Errera's Law, e.g. those who point out that cambial cells evidently do not divide by walls of minimal area, seem to me

to view the matter from too narrow a basis. In essence, Errera was calling attention to a fundamental physiological and physical relationship, the relationship so aptly described in the title of his classical paper—a fundamental condition of equilibrium in living cells. In this general approach we are reminded of the writings of Plateau, from which Errera's idea was derived; for Plateau stated quite clearly that a surface of minimal area was to be understood as a *relative* minimum, determined as it would be by all the material exigencies. If, therefore, we accept Errera's Law in its general aspect (as he himself clearly intended), recognizing that in different biological circumstances other factors will inevitably become incident and may affect the pattern of wall formation to a greater or less extent, then the value of the Law, far from being diminished by seeming exceptions, is enhanced by them. This certainly has been my experience in my studies of embryogenesis in plants; for plant embryos not only afford beautiful examples of the application of the Law in its simplest form, but also many seeming exceptions which serve to intensify our attempt to understand developmental processes. Thus, in the simplest case, some plant embryos illustrate the equal division of a spherical or ellipsoidal zygote by a median partition wall, i.e. by a wall of minimal area; and the further divisions, up to a point, are also in conformity with Errera's Law. On the other hand, unequal division of the zygote by a transverse wall is also of very general occurrence. If, however, we accept the Law as being applicable, we are led to consider other factors that may be involved in the developmental process. If, for the moment, we assume that an ovum and the ensuing zygote have initially a homogeneous cytoplasm, then we may recognize that one of the earliest embryological developments consists in a drift towards a heterogeneous condition and the inception of polarity. The distal pole of the zygote becomes densely protoplasmic and the main seat of protein synthesis, whereas the basal pole is often characterized by an accumulation of osmotically active substances and by enlargement. The forces generated during the very different metabolism at the two poles must affect the position in which the first partition wall will be laid down. This, in fact, is usually such that a small embryonic cell is separated from a larger basal cell, i.e. the first partition wall is not one of minimal area. But, having regard to all the circumstances, Errera's Law, in the precise form stated by him, will still apply, the unequal cell division calling attention to the fact that certain factors, which have become incident in development, have modified the working of the Law as envisaged in an idealized situation. The developing embryo as a whole is in a state of dynamic equilibrium, and it was with the maintenance of this state that Errera was so profoundly concerned. In the encapsulated embryos of archegoniate and seed plants, the pre-fertilization ovum has already been exposed to various factors, e.g. physiological gradients in the maternal environment, and if its polarity has not already been determined, this takes place at a very early stage in the zygotic development. In these embryos, the inception of polarity is almost always, and sometimes demonstrably, attended by metabolic differences at the poles, and hence the first transverse partition wall is unlikely to be one of minimal area. The first longitudinal division of the distal embryonic cell, on the other hand, usually exemplifies Errera's Law.

Because the Law deals with a fundamental relationship during the growth and development of organisms, it has its place in contemporary studies of morphogenesis and physiological genetics. In the embryonic development of any particular species, the division of the zygote and the ensuing cell divisions take place with great constancy and fidelity. The characteristic physiological processes in the different regions of the embryo are gene-determined. Accordingly, the position of the first and later formed partition walls reflect both the gene-determined metabolism of the organism and the working of Errera's Law. The fact that the early segmentation of organisms of quite different taxonomic affinity is very much alike emphasizes the importance of the physical forces at work.

The historical setting of Errera's concept deserves further mention. His classical paper of 1886 appeared at a time when many botanists were preoccupied with evolutionary studies, following the publication of Darwin's *Origin of Species*. The phylogenetic relationships of living species were based on comparative morphological studies. Some features were regarded as primitive, others as being relatively advanced. Although some of the conclusions reached could be checked against the fossil record, some departure from scientific objectivity was hard to avoid. The cellular construction of embryos, of apical growing points, of fern sporangia, etc. tended to be considered almost entirely from the evolutionary point of view, i.e. as exemplifying a primitive or derivative condition. There were, however, percipient investigators such as Hofmeister (in his *Allgemeine Morphologie*) who saw that although all formal and constructional features are manifestations of the potentialities of the specific hereditary substance, they must also originate and develop in accordance with physical laws. Léo Errera belongs to this distinguished group whose work did so much for the advancement of scientific botany. As the present writer has emphasized (*Phylogeny and Morphogenensis*, 1952), both the comparative and the causal aspects of morphology have their justification and their place and both are essential to an adequate understanding of form and structure. A contemporary view might be stated as follows: (i) a developing zygote, in its normal environment, usually follows a characteristic course of development; (ii) factors in the genetical constitution determine and control the specific metabolism of the zygote and of ensuing embryonic and differentiating cells; (iii) physical factors and processes which become incident in relation to metabolism and growth are the proximate causes of the shape and mode of division of the developing zygote and young embryo. Thus we begin to see, at least in a general way, how Errera's Law, properly understood and judiciously extended, affords a point of contact for causal and phylogenetic investigators. For this, and for other reasons, his may truly be regarded as a lasting contribution to biological science.

2 1

The inception of
shoot organization*

The vegetative sporophyte of vascular plants consists essentially of an axis, or shoot, and its lateral appendages, chiefly leaves and branches. Differences in the relationships of the organs of the leafy-shoot admit of great morphological diversity and raise innumerable individual problems of morphogenesis. Thus the shoot may be short or tall, simple or much branched; its lateral members may be very varied in their size, shape and position on the axis; and some or all of the component parts may be more or less extensively modified in form and function. But under all this diversity, there is unity: in the developing leafy-shoot the dynamic structural plans, or major organizational features, are common to all. We may, indeed, speak of the *leafy-shoot type of organization*. It is the essence of botanical science to investigate such *general phenomena*, in the hope that we may in time be able to make simplified but comprehensive statements regarding the factors that are involved. The aim in this paper is to deal with some aspects of this organization.

The distinctive organization of the individual species has long been recognized, but what constitutes this organization, and how to define it, are very difficult problems. In vascular plants, a considerable part of biological enquiry is to discover what factors determine the leafy-shoot type of organization, and, in particular, its pervasive unity and harmonious development. I am well aware that organization in plants is one of those topics that we tend to discuss in general terms only, more particular concepts being not only difficult to formulate but seemingly incapable of being validated by observation and experiment.

* A paper presented at the 9th International Botanical Congress, Montreal, 1959. *Phytomorphology*, Vol. 10, No. 2, July 1960.

But should we accept this view? I, personally, do not think so. After all, specific organization is the most distinctive feature of living things, and, as we have seen, certain organizational phenomena appear to be common to all vascular plants. In the investigation of morphogenesis, the formulation of concepts or ideas about organization is not only essential: it is the very essence, the summation, of what we are trying to do. The complexity and difficulty of the task need no emphasis; but what an exciting challenge it presents, and what interest and delight it can afford those who are prepared to accept it.

Apical constitution and unity

Every developmental situation in plants, from the fertilized ovum onwards, yields evidence of organization, particularly in the embryonic regions. The very young embryo affords evidence of polarity and the irreversible establishment of a distal, self-maintaining, growing point. In fact, it shows nascent, or incipient, axial construction. From the outset lateral organs are formed at the apex in characteristic positions. On the evidence, the characteristic organization of the leafy shoot can be ascribed to: (i) the genetically-determined organization and morphogenetic activity of the *apical meristem*, which yield the *primary* or *basic pattern*; (ii) the elaboration of this pattern in the *subapical regions*; and (iii) the action of organismal factors, e.g. those responsible for correlative developments. This view evidently invokes a consideration of the constitution of the apical meristem and its pattern-forming activities as an essential initial stage. Here the *shoot apex* is to be understood as comprising both the apical meristem and the subapical regions.

At different times, special importance has been attributed to different regions of the apex. The early investigators were much occupied with the extreme tip, questions of paramount importance being whether one or several initial cells were present there, their mode of segmentation, and the cell lineages to which they gave rise. Some time later, the emphasis was on apical constitution and growth in terms of detectable histogenic layers, or histogenic regions; and so on. In recent years, we have witnessed the birth and development of a new concept proposed by our French colleagues, in which the important region of the vegetative apex is considered to be the *anneau initial* (or

initial ring) situated some distance below the tip. This band of
tissue is held to be the region of active growth and morphogenetic activity, responsible for the maintenance of the apical meristem and giving rise to leaf primordia: the extreme tip, on the
other hand, is virtually inactive, a resting meristem which only
becomes active at the onset of flowering. Now all the very
different conceptions of the apex have one point in common,
namely, that a particular region, or a particular aspect of apical
activity, is emphasized at the expense of others; in each view,
a part is abstracted from the whole. It now seems evident, in the
light of many experimental investigations, that the *shoot apex*,
comprising both the apical meristem and the subapical regions,
must be regarded as an integrated whole; for only a conception
of this kind is capable of incorporating all that has been learnt
about apical growth, self-maintenance, response to stimuli,
morphogenetic activity and genetical expression.

The primary morphological pattern
With direct reference to the shoot and its appendages, experimental observations suggest the following conclusions regarding
the apical meristem: (i) it consists of embryonic cells and is an
organized and biochemically differentiated region; (ii) the
physico-chemical reactions which take place in it give rise to a
pattern of growth centres, which usually but not invariably
become leaf primordia; (iii) in this apical reaction system,
genetically-determined substances control the characteristic
allometric growth pattern and organogenesis; (iv) any significant change in the morphogenetic activity of the apical meristem must be mediated through its organization and reaction
system. (In parenthesis, it would appear that some of the
numerous attempts to explain correlative inhibition and similar
phenomena have failed because they have not recognized that
when specific substances, e.g. auxin, act, they do so in organized regions.)
    The pattern-forming property of apical meristems is an ever-
present fact. The regular pattern of primordia, which becomes
apparent when the apex and youngest leaf primordia are laid
bare, is characteristic and distinctive for the species, varying
usually within narrow limits, or in a regular manner, as the
apex undergoes its ontogenetic enlargement, or as it diminishes

in size in relation to decreased nutrition. As closely comparable patterns of primordia occur in taxonomically quite unrelated species, it may be inferred that this pattern-forming activity is a general and characteristic feature of organized meristem cell masses. This remarkable pattern-forming property of apical meristems is further exemplified in certain species once they have entered upon the flowering phase. Thus in *Nuphar lutea*, flower buds typically originate in leaf sites, and in *Victoria* spp. the flower buds arise, with great regularity, just adaxial to one of the stipular flanges. In each instance a growth centre originates in a precise position in the apical meristem; and whether or not we have yet an adequate theory of physico-chemical patternization in embryonic regions, such as that advanced by Turing in 1952, the fact itself is indisputable. In any consideration of the shoot and its appendages, the starting point is the reaction system in the apical meristem. For example, if we assume (i) the inception of a decussate, alternate or spiral pattern of leaf growth centres in the apical meristem, and (ii) specific differential or allometric growth in the subapical regions, this affecting the nature of the internodal development, then the main architecture, or structural plan, of the leafy-shoot is determined. This, of course, is a great simplification: buds, or bud rudiments, also originate in the apical meristem in characteristic positions. In ferns, such as *Dryopteris*, bud and leaf primordia originate from identical groups of meristem cells; and, as we know from experimental studies, young leaf primordia, or primordium sites, can be transformed into buds, and bud sites can be made to develop as leaf primordia. In the formation of buds we have further evidence of the patternizing activity, or property, of the apical meristem; and the morphological differences between leaf and bud primordia, whether they are due to qualitative or quantitative metabolic differences, or to both, have their inception in the apical reaction system. In the Nymphaeaceae, where both leaf and flower-bud primordia are components of the apical pattern, the complexity of the underlying reaction system raises problems of which the solution is not yet in sight. We can, however, see clearly where the problem of shoot organization ultimately lies.

Two general observations may be made at this point. The

view has long been held that one of the main differences between plants and animals is that the major organs in the latter are defined once and for all in the very young embryo, i.e. in embryonic tissue showing little evident differentiation. But, in the light of the foregoing observation on the apical meristem, is organ inception in plants so different after all? Both chemical and surgical experimental treatments indicate that the apical meristem possesses considerable stability: its fundamental organization is not readily modified, e.g. by direct or indirect applications of auxin, gibberellic acid, etc. It may perhaps be inferred, therefore, that it is controlled by a very stable part of the genetical constitution, or that the protoplasts in the meristem cells do not respond readily to some of the more common morphogenetic stimuli. They are not, however, inert or non-reactive.

The elaboration of the primary pattern
Relative to the apical meristem, the subapical region is one of active enlargement, the transverse and vertical components of growth being involved to different extents according to the species, its stage of ontogenetic development, and the impact of environmental factors. If the vertical component of growth is the more active, an elongated axis is formed, with separation of the leaves, or leaf groups, by more or less conspicuous internodes: if the transverse component is the more active, a condensed axis, closely beset with leaves, is formed, as in many ferns, cycads, rosette plants, etc. Accordingly, it is in the subapical region (or regions) that the action of certain genetical and regulative factors becomes most evident. In this region, also, the primordia of leaves, buds and other lateral organs grow and enlarge conspicuously: their primary patterns are elaborated in characteristic ways and their distinctive morphological features become manifest. A simple example will illustrate this point. At their inception in the apical meristem, leaf primordia are very much alike in all classes of plants. But, as they enlarge in the subapical region, it soon becomes apparent whether the petiole-base is going to be small, simple and discrete, or large, clasping and stipulate; whether the primordia will develop as microphylls or megaphylls; and so on. Now these developments, including the main constructional

features of the lamina, are recognized later as genetically-determined, taxonomic characters. It follows that, by precise observation of axis and primordium development, it is possible to indicate the particular stage at which the genes determining or controlling major leaf and stem characters begin to exercise their specific effects, e.g. in a comparison of mutant and parental materials of known genetical constitution. This morphological information will obviously be essential if the aim, as in physiological genetics, is to ascertain which particular substances and reactions, considered to be gene-determined, are involved in the inception of some particular major morphological character. In brief, if our primary interest is in organisms as organisms, as distinct from the analysis of particular processes, a synthesis of genetics, physiology and morphology is not merely desirable: it is unavoidable. There are indications that this desirable state of affars is no longer at the stage of being merely a possibility: it is becoming a reality.

What has been said of the vegetative apex may, with appropriate modifications, be applied also to inflorescence and floral apices.

The investigation of apices
Morphogenetic investigations of shoot organization, being concerned as they are with the inception of the major morphological features of species, are rightly accorded a central place in contemporary investigations; and the more morphologists can explore the same materials in collaboration with physiologists and geneticists, the more satisfying will the result be. With these aims in mind, the study of apices should be essentially of a comprehensive nature, including observations on: (i) the size, shape, and histological pattern of the apical meristem and subapical regions, throughout ontogenesis, this affording evidence of the basic allometric growth pattern in the axis and primordia; (ii) the size and shape of embryonic cells, and the nature and rate of their differentiation; (iii) the position, size and shape of leaf primordia, buds, etc. in the apical meristem and the manifestation of genetical characters during their further development in the subapical regions, together with the related anatomical developments. These and other relevant observations should, if possible, be made on growing apices

kept under regular scrutiny, and materials collected at critical stages for cyto-chemical study or micro-chemical analysis. Investigations along these lines of apices of related plants of known genetical constitution, e.g. related genera, species, varieties, induced mutants, etc., in which there are morphological differences of some magnitude, or of apices which have shown responses to chemical treatment, e.g. gibberellic acid, seem likely to be attended with interesting and important results. A modest but growing literature already indicates the value of this procedure, but a vast and fascinating field awaits exploration both in the interests of academic and of applied botany.

R

22
Morphology

These discourses on *Contemporary Botanical Thought* immediately commend themselves as an excellent idea: they are timely, provocative and challenging. And the more one lends oneself to the theme, the greater is the reward, not least because it invokes philosophic reflection on the present state of our science.

It is impossible to think of plants without considering some aspect of their form or structure; hence morphology is a very ancient and pervasive subject. To discuss what is new in it is therefore a considerable test of scholarship. Thus, while my main task is to discuss what has been going on in recent years, and to show how new ideas and new techniques are advancing the subject, it is no simple matter to indicate what is really new in morphology. For although we owe the term *morphology* to Goethe, the study of form in animals and plants was already of acute interest to Aristotle and his contemporaries. Moreover, as Dr. Arber has pointed out in her excellent book *The Natural Philosophy of Plant Form* (1950), the term *form* has undergone a considerable degradation and limitation of its original fullness of meaning. To Aristotle, according to Dr. Arber, 'the scope of "form" was wide enough to cover the whole of the intrinsic nature of which any given individual was a manifestation'. The contemporary state of this branch of botany justifies the view that the original fullness of meaning should now be restored to the term. That at least is the sense in which I use the word *morphology* in the present context.

Because so much attention was devoted to plant morphology

'Morphology'. *Contemporary Botanical Thought* (Oliver & Boyd), 1–26, 1961.

and anatomy during the nineteenth century—i.e. since the regeneration of botanical science at the hands of Schleiden, Naegeli, Von Mohl and others—it is difficult to point to recent general ideas or discoveries that, in a strict sense, are absolutely new. With that reservation, however, recent decades have undoubtedly witnessed a great output of new work, the development of new ideas, and extensions and critical reappraisals of old ones. Perhaps not least significant, there have been important changes of attitude and outlook. To cite a single example, contemporary investigators of the physiology of development—and there are many of them—must be both morphologists and physiologists if their work is to have a desirable fullness.

What is going on in contemporary morphology, then, had its roots in the nineteenth century. All the major trends, whether in comparative morphology, physiology of development or experimental morphology, had already received attention at the hands of the great nineteenth century continental botanists (*see* pp. 2 *et seq.*; 19 *et seq.*; 40 *et seq.*).

In this paper I propose to discuss briefly some developments in morphology from about 1910 onwards, to indicate the outlook of contemporary workers and to speculate, without too much reserve, on what the future may, or could, hold for botanists.

A brief survey of some developments

Comparative Morphology There can be no question here of dealing with the whole development of plant morphology, and the year 1910 may be selected as a convenient if arbitrary starting point for my purpose. Some two years earlier Bower (1908) had published his *Origin of a Land Flora*. This book is sometimes described as marking the end of the great post-Darwinian period of Comparative Morphology—the so-called Phyletic Period. In 1915, in his Presidential Address to Section K of the British Association, Lang compared the claims of *General* or *Causal Morphology* with those of *Comparative Morphology*, indicating that the latter had failed to achieve its major aim, namely, the establishment of a 'monophyletic genealogical tree' of the Plant Kingdom—it was more like a bundle of sticks (he said)—and emphasizing the great scope which the former approach, pioneered by Hofmeister, Sachs, Goebel and others, offered to contemporary botanists. Again, in another Presidential Address, Tansley (1923) recorded the view that the investigation of phylogeny was virtually at an end, that 'it left the majority of the younger generation of botanists cold', and that they would do well to direct their attention to the process of development. However, it seems that there are dangers in prophecy. As we now know, although comparative and phylogenetic investigations no longer occupy the centre of the botanical stage, they are still of lively interest to many botanists throughout the world, as judged, for example, by the number of contributions made at sessions of the International Botanical Congresses, held in Stockholm, 1950, Paris, 1954 and Montreal, 1959. Moreover, there is scope for reflection on the fact that the great problems of phylogeny are still largely unresolved.

The Rhynie fossils described by Kidston and Lang (1917–21) from Middle Devonian horizons in Scotland, and related materials from other parts of the world, raised new expectations. They appeared to afford a basis, or starting point, for the evolution of vascular plants, and suggested that the establishment of a land flora might have taken place during this great geological epoch. The simple, leafless, rootless, bifurcating axis of *Rhynia* and other Psilophytales also provided the materials and inspiration for new morphological theories of the origin and nature of the leafy shoot, such as the telome theory of Zimmermann (1930, 1938), and the enation concept of microphylls and the cladode origin of megaphylls by Bower (1935). The discovery and investigation of the fossil *Baragwanathia longifolia* from Australian Silurian strata by Lang and Cookson (1935) showed, however, that the establishment of land plants must be sought in geological periods much earlier than the Devonian. Contemporary investigators have now shown that vascular land plants existed in Ordovician times, and possibly even earlier, and that collectively they were probably considerably more numerous in species and morphologically diverse than had been inferred from investigations of the Psilophytales.

The impact of palaeobotanical observations on morphological thought was discussed in 1932 by H. Hamshaw Thomas in a paper entitled 'The Old Morphology and the New'; and

more recently Lam (1959) discussed 'Some Fundamental Considerations on New Morphology' before this Society. While one
may doubt the validity of some of the morphological theories
which have been based on the study of fossil materials, it
becomes increasingly evident that, for an adequate understanding of the course of the great historic evolutionary trends, there
can be no substitute for an enlargement of our knowledge of the
fossil record.

A brief reference should be made to the effect of recent cytological work by Manton (1950) on phylogenetic conclusions
reached by comparative morphologists. Her investigations show
that some of these conclusions are valid, that others are definitely wrong and stand in need of revision, and that, because
of the prevalence of parallel evolutionary development, which
had also long been recognized by morphologists, close similarity
of form may sometimes be misleading as evidence of close
genetical affinity. But the cytological investigation of Pteridophytes has raised its own difficulties: in view of the high chromosome numbers found in plants of ancient affinity, such as lycopods, *Psilotum* and *Tmesipteris*, and in eusporangiate ferns such
as *Ophioglossum*, it is surely an odd and perplexing fact that the
species of *Selaginella* and *Isoetes* which have been examined, and
which are also plants of ancient ancestry and exemplify the
advanced heterosporous condition, have very low chromosome
numbers.

Morphogenesis   Comparative, phyletic, taxonomic and anatomical investigations apart, what major new developments
have taken place in morphology since 1910? The short answer
is that there has been a considerable advance in our knowledge
of morphogenesis, especially during the past two decades.
D'Arcy Thompson (1917, 1942) discussed selected aspects of
the form and structure of plants and animals in terms of their
physical and mathematical relationships under the title *On
Growth and Form.* In 1932 J. S. Huxley took the mathematical
aspect further in his *Problems of Relative Growth.* Experimental
investigations of plant form and structure have been variously
described: when undertaken by professed morphologists, they
are sometimes referred to as *experimental* or, less frequently
nowadays, as *causal morphology* (the *general morphology* of Hofmeister). Goebel had published some of his investigations under

the title of *Introduction to Experimental Morphology*, and in the 2nd
and 3rd editions of his *Organographie* many illuminating observations of the same kind are recorded. When investigations of
morphogenesis are carried out by physiologists, usually but not
invariably with the emphasis on the physiological mechanism,
they tend to be described under the heading of *physiology of
development*. Bünning (1948, 1953) has described and discussed
some of the main results in his book *Entwicklungs- und Bewegungs-
physiologie der Pflanze*. Here, too, it may be noted that a volume
of the new *Handbuch der Pflanzenphysiologie* is being devoted to
developmental physiology. This, it need hardly be said, is a
vast field. Some contributions have been almost entirely analytical and biochemical in conception and outlook. In other
investigations, the selected plant has been given some experimental treatment, often on a purely trial-and-error basis—the
direct or indirect application of physiologically-active substances to growing regions, irradiation, exposure to different
day-lengths, temperatures, etc.—and the induced morphological and structural changes have been recorded and illustrated, if not always explained. Between these extremes there
have been many carefully planned experimental and analytical
studies in which the aim has been to discover what the physiological factors are, and how they work in the embryonic tissues
of the apical and subapical regions of shoot or root, determining
or regulating the growth, morphological development and
tissue differentiation in these important formative regions. The
interesting and important results obtained by Robert Brown
and his co-workers, using techniques of great ingenuity and
delicacy, show what can be achieved in a well-conceived and
well-executed programme (Robinson and Brown, 1952, 1954;
Brown and Robinson, 1955; Sunderland and Brown, 1956;
Sunderland, Heyes and Brown, 1956, 1957).

In investigations of the onset of the reproductive phase and
of flower development the aims have sometimes been almost
entirely physiological, as in many of the published works on
photoperiodism; they have sometimes been almost entirely morphological, as in ontogenetic investigations; and they have
sometimes combined both physiological and morphological
ideas and observations. In the third category, information of
considerable interest not only to morphologists but to botanists

in general has been obtained, for example, by J. and Y. Heslop Harrison and co-workers (1958, 1959) by the judicious use of both disciplines.

Tissue Culture   Although the idea of growing plant tissues in pure culture is not new, the discovery of a satisfactory technique independently, and more or less contemporaneously in 1939, by White, Gautheret and Nobécourt, is indeed a new achievement. Progress has also been made in the *in vitro* culture of organs, including roots, shoot apices, excised leaves, floral organs and embryos. However, of the vast literature on tissue culture, the contribution to physiology, that is to the nutritional requirements of tissues and organs, has greatly exceeded the contribution to morphology. During the past ten years or so, however, important discoveries, contributing to our knowledge of organogenesis and histogenesis, have been made; e.g. by Wetmore (1953-9), and Wetmore and Sorokin (1955), on the differentiation of xylem and phloem in previously undifferentiated lilac callus by grafting in buds and by applying auxin (IAA); by Skoog and his co-workers (1948-57) on the induction of buds in undifferentiated tobacco tissue by using combinations of kinetin and auxin; by Ball (1946), Wetmore *et al.* (1953-9), on the culture of excised shoot apices and their growth into viable plants; by Steeves and Sussex (1957) on the culture of excised leaf primordia; and by Steward *et al.* (1958) on the culture of single cells and small cellular masses and their growth and differentiation into viable plants.

Physiological Genetics   Our knowledge of morphogenesis has also been advanced by investigations described as physiological genetics. Here the aim has been to discover how and when, in embryonic regions, particular genes, through the chemical action of their products, determine or contribute to the primary morphogenetic pattern and its subsequent amplification.

Embryology   The past four decades have seen advances in plant embryology, especially that of seed plants, at the hands of Souèges and his school in France, Maheshwari and his colleagues in India, and Buchholz in the United States, to mention only a few of the many contributors. Much of this work is almost entirely descriptive of the stage-by-stage development of the embryo from the fertilized ovum onwards. During recent years, however, new experimental approaches, including the excision of embryos and the use of tissue culture technique, the application of physiologically-active substances to very young embryos *in situ*, and the investigation of physiological-genetical relationships of embryo and endosperm, have advanced and brought new interest to this branch of morphology (Wardlaw, 1955, 1960).

Investigation of Apices   The morphologist's contribution to morphogenesis has consisted largely in the investigation of the apical growing points of shoots, leaves, buds and roots of vascular plants in general, including also those of inflorescences and flowers in the angiosperms. While many of these investigations have been along classical lines and largely confined to histological and anatomical accounts of apical growing points, some have attempted to break new ground by mathematical analysis, by using cytochemical techniques, by the comparative study of related materials of known genetical constitution, by surgical and chemical treatments of the apical meristem, and by the use of radio-isotopes. Also, in conjunction with physiologists, a beginning has been made on the micro-chemical analysis of the apical meristem and its subjacent regions. (For reviews and references, *see* Wetmore and Wardlaw (1951) and Wardlaw, 1952, 1957, 1959, 1960.)

All of the above-mentioned investigations have, in varying degrees, contributed to our better understanding of plant morphology, though some of the workers would almost certainly not describe themselves as morphologists. Yet it is evident that the kinds of inquiries which I have briefly surveyed were very much those which Hofmeister—surely a brilliant morphologist—had in mind when, in 1868, he published his *General Morphology of Growing Things*. For in this book he asked: how, during growth, does the observed form or structure come to be as it is and what factors are involved? In his analysis of morphogenetic development, Hofmeister recognized that both extrinsic and intrinsic (i.e. genetical) factors were at work and that, of these two kinds of factors, the latter were of paramount importance in the primary morphogenetic development, i.e. the inception of pattern.

New conceptions and outlook

Some general results   We may now consider what general results, new thoughts and outlook have emerged from recent work on morphogenesis. We may recognize, I think, that there has been some convergence of genetics, physiology and morphology, though a much greater measure of integration of the data derived from these branches will evidently be required if we are to arrive at any adequate understanding of the unique kinds of organization that characterize living things. Our appreciation of what is involved in ontogenesis has certainly been deepened and broadened, though it is still very elementary and inadequate, if not, indeed, naïve: each new discovery brings in its train a fuller recognition of the complexity of the process of development. Information on the facts of development has increased greatly as the result of experimental, analytical and observational investigations, and our understanding, at least in a general way, of the mechanisms determining the formation of leaves and buds, phyllotaxis, the differentiation of tissues, the reciprocal relationships of parts, correlation and regulated development, has advanced, though it is recognized that further significant progress must await new discoveries and refinements in the techniques of cytochemistry, microchemistry, the use of radio-isotopes and biophysics. I think, too, that there is a growing recognition of the need for integrating and synthesizing the data that accrue from the several branches of botany, towards a better understanding of what constitutes organization in plants (Wardlaw, 1952, 1955, 1960), though there is probably no very clear idea as to how we should set about this task.

The shoot apex   To give an indication of where at least one morphologist has got to in the matter of new thinking, I propose to set down some conclusions and conjectures based on my study of the shoot apex in vascular plants. The starting point was the rather simple and limited kind of information that was available in standard texts and monographs some 20–30 years ago, bearing in mind, however, that some very percipient biologists had worked in this field since the days of Naegeli, and not forgetting that, in 1759, Kaspar Wolff had already stated in his *Teoria Generationis* that he saw nothing in the plant but *leaves* and *stem*, that the growth of the shoot proceeded from an undifferentiated apical growing point, and that the new organs originated there successively, thus exemplifying a true epigenesis as distinct from some hypothetical phenomenon of preformation. One should perhaps note that, by the 1930's, the literature of phyllotaxis was probably already sufficiently abundant and complex for most tastes. At the purely descriptive level, in the aforementioned standard texts, the main facts about the shoot apex were easily stated and apprehended: it was a region, consisting of delicate embryonic cells, exemplifying the phenomena of *continued embryogeny*, as Bower (1919) described it, the formation of leaf primordia in an orderly and characteristic phyllotactic sequence and their associated buds, and the initial differentiation of the several tissue systems. The more searching inquiries of the past 30 years into the varied histological constitution of apices* in vascular plants—their growth, morphogenetic and histogenetic activities, their physico-chemical properties, the factors determining phyllotaxis, the interrelationship of parts, and the harmonious development of the leafy shoot as a whole—have vastly changed our earlier ideas and made new thinking unavoidable.

Histological and experimental observations indicate that the apical meristem is not simply a distal region consisting of identical and physiologically uniform embryonic cells. We are forced to recognize that it is an organized and biochemically differentiated region, that it is the seat of complex physicochemical reactions, and that this reaction system typically gives rise to a pattern of *growth centres*, which usually, but not invariably, become leaves. Following Schoute (1913), we see that the apical meristem probably consists of a system of three interrelated zones, the growth centres having their inception in the *sub-distal* zone of the meristem, while the actual appearance of the very young primordia, as discrete, flattish, rounded outgrowths on the smooth surface of the meristem, takes place in the next zone below, i.e. the organogenic zone or region. The primary organogenic pattern, and also the associated histogenic pattern, is thus established in the apical meristem and the elaboration of this pattern, the enlargement of the axis and its

* The term *shoot apex* as used here is intended to denote the *apical meristem* together with the subjacent regions which constitute the terminal bud.

organs, and the further differentiation of the tissue systems take place in the subapical regions. The characteristic distribution of growth, i.e. the allometric growth pattern in the apex, and organogenesis are, presumably, primarily determined by genetical factors, though environmental factors and the stage reached in ontogenesis are also involved.

Growth centres   However little we may be able to explain the pattern-forming property of the apical meristem, i.e. the inception of growth centres in regular positions, it is an ever-present and unavoidable fact. If apical meristems are laid bare, it can be shown that the youngest leaf primordia originate with great regularity in predictable positions and are of a shape and size characteristic for the species. In brief, growth centres, though invisible and not readily demonstrable by existing cytochemical means, have a real existence. Van Fleet (1959) has been able to obtain some evidence that growth centres are loci of special metabolism, differing from the adjacent tissue of the meristem. The morphologist is thus led to a lively appreciation of the further view, long since propounded by Sachs, that, in every organogenic situation, a biochemical phase, involving a patternized distribution of metabolites—including general metabolites, growth-regulating substances, enzymes, etc.—precedes, and is the basis of, the visible morphological and histological developments (Wardlaw, 1955).

Since closely comparable patterns of growth centres, and of the primordia formed from them, occur in taxonomically unrelated species, it may be inferred that this pattern-forming activity is a very general property of organized embryonic cell masses, i.e. of the kind that we find in shoot apical meristems. All the growth centres formed during the growth of an apical meristem have presumably much in common, but they need not be identical in their metabolism. Accordingly, in particular circumstances, the organs formed from them need not be identical—they may even belong to different morphological categories. Foster (1929, 1935), for example, has shown that, in the formation of cataphylls as compared with foliage leaves, differences in the distribution of growth can be detected in the very young primordia. In the ferns, by simple surgical and chemical techniques, it has been possible to cause buds to be formed in leaf sites; and in species of *Nymphaea* and *Nuphar* the flower buds typically orginate in normal leaf positions. The concept of metabolic differences between growth centres is of special value in attempting to account for the heteroblastic leaf formation which we see in the normal ontogenesis of many species of both ferns and flowering plants and which typically attends the transition from the vegetative to the flowering phase in angiosperms. It may also have an apt application to the formation of the several helices or whorls of floral organs (Wardlaw, 1957). In the Ranunculaceae, for example, recent investigations (e.g. Tepfer, 1953) have made it increasingly evident that sepals, petals, stamens and carpels are homologous with vegetative leaves, as judged by the basic criteria of position of origin and the associated histological details. But we need no longer be troubled by the old morphological difficulty as to how one organ could originate by the 'transformation' of another organ, e.g. how a stamen could be transformed from a petal or *vice-versa*. As Engard (1944) clearly pointed out, there cannot be transformations at the morphological level, but there can be 'physiological transitions'. With the development of the growth centre concept, we can now see how there can be differences in homologous organs, i.e. they are referable to physiological differences in the growth centres (Wardlaw, 1957). It may perhaps appear to readers that these considerations pertain to physiology rather than to morphology. But as they certainly pertain to morphogenesis, so they also do to morphology, constituting, as it were, part of that fullness of content and meaning which the term once possessed.

A diffusion reaction theory of morphogenesis A provisional acceptance of the growth centre concept brings us to what is perhaps *the* most fundamental problem in morphogenesis, namely, how in an embryonic tissue in which the metabolites involved in organ formation are initially more or less homogeneously distributed, the accumulation of particular substances in regularly positioned loci is brought about. To date, the theory most closely directed to the solution of this all-important problem is that of Turing (1952). It is described as the Diffusion Reaction Theory of Morphogenesis; it is mathematical in its inception and is based on well-known laws of physical chemistry. What Turing has done is to indicate how an initially homogeneous diffusion reaction system, such as might be present in various morphogenetic situations, for example, in the sub-distal region of the apical meristem (Wardlaw, 1953, 1957), may drift towards a heterogeneous condition and eventually give rise to the regular, or pattern-ized, distribution of metabolites, this being the basis of the ensuing organogenic or histogenic developments. For example, in different species of *Pinus*, and within particular species, the number of cotyledons in the embryo may be very variable. Now, at a stage before the cotyledon primordia appear, the embryo has a rather massive, rounded apical meristem. It is at this stage that the growth centres, few or many, presumably have their inception. Chamberlain (1935) has illustrated a group of youngish embryos in which there is an evident pro-portionality between the diameter of the axis and the number of cotyledons. Further information of great interest and rele-vance has been given by Bucholz (1946) on the size and mor-phology of the embryo in *Pinus ponderosa*. By making measure-ments of embryos, endosperms and seeds he has shown (*a*) that the embryos of large seeds, at the time of cotyledon inception, are much larger than those of small seeds, and (*b*) that there is a direct relationship between the size of the seed and the number of cotyledons in the contained embryo. Embryos in small seeds may have only 6 cotyledons, whereas those in large ones may have 15 or 16. There is evidently a close correlation between embryo size and the number of cotyledons. As I see it, the provisional acceptance of Turing's theory, or of some other physico-chemical theory, possibly along similar lines, which would account for the patternized distribution of metabolites, and attempts to find supporting evidence for it, are essential if our knowledge of plant morphogenesis is to advance. In passing, an important feature of Turing's theory is that differently con-stituted systems may nevertheless yield closely comparable reaction patterns.

Organ differentiation in endogenous buds   It is not only in apical meristems that we see the need for a theory to account for the patternized distribution of metabolites. In *Ophioglossum vulgatum*, for example, the formation of endogenous buds can be in-duced in the middle cortex of isolated fragments of roots and in the pith of decapitated shoots (Wardlaw, 1953, 1954). In both cases the morphogenetic process begins by the repeated division of a parenchyma cell or a group of such cells. After some time the daughter cells, which become densely proto-plasmic, constitute an ellipsoidal or spheroidal mass of embryonic tissue. Within this tissue mass, the differentiation of a shoot apical meristem and first leaf takes place almost simultaneously, and soon afterwards, in a morphologically lower position, a root initial is formed. In short, a nascent plantling has become organized. If we follow Sachs in the view that in the formation of the several organs, viz. apical meristem, leaf and root, different specific substances are necessarily involved, or that qualitatively and/or quantitatively different assemblages of substances are involved, then the question arises as to how the characteristic distribution of these substances is brought about in the developing ellipsoidal or spheroidal mass of embryonic cells. The working of some kind of diffusion reaction system, it would seem, must underlie and precede the visible morphological differentiation. Comparable developments take place during the organization of endogenous buds in certain dicotyledons and monocotyledons, exemplifying yet another remarkable homology of organization, and suggesting that, notwithstanding the wide differences in genetical constitution, general factors are at work.

Diversity and similitude

The abundance and diversity of materials   As a result of the work done during the great periods of systematic, anatomical and phylogenetic investigation, botanists have come to realize how very numerous, varied and complex are the materials of the Plant Kingdom. Contemporary taxonomists, for example, recognize that there may be as many as 250,000 species of flowering plants, and other major phyletic lines are not lacking in species. If botanists at times view this situation with some dismay, and tend to confine their work to more restricted fields, is it after all very surprising? Still, these materials do exist, their super-abundance is itself a natural phenomenon, and one cannot escape the feeling that something ought to be done about them, in particular by morphologists. We cannot just stay feeling lost in this multifarious mass of materials. One feels that some drastic but tenable simplification should be sought; that morphological variation may eventually be attributed to a relatively small number of general relationships pervading all development, e.g. polarity, the effects of differential growth, increases in size, and so on. In brief, in the interests not only of morphology but of botany as a science, some sustaining aim of a more general kind, some unifying philosophy, seems desirable. In the Phyletic Period, for example, contributing to the theory of evolution and attempting to construct a monophyletic tree of the Plant Kingdom provided comparative morphologists with a major aim and a unifying philosophy. These important topics are by no means exhausted. After all the effort, our knowledge of phylogeny is still far from complete—phyletic gaps confront us at virtually every critical point—and it may be that what is needed is an approach from a new angle. A line that offers abundant scope for new work, and one which should be of interest to both phyletic and causal investigators, is the intensive study of some of the numerous parallelisms of development in unrelated organisms, i.e. homoplasy. This suggestion is not new. Homoplastic development was recognized by Ray Lankester (1870) and the general concept of *homology of organization*, which comprises both genetical homology and homoplasy, has been discussed from time to time by eminent botanists. Thus Lang (1915) pointed out that the study of homoplastic developments or, more generally, of homologies of organization, might lead to the discovery of common causes underlying the assumption of form in plants—but, in actual fact, little has been done about it. Morphogenetic developments are attributable to physiological-genetical, physico-chemical and environmental factors, but it is important that we should ascertain more precisely the contribution made by each of the several kinds of factors. There are probably few contemporary biologists who dissent from the view that genetical factors pervade all development and it is generally accepted that specific gene action is primary and fundamental in determining morphological and structural characters. Yet, when we bear in mind the prevalence of homology of organization, especially in quite unrelated organisms, the too-ready or outright acceptance of these genetical views may be an over-simplification which obscures the truth. Certainly, in botanical science, geneticists have had very little to say about parallel developments, or, indeed, about gene action in major morphogenetic developments. In some morphogenetic developments, although gene action must obviously be involved (for indeed it cannot be excluded) gene products may not be proximate factors determining particular conformations. Factors of a general biophysical kind may be more directly involved. Thus Turing (1952), in his theory of morphogenesis, has pointed out that differently constituted reaction systems, in which gene-determined substances are the component factors, may nevertheless yield the same pattern. Heteroblastic development, i.e. ontogenetic changes in leaf shape, which may be very extensive and conspicuous in some species, is known in both ferns and flowering plants, and is not directly attributable to the action of specific genes. In numerous experiments, nutritional and various extrinsic factors have been shown to be directly involved. In a survey of early embryogenesis in all classes of plants, from algae to angiosperms, I have shown that the same morphogenetic developments are common to all, e.g. the early establishment of polarity, the regularity of the segmentation pattern, the establishment of histological and other gradients, the organization of a distal formative meristem, and so on (Wardlaw, 1955). And in floral morphogenesis it appears possible that the same, or closely comparable, growth mechanisms may determine such features as zygomorphy, meiomery and pleiomery, etc., in different families which are not closely related. In brief, intensive investigations of homologies of organization will not only carry their own interest but should enable us to test the hypothesis which I have proposed, namely, that notwithstanding the great diversity in plant form and structure, only a relatively small number of developmental mechanisms (or basic processes with their concomitant organismal relationships), of high generality, are involved in determining the major formal developments. If one regards a growing plant as a dynamic geometrical system, one might say that, underlying the evident diversity of form, there are only a few basic geometric patterns. Or, in quite general terms, one might say that substantial unity underlies the evident diversity, that morphology consists largely of variations on a small number of themes, and that many seemingly unlike morphological and structural features may not, after all, prove to be so very unlike, the differences being of degree rather than of kind. Of course, the same general idea underlies D'Arcy

Thompson's fascinating chapter in which he deals with theory of transformations and the comparison of related forms. There he has shown how, by simple modification of the Cartesian co-ordinates, seemingly very unlike adult forms can be brought into a close morphological relationship. Today we recognize that the comparisons ought to relate to ontogenetic developments and not to adult forms; but we may note that changes in the allometric growth pattern, of the kind which D'Arcy Thompson had in mind, are now known to be determined by, or associated with, changes in the genetical constitution.

No doubt all this is a vast over-simplification, but it does suggest a guiding line for thought and action which will contribute to both causal and comparative morphology. An example may be helpful at this point.

The leafy shoot All vascular plants, with some very rare exceptions, consist of a root system and an axis or stem. The latter has a distal apical growing point which gives rise to lateral members, typically leaves and associated buds, in some characteristic phyllotactic sequence. Associated with the apical meristem and leaf primordia there is an incipient vascular system. Although, according to some contemporary views, the major classes of plants have probably evolved as independent phyletic lines, and show considerable histological diversity in their apical meristems, the major and conspicuous features of the vegetative phase are common to them all: that is, the leafy shoot in the several classes affords a remarkable homology of organization. Further, we may note that closely comparable phyllotactic systems occur in ferns and in dicotyledons. Moreover, as the conspicuous differences in the vegetative phase of vascular plants are largely due to leaf size, shape and phyllotaxis, we may note, as Richards (1951) has shown, that many of the different phyllotactic systems are not, on close mathematical analysis, so very different, and that one may readily pass into another. In my own laboratory, Cutter and Voeller (1959) have shown that, in individual apices of *Dryopteris dilatata* kept under close observation over a period of time, the transition from one phyllotactic arrangement to another can be followed; and we may also recall that M. and R. Snow (1931–5) induced changes in leaf position by incising the apical meristem. In one experiment (1935), a spiral phyllotactic system was obtained by diagonally bisecting the apical meristem of *Epilobium hirsutum*, in which the leaf arrangement is normally decussate. Once again we may note that although genetical factors cannot be excluded from the inception of these phyllotactic arrangements, other factors or relationships appear to be more directly involved.

Within the leafy shoot type of organization, there is scope for almost unlimited diversity. The axis may be long or short, simple or much branched; the leaves may originate in various phyllotactic patterns; and they may be large or small or show some characteristic heteroblastic development.

Again, although seven or more types of apical histological pattern have now been described, their major growth, organogenic and histogenic activities are all closely comparable. The general property of meristems, apparently with very rare exceptions, of giving rise to growth centres, which usually become leaves during the vegetative phase, has already been noted. All leaf primordia have their inception in the apical meristem and have their characteristic symmetry and orientation determined, usually irreversibly, in that region. They all result from the active growth of a localized group of superficial and subsurface cells in characteristic positions on the lower flanks of the meristem, the very young primordium becoming visible as a small outgrowth, of a shape and size that is typical of the species, on the surface of the meristem. This account of leaf inception applies equally to microphylls and macrophylls (Wardlaw, 1957). Also associated with the activity of the apical meristem and its growth centres is the inception of vascular tissue. Variations in the amount of primary vascular tissue formed and its pattern of distribution can be referred to the respective physiological activities, of some particular kind, of the meristem and primordia.

The completely leafless condition is very rare. *Psilotum*, which has only very minute scales, often without a vascular strand, is sometimes described as a 'leafless' plant. At their inception in the meristem, however, these scales are quite massive multicellular structures, generally comparable with the primordia of megaphyllous pteridophytes (Wardlaw, 1957). Again, in virtually 'leafless' parasitic flowering plants, such as *Cuscuta*, the young primordia are closely comparable with those of other members of the family which have normal foliage leaves. Whether the classical leafless species of *Rhynia* were primitive leafless plants, or leafless by reduction as some investigators aver, must remain an open question. But we may note that they are almost unique among vascular plants in being leafless, and that both contemporary and earlier species of vascular plants possessed either well-developed microphylls or megaphylls. Leaf inception in all classes of vascular plants thus affords a further remarkable example of homology of organization, and suggests that the same physico-chemical factors, or generally comparable reaction systems, may be present in all shoot meristems.

Lateral buds In leptosporangiate ferns such as *Dryopteris*, *Onoclea*, *Matteuccia*, etc., buds, which will become lateral branches, are not usually formed on the apical meristem, though their formation can be induced there. They originate from *detached meristems*, i.e. small discrete portions of the apical meristem, occupying characteristic interfoliar positions, which become detached during growth but remain in the embryonic state. Some distance below the apical meristem, detached meristems may undergo a renewal of growth, and form buds of characteristic radial symmetry which subsequently give rise to lateral branches (Wardlaw, 1952). With modifications of the histological detail, the principle of branch formation from detached meristems has been recognized in angiosperms; and in the brown alga, *Ascophyllum nodosum*, the apical meristem of the strap-like thallus has a histological organization not unlike that of a fern and its lateral branches are formed from detached portions of the apical meristem. In these details of bud inception we encounter yet another quite remarkable homology of organization.

Inferences from causal and comparative investigations Morphogenetic investigations of apical meristems have yielded information and inferences that are at variance with certain views based on comparative studies. To the phyletic morphologist, for example, microphylls and megaphylls are foliar members of different evolutionary origin and morphological category. On the basis of comparative studies, Bower (1935) described microphylls as *enations*, i.e. scale-like outgrowths from 'untenanted' regions of the apex, whereas he considered megaphylls to be of cladode origin, i.e. modified dichotomizing

branch systems, and he argued that lateral branches resulted
from modified dichotomy. To me, at least, these accounts of
the inception of leaves and buds no longer seem probable or
tenable; and, if this is valid, the systematic treatment of the
several classes of vascular cryptogams, formerly regarded as a
coherent group—the Pteridophyta—but now held to be independent subdivisions of long standing (Lycopsida, Fili-
copsida, etc.), may require yet further revision.

The views which I have expressed are, of course, in no sense
final. They may be quite wrong. But they show that, quite
apart from the inherent interest of contemporary morphogenetic research, comprising as it does so many confluent
branches of botany, the results obtained from it may afford
new criteria in phylogenetic studies (Wardlaw, 1952).

Floral morphology Comparable general ideas can also be
applied to the development of inflorescences and flowers
(Wardlaw, 1957, 1961). In some families, e.g. Ranunculaceae,
contemporary histological investigations show that the floral
meristem has many of the properties of the vegetative meristem;
thus, it gives rise to regularly positioned growth centres and to
lateral members. These are essentially foliar in nature at their
inception, but under the impact of specific genes, which, we
may suppose, become successively evoked, they become modified in characteristic ways and yield the several floral organs.
Moreover, within limits, as experimental investigations have
shown, the normal floral development can be modified by
modifying external factors, light, temperature, etc., and by the
timely application of auxins, etc. (Heslop-Harrison, 1959).

In their vegetative development and in their inflorescences
and flowers, the angiosperms confront the botanist with a truly
formidable array of material. The work of taxonomists in
bringing order into this complex mass has been a remarkable
achievement, though this is perhaps not always sufficiently
realized. Nevertheless, we still know all too little about floral
morphogenesis and about the major phylogenetic relationships
within the group. The very abundance of materials may well
have deterred many potential investigators from engaging in
work on floral morphology. Yet, in every way, the flowering
plants are the greatest group of organisms known to us, comprising as they do the most highly evolved and highly adapted

species. Of course, there have been some very interesting investigations of floral morphogenesis—witness the beautiful and remarkable work of Payer (1857). Contributions have been based mainly on the ontogenetic method, usually in the interests of evolutionary theory. A more considerable attack on the problems of floral morphology would now seem to be timely. Here again, because of the prevalence of parallelisms of development, a sustained search for common factors in development is likely to be rewarding. In general terms, the hypothesis to be tested is that such factors do exist and that, accordingly, a considerable measure of unity underlies and pervades much of the evident diversity.

A basic assumption in studies of floral morphogenesis is that the floral meristem has many of the properties and morphogenetic activities of the vegetative meristem and is derived from it. If, now, we have in mind some hypothetical prototypic flower, perhaps not unlike one of the Ranunculaceae, then it is not difficult to see that, by assuming certain characteristic gene-determined changes in the distribution of growth in the meristem, we have a means of accounting for such features as the perigynous and epigynous conditions, zygomorphy, and so on; other changes in the apical reaction system could account for increases or decreases in the number of growth centres, or changes in their size, and consequently changes in the floral organs. In brief, it is conceivable that a relatively small number of growth changes in the developing floral meristem, together with correlation and other relationships that pervade development, could account for many of the varied kinds of floral construction. Taxonomists have shown that, in floral evolution, certain characteristic developments have apparently taken place quite independently in different families, e.g. zygomorphy, meiomery, etc. In many zygomorphic flowers the plane of zygomorphy is that of the axis and subtending bract, suggesting that unequal distributions of growth in the floral meristem in this plane tend to be of general occurrence. Other aspects of floral construction lend themselves to similar analyses in terms of the distribution of growth and the factors that control it. Cumulatively, there are strong indications that the same, or closely comparable, processes are of common occurrence in floral ontogenesis. If this view can be confirmed, it may in time be possible to refer the diversity of floral structure to a relatively small number of principles of development and thereby to simplify and unify our conception of the angiosperms.

The shoot and the flower   Reflection on vegetative and floral meristems leads one further along the endless, if sometimes tortuous, pathway of new thinking. The flower is usually regarded as being something rather different—by some morphologists as being utterly different—from the antecedent vegetative development. And so, indeed, as a visible manifestation, it is. I now suggest that, fundamentally and inherently, much of this difference may be more apparent than real. Some such idea is implicit in the classical view of the flower, i.e. that the flower is a condensed shoot axis and that its organs are homologous with the vegetative leaves. Obviously, I do not mean to suggest so untenable a view as that there are no differences between the vegetative shoot and the flower: there are very evident and important differences but these should not make us overlook the less evident but perhaps more important similarities. In both, the major constructional feature is that there is an axis which gives rise to regularly spaced, dorsiventral, lateral members in an orderly sequence. After all, is it reasonable to expect differences of a fundamental kind, in that the same genetical system determines both phases of development, the floral phase being in close organic continuity with, and physiologically dependent on, the vegetative phase? If we follow this line of thought, perhaps in a somewhat radical fashion, we may note that in some genera and families the floral construction shows a high degree of constancy, whereas the vegetative features may be rather diverse. Since all of the related species must share the same genetical constitution in a very considerable measure, one may enquire how fundamentally different the vegetative features really are; in other words, what magnitude of genetical difference is involved. As we know, single genes, or small groups of genes, sometimes determine morphological differences of very considerable magnitude. Since in comparative morphology, morphological features are sometimes considered to be as fundamentally different from each other as they appear to be, there has perhaps been an understandable tendency to emphasize the evident differences rather than to look beyond them to the less evident common basic features.

Conclusion

However valid or otherwise these ideas may be, they show that broadly-based morphogenetic inquiries can lead to new points of view on morphological phenomena of classical interest. From time to time morphologists have been inclined to think that the cream had been scooped off their subject by the previous generation of investigators. Nothing could be more erroneous. The scope for constructive thinking and for new investigational work is probably greater than ever before, with who can tell what important effects not only on the further development of botanical science but on its underlying philosophy. New discovery, however it comes about, is the very essence, the *sine qua non*, of scientific advance. In the biological sciences, however, isolated new discoveries do not take us very far unless they are integrated into the general corpus of knowledge. Every living plant is a complex organization in which many factors, extrinsic and intrinsic, are involved; it is a unique physical system, holistic and harmonious in its development. Because this is so, it is not enough to give a factorial account of the numerous processes that take place during the individual development: there must be synthesis, if our understanding of plants as organisms is to advance. To those who entertain this concept, it appears that morphology—which is concerned with the visible manifestation of specific developmental processes, and the factors involved in them—must continue to occupy a central position in botanical science.

23
'On Growth and Form'

In 1917 a remarkable work of scholarship and biology was presented to the scientific world. This was the first edition of the late Sir D'Arcy Thompson's book *On Growth and Form*. It was a relatively small edition by modern standards but, over the years, the book became widely known among biologists, exercising a real if perhaps somewhat intangible influence. With some exceptions, indeed, its effects were probably general and pervasive rather than direct and specific. The book did not escape criticism, but it can rarely have been read without evoking some measure of delight and an awareness of the integrity, and of the deep and abiding sense of scholarship, of its author in his search for simplifying concepts and general truths in the varied and complex world of animate things. For, in memorable phrases, in a book which many must appreciate and perhaps envy for its reasoned arguments in distinguished and gracious prose, D'Arcy Thompson told his readers that '. . . Cell and tissue, shell and bone, leaf and flower, are so many portions of matter, and it is in obedience to the laws of physics that the particles have been moved, moulded and conformed' . . . and that '. . . In general, no organic forms exist save such as are in conformity with physical and mathematical laws'. In short, in this singular, indeed unique, book, D'Arcy Thompson showed, by reference to many fascinating examples drawn from plants and animals, that some knowledge of physics and mathematics may contribute to a more adequate understanding of the

A Review of *On Growth and Form*. By the late Sir D'Arcy Wentworth Thompson. An abridged edition, prepared by J. T. Bonner. 8½ × 5½ in. Pp. xiv + 346 with 181 text-figures. Cambridge University Press. 1961. *New Phytologist*, Vol. 61, 226–8, 1962.

development of form and structure. Today, this may seem evident and commonplace, but the situation was very different in 1917 when botany, for example, was just beginning to emerge from the long phase during which comparative morphology, undertaken in the interests of evolutionary and phyletic theory, had been the prevailing discipline. As a very young botanist, the present writer still recalls the thrill with which he dipped into that first edition—a handsome book by any standards—and the profound interest he experienced in D'Arcy Thompson's suggested causal explanations, or interpretations, so singularly lacking in many of the morphological works available at that time. In parenthesis, it strikes the writer as curious that little or no mention is made of D'Arcy Thompson's ideas even in some contemporary standard texts.

The first edition of *On Growth and Form* was a wonderful and stimulating book but, let us be clear, it took a deal of reading! It was so full of novel ideas and demonstrations, erudite footnotes, often with pertinent quotations in several different languages on a single page, and elegant scholarly expositions, which one wished one had been able to write oneself; in short, it was so full of substance that it just had to be studied, closely and consistently, over a considerable period of time. For D'Arcy Thompson had a superb knowledge of the literature and he was able to present, with refreshing clarity and cogency, many valuable ideas on factors in morphogenesis which were in danger of drifting into total obscurity.

Biological books of this fullness of content and level of scholarship are rare and it is little wonder that, over some four decades, *On Growth and Form* has exercised at least some effect on so many different minds and pervaded so many branches of biology. Of course, D'Arcy Thompson was in no sense the originator of all the ideas and concepts round which the book was written. Nor was he much of an original investigator. But he had a prime appreciation of the value and relevance of mathematical and physical concepts in biology and a capacity, not merely for bringing different fragments of biological information together, but for integrating them so that aspects of organization in plants and animals stood forth with a new clarity and meaning.

The first edition, which comprised some 793 pages, was

evidently a very substantial, not to say 'meaty', work of scholarship. Then, in 1942, a new edition of 1116 pages was issued.
Much of the text was precisely as before, but there were some
extensions and additions, notably in the chapter on 'The rate
of growth'. Here, D'Arcy Thompson greatly extended his
original treatment, in fact, to the tune of 208 pages! But there
were also some very curious omissions. For in the new edition
he more or less completely omitted any reference to the considerable recent and contemporary advances in biochemistry
and genetics, either in their own right or, more particularly, as
they may be involved in the development of form and structure;
and even certain mathematical treatises, which presumably
had been inspired by the earlier volume, were given the most
inadequate treatment or totally neglected. Some 3 years later,
as a tribute to the 'grand old man'—now known to a wide
circle as D'Arcy—on the occasion of his completing 60 years
as a professor, a volume entitled *Essays on Growth and Form* was
prepared under the editorship of Professor W. E. Le Gros Clark
and P. B. Medawar. While many of these essays are critical,
and sometimes severely so, of D'Arcy Thompson's ideas and
interpretations of form and structure, they reveal to what an
amazing extent his attitude of mind had pervaded different
branches of biology and how great was the stimulus his book
had exercised on more than one generation of botanists and
zoologists.

It is against this background that we come to consider the
merits of the current abridged edition prepared by Professor
J. T. Bonner, himself a notable exponent of morphogenesis. Of
course, it is understandable that the reviewer has proceeded to
his task of assessment quite simply and naturally by asking if,
in the abridged edition, what he regards as the 'right things'
have been retained and other, less desirable or less essential,
matter omitted. There can be no doubt that Professor Bonner
has set about his task with devoted care and discrimination and
that, in some 327 pages, he has succeeded in making more
readily available than heretofore most of what is best and still
significant in the great book. This he has achieved partly by
curtailing some of the less essential parts of the book, partly by
omitting sections which are now out of date and partly by
limiting the numerous examples in which the full edition

abounds. In an admirable introductory essay, Professor Bonner has explained in sufficient detail how he set about his task and the criteria which he used for retention or rejection in different instances. For, as he has rightly said, 'it is a grave responsibility tampering with a classic'; and, recognizing the inimitable polished style of the original, he has been at pains 'never to condense, rewrite, or digest his [D'Arcy Thompson's] materials; the words are all his own'. He has also recognized that in the process of condensation something of importance will inevitably be lost. In the present writer's view, it seems probable that Professor Bonner's reasons will also seem good and sufficient to those who, over the years, have made a close study of the first and second editions.

Among others that have been omitted is the large chapter on 'The rate of growth' (208 pages); also, for evident contemporary reasons, the chapter 'On the internal form and structure of the cell'; the short chapters under the headings of 'A note on adsorption' and 'A parenthetic note on geodesics'; and the considerable, but today rather incomplete, dissertation 'On leaf arrangement or phyllotaxis'. Notwithstanding these excisions, a rich and varied feast, elegantly served, remains and Professor Bonner and the Cambridge University Press are to be thanked for making available to students, research workers and general readers, the essence of a very exceptional book which, for many a year, will persist as a unique testimony to scholarship in biology.

24
A surfeit of species

The world is so full of a number of things,
I'm sure we should all be as happy as Kings.
R. L. STEVENSON.

In this lecture, in which we honour the late Professor Sir William Wright Smith, F.R.S., we not only remember the man himself as a distinguished and indefatigable scholar in the field of taxonomic botany: we also commemorate the long line of devoted herbalists, naturalists, collectors and systematic botanists to whom, collectively, we owe our knowledge of the great floras of the world and much of our understanding of the relationships of plants. In this work Sir William Wright Smith was an active participant. I still recall how, as a young man, I was thrilled and, indeed, stirred to envy, by his tales of travel and botanizing in lofty and remote Himalayan regions. At that time, to follow in the footsteps of Sir Joseph D. Hooker, the author of *A Himalayan Journal*, seemed almost more than any young botanist could reasonably hope. It would be quite out of place, especially before this Society, for me to attempt to catalogue Sir William's many contributions to taxonomy, or his numerous introductions of rare plants to the Edinburgh Botanic Garden (*see* Matthews, 1957). May I, therefore, say, quite simply, how honoured I am to have been asked to give this discourse.

At this time of varied and specialized activities, it is important, in the interests of botanical science and scholarship, that we should not forget how much the overall development of the science is, and has been, due to collectors and taxonomists. The

The Wright Smith Lecture, 1962. A *surfeit of species*. Reprinted from *Trans. Bot. Soc. Edinb.* Vol. 39, 1962–3.

order, which we hold to be pervasive in Nature, has largely become evident because taxonomists have reduced the seeming chaos of multifarious plant life to orderly, if still elaborate, systems.

It is a simple truism that, if we are to make sense of the vast and complex array of plant materials, there must be collecting, identification and classification. A great deal of this work has now been accomplished, so much so, indeed, that specialists in new fields tend to regard classical taxonomy as a rather static subject, belonging to the past rather than the present. But is such a view really tenable? Have we really achieved an adequate understanding of the major phenomena of evolution, involving, among others, the differentiation of the all too numerous species? One may well have reservations on this point. And here it is relevant to recall how much the theory of evolution, in the earlier stages of its formulation, owed to taxonomic studies. Sachs, in his *History of Botany*, recorded that, during the period of 1825–45, some twenty-four natural systems of classification, in a high degree of completeness, had made their appearance. Thus, before the publication of Darwin's *Origin of Species*, the materials of the plant and animal kingdoms had already been surveyed and assembled into orderly groups, merely awaiting, as it were, the catalytic action of a general theory to make them scientifically meaningful. We may also recall that in such great works as Sir Joseph Hooker's *Tasmanian Flora*, the floristic data were not merely recorded but were considered in relation to the factors which may have determined the distribution of species and their isolation and evolution. Information of this kind gave invaluable support to the edifice of evolutionary theory. The compilation of the great regional floras has not only provided very essential information as to what actually exists: it has stimulated many kinds of new interests; so much so that, as some contemporary taxonomists have rightly reminded us, there is virtually no branch of biological science which does not make at least some use of the fruits of these studies.

A surfeit of species

Having spoken in honour of those who have devoted themselves to works of classification, it may well seem to you that the title of this paper carries undertones of a somewhat querulous nature. What I have in mind when I refer to a 'surfeit of species' is simply that, for most of us, there are far more species than we can cope with. Even for the most gluttonous of taxonomic appetites, evolutionary processes have produced a vast super-fluity of species. At one end of the evolutionary scale, we are confronted with the great and varied array of the flowering plants, up to quarter of a million of them, according to some authorities: at the other end, there is the populous world of micro-organisms, viruses, bacteria, flagellates, algae, fungi and so on. And between the two there is a substantial assemblage of archegoniate plants. With some exceptions, there is a sur-prising number of species at every evolutionary and organiza-tional level. And the taxonomists are still hard at it! Every year sees new accessions and records, from viruses to angiosperms.

Although it may bring them little comfort, botanists are by no means alone in this *embarras de richesse*. They are not even ahead of the field! Upwards of three-quarters of a million insect species have been described and named. The yearly record of new accessions exceeds six thousand species and the end is not even remotely in sight. The total number of named animal species is roughly estimated at a million. Again, if we leap into quite a different realm of order and magnitude, we find that the astronomers, with their new techniques for reaching out into space, are finding that there is not only a surfeit of individual heavenly bodies, but also of the assemblages of these bodies. One might also cite the vast and ever-increasing encyclopaedia of compounds with which the organic chemist has to cope. In short, a superabundance of materials for study exists in virtually all branches of science. As scholars, we cannot escape the fact that, in different material categories, the num-bers of specific entities are very, very large.

A personal note: humiliation by numbers
There must be many who, like myself, are not taxonomists but who enjoy field botany and have a lively, general interest in their own regional flora. Thus, as a young man in Britain, I knew most of the species in my own immediate district and also in the neighbourhood within reasonable reach. One also mastered, to some extent, the flora of more distant localities visited during holidays, and so on. Perhaps, in all, one could

readily have named some seven or eight hundred flowering
plant species. Then I went out to the West Indies and Central
America to do specialized work in applied botany—and my
humiliations began. For it is a very big step-up from our modest,
though very interesting British flora, of two to three thousand
species, to one which is probably ten times as large. During my
evening walks, and on other occasions, I began to be interested
in the Melastomaceae. They were among the common wayside
plants and easy to recognize because of their flowers and distinctive leaf venation. As a professional botanist, I thought I
really ought to be able to name some of them, if only as a kind
of hobby interest. Then my difficulties began. I found that the
family consisted of some 200 genera and about 2,500 species.
What with preoccupation with my work and the rather inadequate works of reference at hand, it soon became evident
that, for me, there could be no deep excursions into the
Melastomaceae. Then there were the Bromeliaceae, occurring
abundantly everywhere in the moist Caribbean islands and
mainland. These epiphytes are most attractive, varied and interesting. To see them is to wish that one could know them
better. But in Central America and Colombia alone, some
1,400 species have already been named and classified; and new
collections are daily coming to hand. So, too, with the Orchid-
aceae. This is a pantropical family, rich in species. In Guate-
mala alone, some 700–800 species have already been enumerated. And so the sad tale of the interested, but overwhelmed,
botanical neophyte in the tropics could be continued almost
indefinitely. In Malaya and the East Indies I had looked
forward to seeing some of the fascinating members of the Zingi-
beraceae. I was not disappointed; but, with other tasks on hand,
I could not hope to cope with their numbers and morphological
diversity: forty-five genera and 800 species take a bit of know-
ing. Then there are the many other attractive Scitamineae
which one sees everywhere in the moist tropics. These are
merely samples of personal experience for I still have made no
reference to some of the really big tropical families, such as the
Rubiaceae, with some 450 genera and more than 5,500 species.
But I need not elaborate further: we are indeed confronted
with a surfeit of species. Even if one's scholarship were encyclopaedic, one could not hope to know well all the families of

flowering plants, let alone their botanical significance. What I have said of the flowering plants would also apply, to some extent, to other groups, e.g. the micro-fungi. In discussing the classification of the ferns, which comprise some 300 genera and 10,000 species, a leading authority has recently indicated that no one person could hope to have a critical knowledge of all of them, even in the matter of assigning some of them to their proper genera (Holttum, 1949).

An imaginary colloquium

As scholars, scientists and teachers, what are we to do with these great masses of species? (In the discussion that follows I shall confine myself to the flowering plants.) How much systematic botany should we ourselves know? How much should we try to teach our students? And, not least important, what are the essential things that they should be taught? These questions have often been discussed. It is not difficult to imagine colloquia between professional botanists, of different outlook and interests, in which very different views would be expressed; such as (1) that, whatever our special interest may be, we ought to have a general knowledge of the Plant Kingdom, implying at least a reasonable working knowledge of systematic botany, in much the same way that a biochemist must have a good knowledge of systematic organic chemistry; (ii) that, as specialist workers, fully occupied with our own particular problems and materials, we just do not have the time, nor perhaps the energy or inclination, to widen our general knowledge of plants; but, nevertheless, we ought to know 'some' plants; or (iii) that a knowledge of systematic botany is quite unnecessary: all we need are appropriate labels for the species we are working with; and, in any case, is not systematic botany dead, dry-as-dust, and generally deplorable stuff?

Each of these views can no doubt be defended, if only on the grounds that, in a free society, scholars are free to do what they specially want to do. If you happen to have a distaste for floristic or systematic botany, nobody can make you like it. Moreover, intensive and specialized work in particular branches is quite essential if botanical science is to advance and, as we all know, such work can be very exacting, completely absorbing and satisfying. It could also be argued that the day of the general scholar is past; and it is evident that no one can now cover the whole field of botany in the way that was still possible during the latter part of the nineteenth century.

The flowering plants : a major phenomenon

All these, and other arguments, may be admitted. Specialization is the order of the day: the specialist is entitled to live at peace in his own little niche. But, viewed philosophically, in terms of the wider aims of botanical science, is this good enough? This great mass of taxonomic material, like Mount Everest, is there; it exists. The very existence of the numerous flowering plant species is in itself a major natural phenomenon. It is one of the greatest phenomena with which the botanist has to deal. Surely, in some way, we must try to come to terms with it, ponderous and intractable though it may seem.

That flowering plant species show inter-relationships in different degrees, and that, collectively, they exemplify the ever-present, pervasive phenomenon of evolution, is now beyond dispute, even though many major problems still await elucidation. Accordingly, if botanical science is to develop a coherent philosophy, based on the cumulative experience of its adherents, the facts they have assembled, and the concepts and theories they have evolved, botanists cannot afford to neglect this wealth of material. To do so would be to indulge in a kind of ostrichism, an attitude that surely falls well below the scholarly attainments of the past. So, recognizing the magnitude of the problem, we must try to come to terms with the surfeit of species.

Major biological themes and specialization

In referring to a coherent *philosophy* for botanical science, I am using the word in one of the commonly accepted senses (*Oxford Dictionary*); i.e. the knowledge or study of natural objects and phenomena and their causes—now usually called science; also that department of knowledge or study which deals with the most general causes and principles of things.

The main biological themes, expressed in very general terms, include (i) the inception of 'living' from 'non-living' substance

T

and the organization of primordial organisms; (ii) the adaptation, survival, reproduction and dispersal of primitive organisms, and their progressive genetical and somatic elaboration and organization; (iii) the characteristic growth, morphogenesis and functioning of different kinds of living entities; and (iv) the differentiation and diversification, during vast spans of time, of the numerous varieties and species which the taxonomist groups into higher taxa, i.e. genera, families, orders, etc.

These several topics are all inter-related, our information concerning them, constituting, as it were, a kind of *continuum*. But each aspect has its own particular interest and attraction, with the result that, allowing for the diversity of the human mind, more or less acute specialization is inevitable. We may recognize that this is itself a natural phenomenon, though we may sometimes deplore its consequences. Where there is acute percipience there tends to be strong conviction. We have probably all met geneticists who feel that they are the leaders, not to say the saviours, of biological science. But then, one has also met physiologists, who are no less percipient and active in their own chosen fields, who remain singularly unmoved by the pronouncements of geneticists; or, for that matter, of morphologists or ecologists. In brief, we may recognize that, in an advancing science, specialization is at once natural, desirable and unavoidable, but potentially disintegrative. There must be many, I am sure, who share my view that we should try to keep botanical science coherent and unified; i.e., that we should make a profound effort to develop a philosophy of botany.

Notwithstanding all the wonderful cosmic explorations that are now afoot, the tremendous achievements in telecommunication, the miracles of chemical synthesis, and the almost inconceivable resources of nuclear energy that, for good or ill, are daily being released, all flesh is still grass and it seems probable that *Homo sapiens* will have to occupy, live on and in, and try to understand, his own green world for some very considerable time to come. That being so, it must surely be a sustaining aim among botanists to make their science grow in strength and dignity and take its proper place in the public regard, no less than in what Francis Bacon long ago described as the 'Advancement of Learning'.

Towards coherent scholarship

How is this desirable level of scholarship to be achieved? In each major branch of the science, we must consciously work towards the formulation of simplified, coherent, general concepts, laws or principles, especially those which relate to converging or closely related branches. Among other tasks, this means coming to terms with the 'surfeit of species'. For, unless there is due appreciation of the fact that the abundance and variety of plant life are major and ever-present phenomena, the underlying causes of which must be continuously explored, the general laws which we try to formulate will tend to lack full validity.

In fact, over the centuries, taxonomists have been gradually coming to terms with the numerous flowering plant species, i.e. by identifying them and arranging them into hierarchical groups which, on the evidence, probably indicate genetical affinities and evolutionary trends. They have brought at least preliminary order into a vast and complex mass of materials, though, as is proper in any lively branch of science, plant taxonomy is constantly undergoing revision.

Using a different approach, involving the cytogenetic analysis of small groups of selected materials, experimental taxonomists and geneticists have been exploring the nature of the processes that are involved in the origin of species. The many contributors to what has been described as *The New Systematics* (Ed. J. S. Huxley, 1940; *see also* Turrill, 1938; Stebbins, 1950; Heslop-Harrison, 1953, etc.) have now been able to advance a reasoned if still incomplete account of the differentiation of varieties and species and to extend their findings to higher taxonomic categories. The factors which induce gene mutations and make for their establishment in natural populations are held to be of primary importance, selection being pervasively operative throughout development. In the distinctive development of the individual plant, the collective effects of the genetical constitution, of growth substances, correlations, allometric (or differential) growth, environmental factors, etc., are all involved.

It is held that the same evolutionary processes which resulted in the differentiation of varieties and species, have also been responsible, over great spans of time, for the differentiation of the higher taxa, i.e. genera and families. Along these lines, and with a wealth of interesting observations, which cannot be dealt with here, Stebbins (1950), for example, has given a reasoned account of how the contemporary families and orders of flowering plants may have evolved from the new primitive angiosperms of Jurassic times. Of the possible evolutionary advances or changes, e.g. from actinomorphy to zygomorphy, hypogyny to epigyny, pleiomery and meiomery, etc., either singly or in one or other of the possible combinations, e.g. polypetaly and hypogyny to gamopetaly and epigyny, some have been realized independently in many different families, whereas others have been but occasional. All this not only contributes to a fascinating general theory which may be tentatively accepted, but in conjunction with other approaches to be discussed below, it may also enable botanists, including taxonomists, to acquire a more adequate understanding of the wealth of species.

Towards a new taxonomy

The importance of the species. In botany, the basic biological unit or entity is the species, connoting a group of plants having certain common and permanent characteristics which clearly distinguish it from other groups. Basically, the word species means the appearance, visible form, or kind (*Oxford Dictionary*). So, in classifying the higher plants, we are essentially concerned with their form, the latter term being used here in its more comprehensive, Aristotelian sense. And, as we have seen, it is the apparently endless diversity of form that constitutes our seemingly intractable problem.

Similitude and the diversity of form Now, there is no denying that flowering plants are exceedingly varied in form and structure; and hence even the simplest classification is still very complex. Hutchinson (1926), for example, enumerated 105 orders (or cohorts) and 322 families. So even at the higher taxonomic levels, we still have much more than we can conveniently bear in mind. Some further simplification is desirable and should be possible. Studies of morphogenesis have enabled us to recognize that many seemingly quite unlike adult forms may fundamentally be essentially alike. As an example from plane geometry, a circle is simply a special case of a whole family of ellipses, some of which could be very greatly extended along one axis. Again, as D'Arcy Thompson (1917) has shown, the same basic form can be greatly diversified if it is projected on to variously distorted Cartesian co-ordinates; and we now recognize that closely comparable primary patterns in embryonic tissues may develop into very different adult forms as a result of genetically-determined differential growth. Some phyllotactic systems are simple, others are more complex and yet others are highly complex. Yet they may all be related mathematically, i.e. in so far as they all belong to the same Fibonacci series. In fact, several of them may occur during the ontogenesis of certain species. Even decussate and simple spiral systems are by no means as unlike as they seem to be: the former may be transformed into the latter by a simple experimental treatment. Here, too, we may note that, in floral morphogenesis, the number of basic patterns is quite small; i.e. the floral organs typically occur in groups of 2, 3, or 5, with occasional instances of higher or irregular numbers. And many other instances of the prevalence of similitude could be cited. It thus appears that if plants show great morphological diversity, they also yield abundant evidence of similitude. But whereas the differences are usually evident, the element of similitude may be elusive and has to be sought.

These and other cognate observations hold out the hope that, eventually, although the need for the full taxonomic treatment of our wealth of angiospermous species will remain, it may be possible to formulate simplified general statements which will enable botanists to gain a fuller insight into the main features that have characterized the evolution of this great group of plants.

Evolutionary trends and parallelisms In any considerable group of related organisms it is usually possible to perceive some characteristic evolutionary trend, or trends. Most of them are upgrade, but some, as it seems to us, are downgrade, though they still possess survival value. Such relationships can be distinguished in the families and orders of flowering plants. In fact, many of the more distinctive trends have already been discussed in phylogenetic studies. Actually, their number is not large, for essentially the same kinds of evolutionary change appear to have taken place independently in different families, not necessarily closely related. Now, even although the causes of these evolutionary trends and parallelisms are still not understood—although, thanks to genetics, we are beginning to know something about the probable mechanisms—it is possible to specify and classify the relevant morphological data; but some new system of nomenclature or codification is needed to give simple expression to them. Of course, it is to be expected that many special cases may remain which will not be readily assimilated to a simplified scheme. Nevertheless, with the help of computing machines, to which taxonomists are already having recourse, it should be possible to formulate general conceptions of what has been involved in the differentiation of the flowering plants and, in time, a system of principles which should enable us to gain a deeper insight into any particular group.

A new taxonomy What I am suggesting, then, is that we should work towards the formulation of a *classification of evolutionary processes*, to supplement and simplify the comprehensive systematic classifications. The latter will, of course, be retained, extended and modified as knowledge increases. In support of the new proposal, it is pertinent to recall that taxonomy in biology is not only concerned with the principles and practice of classification, usually based on morphology, with the aim of indicating the true natural affinities between plants: it is also, by definition, concerned with the arrangement or classification of materials or entities *in relation to general laws or principles* (*Oxford Dictionary*). In a contemporary publication, Linsley and Usinger (1961) in defining the comprehensive objectives of taxonomy, have included: 'to devise and perfect a scheme of classification in which the named taxa can be arranged in a meaningful manner, preferably one which will demonstrate at once the unity and diversity of organic life, contribute to an understanding of organic evolution, and provide a convenient and useful basis for recording, analysing and interpreting the data of the biological sciences'. This definition comes very close to what I have in mind in suggesting a new approach to our manifold biological materials.

In the proposed new taxonomy, the aims will be to specify, classify and, if necessary, to *codify*, not the morphological or other characters, but the major processes that are, and have been, involved in the evolutionary trends and morphogenetic parallelisms in representative taxa. The findings both of classical and experimental taxonomy, and of other branches of botany, will all be contributory. Ultimately we may hope to formulate the laws or principles that determine these processes, though these are, admittedly, ambitious aims. But already in physics, e.g. in the laws of motion, and in chemistry, e.g. the laws of thermodynamics, it has been possible to bring together and express in relatively simple terms many seemingly very different phenomena. (It is appreciated that neither the physicist nor the physical chemist considers that his labours in these fields are in any sense at an end; but the general analogy is still valid.) Biological systems are admittedly much more complex than those with which the physicist or chemist has to deal; but, if we can recognize and state a problem, it is usually possible also to make some progress towards its solution. Like the physicist working towards the formulation of simplified general laws, to achieve what Newton described as the perfection of simplicity, the botanist must also, in appropriate ways, come to terms with his surfeit of species. In fact, a not inconsiderable part of the groundwork has already been done by taxonomists and others. If something along the lines I am suggesting can be achieved, one begins to see the possibility not only of reducing the major phenomenon of species differentiation to manageable proportions, but of effecting a vital integration of taxonomy with other branches of botanical science.

25
Reflections on
organization
in plants

Organization: A continuum

A long time ago Lyell noted that ancient fossils were generally
like contemporary living organisms in their basic organization.
This he described as the principle of continuity. One may now
recognize that some aspect of organization can be perceived in
all the materials with which science is concerned—the elements,
inorganic crystals, organic macromolecules, and progressively
up the scale of increasing biological diversification, from the
amoeba to the most highly evolved angiosperm. In this general
sense, organization is a *continuum* in the physical world. Organization is also a *continuum* in the ontogenesis and reproduction of
the individual organism and of the phyletic line of which it is a
component.

Since metabolism, growth and structural development are
considered to take place according to physico-chemical laws, it
would be surprising if living entities did not manifest organization, since many physico-chemical processes have organizing
effects. In some instances it is possible to indicate the physico-
chemical factors, or the mathematical relationships, that are
directly or indirectly involved in, or indicative of, organizing
processes; but, in others, the developmental situations are so
complex, and our knowledge of the kinetics and reactions of
large organic molecules in the protoplasm and between cells
so inadequate, that reasoned conjecture must still take the place
of validated interpretation. Such difficulties led earlier workers
to formulate vitalistic hypotheses, concepts of the 'emergent'
properties of more complex substances, emergent evolution,
and so on.

Maheshwari Comm. Vol., *J. Indian bot. Soc.*, 1963, **42**A, p. 301.

While some problems of organization are yielding to the application of physico-chemical ideas and techniques, and many others may be expected to do so in the coming decades—especially if biophysics and biochemistry become established in biology departments, as they ought to be—the higher levels of organization call for concepts of an organismic kind. The concepts of 'organismic biology', as distinct from 'molecular' or 'atomistic biology', are not only justifiable: they are essential. The development of an organism as an integrated whole, the relation of its parts to the whole and to each other, its co-ordination and functional unity, and many aspects of its life-history, all require concepts that lie outside the physical sciences as presently formulated. Thus while biologists should use to the full the ideas, results and criteria of the physical sciences, they must go further and formulate their own distinctive and necessary concepts. In fact, the aim should be to work towards a philosophy of biology; i.e. a philosophy of *organisms* as distinct from the analysis of individual processes that take place in them. These views are in no sense new but a re-emphasis of them seems timely.

Organization a heritable character

Organization in plants is a *continuum* in the sense that its basis is heritable. At some stage during the phase of chemical evolution, a persisting arrangement of substances that could be specified as 'living', in contrast to 'non-living', was attained. For convenience, we may accept this stage as the beginning of organismal organization. The essential features were that the 'living substance' had considerable stability, that it could increase quantitatively, and that it could reproduce itself so that its distinctive features, the cumulative result of the previous phase of chemical change and selection, were perpetuated. In other words, the nascent organization was heritable; the 'living substance', or entity, consisted of a compatible system, or 'con-sociation', of molecules, capable of growth and self-reproduction.

In considering the phenomena of organization in plants, we are typically concerned with the development of a zygote, a spore or other reproductive cell, or with the active apices of shoots, buds, leaves or roots. Since there is no evidence that runs contrary to the aphorism: 'all cells from cells, all life from life',

every cell or tissue region possesses some characteristic organization from the outset. Accordingly, all the phenomena of metabolism, growth, development and progressive organization
which may be observed during the ontogenesis of a moss or a
vascular plant, involving sequences of chemical reactions, take
place in embryonic cells in which some organization is already
present.

Integration an essential condition
Integration is an essential condition for organization. It presents
itself as a logical necessity, in that organization could scarcely
emerge from the action of forces causing disintegration. Even in
the simplest living system, certain physico-chemical and biological conditions must be satisfied. Thus the molecules and
macromolecules in the system must tend to cohere rather than
disperse in the particular environment; their reactions must be
such as to impart stability and equilibrium, or steady state, to
the system; and they must be disposed in some characteristic
spatial or geometric arrangement so that they can maintain and
reproduce themselves and react with each other. All this is as
much as to say that, collectively, the molecules must constitute
a system which has functional efficiency. In more complex
organismal systems, it may be assumed that similar physico-
chemical relationships also obtain; for organisms in general are
characterized by functional efficiency, by stable relationships
with the environment, and by the arrangement and compati-
bility of their components, making for harmony of development.

Organismal reaction systems
Theoretical considerations together with observations and experiments support the general idea that the primary organizational features in plants are due to the functioning or *organismal
reaction systems*. These cannot yet be defined with any precision,
but they are held to be present in individual cells, as in zygotes
and spores, and in the integrated groups of cells that constitute
embryonic or meristematic regions, e.g. the shoot apex. In each
instance, an inherited cellular organization determines the
nature and potentiality of the reaction system. The activity of
the reaction system is affected by all its components, i.e., by
the specific physico-chemical properties of its inherent reacting

substances and by external factors such as heat, light or nutrient supply. In fact, the reaction system admits of continuity between organism and environment.

The idea of reaction systems in embryonic cells and regions makes possible at least a tentative approach to some of the more fundamental aspects of organization and evolution in plants. Among the things we can say, at least tentatively, are: (i) that the basis of the reaction system is transmitted in heredity, i.e. in the protoplasmic organization of the ovum; (ii) that individual components of the system include specific gene-determined substances and nutrient and other substances obtained from the environment or from contiguous cells or tissues; (iii) that the reaction system admits of the interaction of genetic factors, factors in the environment, e.g. heat, light, nutrient supply, etc., and organismal factors, e.g. those involved in correlation; (iv) that the reaction system in a shoot apex typically gives rise to a patternized (or patterned) distribution of metabolites, this process preceding and underlying organogenic developments and tissue differentiation; (v) that, since the reaction system is physico-chemically an open system which tends towards a steady state, and is the primary locus for the interaction of the several categories of factors indicated above, the organization which results is typically characterized by unity, integration, specificity and adaptation; (vi) that as the reaction system in a shoot apex becomes more complex during ontogenesis, so also does the pattern of organs and tissues to which it gives rise; but the transition from one morphological 'stage' to another is usually characterized by physical and developmental harmony; (vii) that the specific substance determined by a mutant gene must either be compatible with the other components of the reaction system, or simply non-reactive; otherwise the system may be disrupted to a greater or less extent, i.e. the gene will be lethal; (viii) that, in relation to (vii) above, selection begins in the reaction system of the ovum, zygote, or embryonic tissue; and that, however random gene mutation may be, the selection of mutants is to some extent canalized by the constitution of the reaction system; (ix) that evolutionary concepts such as those of orthogenesis, or of the inheritance of acquired characters of adaptive value, may be re-examined in the light of the reaction system theory; (x) that although each species or variety has its own specific reaction system, which gives rise to many specific characters, the number of basic kinds of pattern in plants is really quite small. Reaction system theory can thus contribute to a better understanding of differences in specific organization on the one hand and of the prevalence of homologies of organization on the other.

The interpretation of physiological processes

In theories of correlation in plant development, e.g. the inhibition of lateral buds by the main apex, it cannot be said that truly satisfying 'explanations' have not yet been proposed, though the reality of the phenomenon itself is not in doubt. The general idea in correlation is that an auxin, or a hormonal substance, moves from one region of the plant into another, e.g. from the shoot apex to an axillary bud or bud site, and there exercises some characteristic effect on growth and organogenesis. It is not unusual in such studies to find that the state of the tissue acted upon is not specified, e.g. in respect to its histology or other aspects of its initial organization. Such information is now seen to be essential. Again, studies of organ formation in undifferentiated tissue cultures, e.g. the induction of buds and/or roots, suggest that certain growth-regulating substances, alone or in combination, have definite organizing and morphogenetic effects. Such substances are certainly essential to the observed developments. Nevertheless, explanations of specific developments in terms of metabolism alone are usually quite inadequate—a view that has long been entertained by Sinnott and other workers in this field. It is now suggested that a growth-regulating, 'organizing' or 'morphogenetic' substance exercises its characteristic effects by becoming a component of the reaction system of the meristematic tissue or nascent organ acted upon. It may then affect the inherent organization in characteristic ways. The development which ensues must accordingly be referred to *all* the factors in the situation: the 'morphogenetic substance', though it may be quite essential, is only one factor in the situation.

External and internal factors

As J. B. S. Haldane once remarked: 'It is hard to disentangle the effects of nature and nurture.' At any particular stage in development, e.g. of an angiosperm, the manifestations of organization are due to genetic, organismal and environmental factors. This is well illustrated by the transition from the vegetative to the reproductive phase in flowering plants: genetic factors fundamentally determine the nature and scope of the general and specific morphological developments; but the stage reached in ontogenesis, and the incidence of external factors such as day-length or temperature, are usually closely involved in the inception of the flowering process. One of the important results of experimental taxonomy has been to show how important both genetic and environmental factors can be in different circumstances.

Specificity and homology of organization
In studying any considerable range of organisms, either within the same circle of affinity or in different major groups, one becomes aware of the dual nature of organization. On the one hand, the evolution of plants, in all groups and at all levels, has been marked by active differentiation of new varieties from the parental stock, yielding abundant specificity of organization. On the other, the Plant Kingdom is characterized by much homology of organization. Specificity of organization, e.g. in related varieties or species, can be referred to genotypic specificity. Homologies of organization, by contrast, raise problems of a different order of magnitude. In the simplest case, i.e. that of genetical homology, comparable features may be expected in plants which have originated from the same ancestral stock. On the other hand, closely comparable organizational features may be produced by quite differently constituted genetical systems. Other homologies of organization may be ascribed to the action of self-organizing factors which may not be closely gene-controlled, to extrinsic factors, to size-and-form correlations, or to other mathematical relationships.

In searching for general ideas relating to organization in plants, it may be emphasized that while the number of specific genetical systems in the Plant Kingdom is very large, the number of basic organizational patterns, or plans, is quite small. For example, viewed as geometrical constructions, all vascular plants are essentially alike, i.e., they consist of an axis with regularly-disposed lateral appendages. In addition to genetical homology, the prevalence of homologies of organization may be accounted for on various grounds: (i) that all heritable organizational changes must obey physico-chemical laws; (ii) that the successive organizational innovations, which mark the course of evolution, were those which were most likely to take place; these may include the evolution of important parallelisms in genetical systems; i.e. some kinds of system will tend to have been perpetuated and others to have been eliminated; and (iii) that environmental factors and certain limiting factors tend to canalize development. In plants, it appears that the number of major organizational innovations has been quite small; but the number of variations of these innovations has been very large indeed.

Since the principal metabolic processes are common to all green plants, their reaction systems may be expected to have features in common and to yield some similarities of pattern. This may account for some aspects of parallel or homoplastic development. In fact, certain general configurations or patterns are common to all the classes of higher plants. But within any one kind of pattern, there may be more or less extensive variations. These may be attributed to specific gene-controlled metabolic substances acting as components of the reaction system.

26
Apical organization
and differential growth
in ferns

The Filicales, though not a large taxonomic group, nevertheless exemplify a wide morphological diversity and afford many
interesting parallelisms with the flowering plants (Bower, 1923).
Accordingly, although this paper will be mainly concerned
with aspects of developmental morphology in ferns, the underlying ideas may well have a more general application. The main
thesis considered here is that, in any selected fern species, the
major morphological developments can be attributed to: (i) the
inception of primary organogenic patterns in the shoot and leaf
apical meristems; and (ii) the elaboration of these patterns by
changes in the intensity and distribution of growth in the subapical and maturing regions, i.e. by differential or allometric
growth (Wardlaw, 1960). In passing, it may be noted that
plants of this affinity are specially useful because of the absence
of the complicating effects of secondary thickening.

Apical organization
In the contemporary view, based on the considerable body of
evidence accumulated during recent decades, the shoot apex,
which consists of the apical meristem and the subapical and
maturing regions, is regarded as an integrated and harmoniously developing whole, or *continuum*. Moreover, the apical
meristem is no longer considered to be simply a region of undifferentiated embryonic cells, but rather one with substantial
specific organization and physiological differentiation; it probably comprises several distinct though inter-related and integrated zones (fig. 1). These may be indicated as the (i) distal,
(ii) subdistal and (iii) organogenic zones, or regions (Schoüte,

J. Linn. Soc. (Bot.), 58, 373, p. 385 (1963).

1936; Wardlaw, 1957*a*). While it is in the third of these zones that leaf primordia can first be observed as localized outgrowths, originating in an orderly sequence, the actual inception of the antecedent growth centres must take place somewhat higher up in the apex. The second, subdistal, zone has accordingly been envisaged as the seat of an important system of physico-chemical activities, i.e. the locus of a complex reaction system (Wardlaw, 1957); and evidence has been obtained that it is a region of distinctive cytoplasmic changes (Brown, 1958). Although the mechanism involved in the distribution of metabolites, and their accumulation in particular positions, i.e. leading to the inception of regularly spaced growth centres and the organs to which they subsequently give rise, has not been demonstrated and, indeed, is not understood, this activity may nevertheless be recognized as being a feature of shoot apices in virtually all vascular plants (Turing, 1952; Wardlaw, 1953, 1957). Information on the physiology and cytochemistry of the embryonic cells in different regions of the apical meristem is still very scanty; but in the subdistal region in particular one may recognize the need for fuller information on: (i) the special properties of its constituent embryonic cells; (ii) the effects of various gene-controlled substances formed in that and in contiguous zones, and of other metabolites translocated to the zone and utilized in it; and (iii) the effects of extrinsic factors.

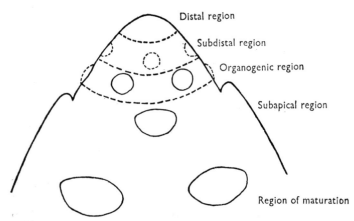

1 Diagrammatic representation of an apex, with whorled phyllotaxis, as a system of inter-related zones.

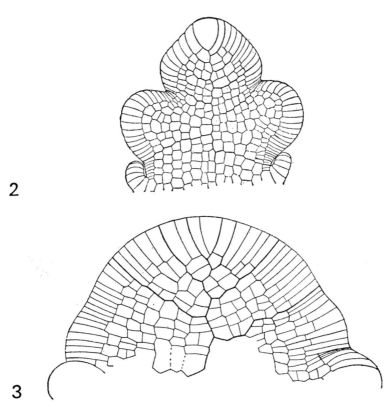

2 *Ceratopteris thalictroides.* Apex of young leaf as seen in surface view, showing the 'two-sided' apical cell and the marginal meristems formed from it. Young pinnae are being formed as a result of growth in localized discrete regions in the marginal meristems. (From Bower (1923), after King.) **3** *Aneimia rotundifolia.* As in fig. 2, the organization of discrete pinna apical meristems from groups of the prism-shaped cells of the marginal meristem can be seen. (From Goebel (1930), after Kupper.)

As apical growth continues, a growth centre occupies positions progressively lower down in the meristem, presumably with changing nutritional relationships at each level, until, in the organogenic zone, it begins to enlarge and become visible as a protuberant lateral primordium. In the subapical regions, the growth and differentiation of primordia continue with increasing acceleration. These and other observations, which need not be given in detail here, support the view that the apical meristem possesses organization, to which the basic or primary pattern of the leafy shoot is due. This concept of apical organization also emerges from a study of the distal and marginal meristems of fern leaf primordia (figs. 2, 3) (Wardlaw, 1958).

The apical organization in shoots and in leaf primordia has considerable inherent stability and, as the experimental evidence has shown, it is not readily modified even by quite drastic treatments. Quite small pieces of the apical meristem of pteridophytes can be excised and grown in aseptic culture to yield normal plants (Wetmore, 1954). Also, as Steeves & Sussex (1957) and Sussex and Clutter (1960) have shown by culturing excised fern leaf primordia in media of known composition, the basic and distinctive features of leaf morphogenesis, which are established in the young primordium in the apical meristem, are retained during the further development of the primordium into an organ of more or less considerable size and complexity.

Types of primary pattern in the shoot

In many procumbent, semi-erect and erect ferns, the leaves are formed on the radially symmetrical axis in a spiral sequence, e.g. *Osmunda, Dryopteris, Athyrium, Matteuccia, Cyathea, Dicksonia*, etc. As Richards (1948, 1951) has pointed out, the changes in spiral phyllotaxis which occur, for example, during ontogeny, are continuous, and phyllotactic systems formerly considered discrete can be linked and also related to rates of growth. We may, in fact, recognize that seemingly unlike morphological features may have their origin in identical or closely comparable primary patterns. In a young sporophyte, or in a lateral bud, e.g. *Dryopteris dilatata*, the phyllotaxis index is low, the second leaf being almost opposite the first. But, with the ontogenetic enlargement of the apex, new primordia are formed in a close sequence and the spiral phyllotactic system becomes progressively more elaborate, with an increasing phyllotaxis index. Moreover, as Cutter (1955) and Cutter & Voeller (1959) have reported, several different phyllotactic arrangements, including spiral arrangements, a bijugate arrangement, and trimerous pseudowhorls, were observed at different times on the same apex of *D. dilatata*. Richards (1951) has provided a method for quantitative comparison of phyllotactic arrangements of diverse kinds, and for relating phyllotaxis to apical growth. In

Dryopteris the different leaf arrangements obtained must be due to growth changes in the apical meristem and are evidently not closely controlled by genetical factors; but these must, of course, be among the components of the reaction system in the subdistal region.

In some fern genera, e.g. *Pteridium, Polypodium, Drynaria, Platycerium*, etc., the primary organogenic pattern in the apical meristem is rather different. From a very early stage in ontogenesis, the axis is procumbent and dorsiventral, e.g. as seen in its cross-sectional tissue pattern, and it gives rise to leaves on the dorsal side only. The roots are mainly, though not necessarily entirely, on the ventral side. In *Polypodium vulgare*, for example, the genesis of these adult morphological features can readily be seen by laying bare the apex. As Hofmeister (1862) and Klein (1884) showed in their classical investigations of dorsiventral ferns, the apical meristem is lens-shaped and horizontally disposed, the leaf primordia originating on the dorsal half, left and right alternately, with a divergence-angle of rather less than 90°. Collectively the leaves are thus formed in two rows on the upper side of the rhizome, while root primordia, as in *P. vulgare*, can be seen emerging in subapical positions on the ventral side. Since, however, the young sporophyte in *Pteridium aquilinum, Polypodium vulgare*, etc., is initially of radial construction with spirally disposed leaves (Hofmeister, 1862; Bower, 1923; Gottlieb, 1958; Dasanayake, 1960), one may perhaps infer that the subsequent dorsiventral leaf arrangement is derivable from the spiral arrangement without any discontinuity of development. (Actually, in *Pteridium*, as described by Dasanayake, the inception of dorsiventrality is first seen in the two bud apices by which the original sporeling apex is replaced.)

Primary pattern in the leaf

The importance of primary morphogenesis in the apical meristem is also seen in the configuration of leaf primordia and the adult organs into which they develop. In studies of cultured, excised primordia of a number of ferns and two species of flowering plants, Sussex & Steeves (1953), Steeves & Sussex (1957) and Steeves (1959) observed that the primordia will usually develop into small but characteristic leaves on nutrient media of relatively simple composition. Steeves (1959) has concluded that 'the factors which control the fundamental aspects of leaf morphogenesis are contained within the organ itself once the primordium has been determined at the shoot apex'. The importance of this process of determination is emphasized by the results of surgical experiments; for example, the findings of Wardlaw (1955) show that the primary or fundamental pattern of leaf morphogenesis, including its symmetry, characteristic orientation, etc., is determined during the early development of the primordium on the meristem, all the subsequent developments being essentially amplifications of this basic pattern (see figs. 2, 3).

Differential or allometric growth in shoot development
With this outline of apical organization and the inception of the primary organogenic pattern before us, and having recognized that the primary pattern may change with the ontogenetic increase in size of the apical meristem, we may now consider how differences in the distribution of growth, i.e. differential or allometric growth, in the subapical regions may affect the subsequent morphological developments (figs. 4–8).

In the ferns the subapical region is characterized by active growth, both vertically and transversely, but usually more conspicuously in one of these directions than the other. As a result, in different fern species, or during different phases of growth of the same species, the axis may show more or less marked differences in its morphological development. Some examples of the effects of differential growth on the eventual morphology of the leafy shoot may now be given.

In the adult plant of many ferns, e.g. *Dryopteris filix-mas*, *D. dilatata* (fig. 5), etc., the vertical component of growth in the subapical region is small compared with the radial or transverse component. As a result, the leaf primordia and young leaves form a close spiral round the apex with a high phyllotaxis index, and the apical cone is seated on a conspicuously broad subapical region (Wardlaw, 1952). This phenomenon is exemplified in a high degree in the cinnamon fern, *Osmunda cinnamomea*, as described by Steeves & Wetmore (1953); for whereas as many as twenty-five or more new primordia may be formed in the course of a year as a close assemblage round the apex, the extent of internodal elongation is very slight, i.e. the rhizome is a short shoot.

In some epiphytic ferns, such as *Polypodium glaucum* (R. & C. Wetter, 1954) and *Adiantum nidus* (Wardlaw, 1956), the narrow axis of the young sporophyte tapers off into the apical cone. But, with the ontogenetic increase in size, radial growth is so relatively great that the apical cone is eventually seated in a saucer-shaped depression (figs. 4, 7, 8). The bases of the leaf primordia, which also participate in this characteristic distribution of growth, become almost circular in transverse section, and they too contribute to the sunken position of the apical cone. This is also seen in some tree-ferns, e.g. *Cyathea manniana* (fig. 6) (Wardlaw, 1948). Similar configurations have been experimentally induced in *Dryopteris dilatata*, in which the apex is not normally sunken, by direct treatment of the shoot apex with aqueous solutions of indoleacetic acid, indoleacetonitrile, and DNA (Wardlaw, 1957).

These characteristic developments are mainly due to the progressive augmentation of the cortical and medullary tissues. In these and other ferns, then, it can be seen that observations of: (i) the steady ontogenetic increase in the size of the apical meristem, with the concomitant elaboration of the primary organogenic pattern; and (ii) the maintenance or experimental modification of allometric growth in the subapical and maturing regions, not only contribute to our understanding of the morphology of the adult plant, but may also be of value in comparative morphological studies.

In contrast to these ferns with short, broad axes, some rather different developments can be observed in *D. dilatata*, *Matteuccia struthiopteris*, etc. (figs. 9–11). In these species the stout, erect leafy plant, which resembles *D. filix-mas* described above, gives rise to more or less extensive horizontal underground rhizomes, stolons or runners. In the latter, the apical meristem, the formation of primordia and the phyllotaxis are basically closely comparable with those in the erect parent shoot. There is, however, one conspicuous difference: the vertical component of growth in the subapical region is large relative to the radial component. As a result the lateral bud develops into a thin, elongating rhizome, with widely spaced leaves, each of the latter consisting of a closely adherent, elongated, narrow petiole with a small undeveloped transitory lamina. Sooner or later, however, the rhizome becomes detached from the parent axis, or ceases to be regulated by it. The rhizome apex then grows upwards, becomes exposed to the light, and in due course gives rise to a typical, compact, erect leafy shoot. In *D. dilatata* (fig. 9) and *M. struthiopteris*, which the writer has examined in detail (figs. 10, 11), the transition region is characterized by: (i) a marked shortening of the internodes so that the leaves form a tight spiral around the apex; (ii) an evident thickening of the

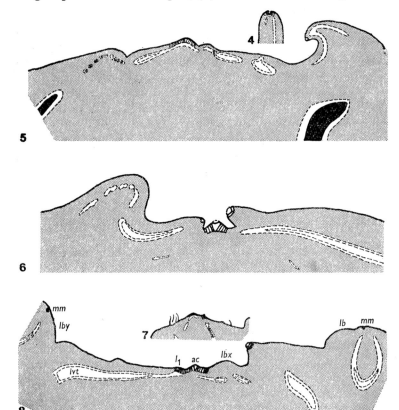

4–8 Longitudinal median sections of fern shoot apices. (Semi-diagrammatic, all ×6) **4** *Asplenium adiantum nigrum*. **5** *Dryopteris dilatata* (*aristata*). **6** *Cyathea manniana*. **7, 8** *Asplenium nidus*, small and large apices. *ac*, apical cell; *l₁*, leaf primordium, recently formed in apical meristem; *lb*, leaf-base; *mm*, marginal meristem of leaf; *ivt*, incipient vascular tissue.

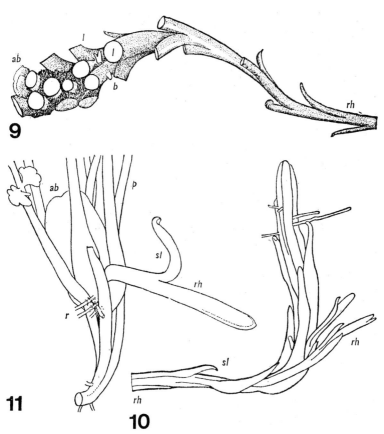

9 *Dryopteris dilatata.* Thin, horizontal underground rhizome, with widely separated scale-like leaves; at the distal end the apical bud (*ab*) has turned upwards and the distal region is becoming broader and bears a close assemblage of leaves (*l*), here dissected off; *b*, lateral buds; *rh*, rhizome. (× 1½.)
10, 11 *Matteuccia struthiopteris.* Two specimens of thin horizontal rhizomes which have turned upwards and are forming erect shoots with a close assemblage of foliage leaves and a stout apical bud (*ab*); new young rhizomes (*rh*) are growing out from the region of transition from the horizontal to the vertical habit; *r*, roots; *sl*, scale-like rhizome leaves; *p*, petiole. (Fig. 10, ×½; fig. 11, ×2.)

stem; and (iii) the development of normal assimilating leaves. The elimination of the regulative effects of the parent stem and exposure to light thus have important effects on the distribution of growth in the subapical regions. In both plants, growth and correlative factors rather than specific genetical ones are evidently the proximate causes of the observed developments.

Comparable contrasted developments were obtained experimentally by growing pieces of *Matteuccia* rhizome on the soil surface and at various depths. At the soil surface the bud rudiments developed into stout, compact, leafy buds, seated directly on the rhizome; but in relation to the depth of planting, the bud rudiments grew out into more or less elongated, ascending, thin rhizomes, these widening out into stout compact leafy plants when the apex eventually became exposed at the soil surface.

As noted on p. 309, the apical meristem in dorsiventral ferns such as *Polypodium vulgare* and *Pteridium aquilinum* gives rise to a dorsiventral pattern of primary morphogenesis. Moreover, in the subapical region, especially of the latter, as many workers have observed, the pattern of differential growth is also dorsiventral with emphasis on the ventral side and on the transverse component. The results are that: (i) the apical meristem occupies a slightly sunken position on the broad, sloping, distal region of the rhizome; (ii) the ventral portion protrudes beyond the dorsal portion; and (iii) the meristem occupies a position facing obliquely upwards. Further conspicuous evidence of the importance of differential growth in *P. aquilinum* is seen in the development of long and short shoots. (For descriptions and references to earlier literature, see Webster & Steeves (1958), and Dasanayake (1960).)

Differential growth in foliar developments

The elaboration and diversification of the primary pattern by allometric growth is seen to advantage in fern leaves. As Bower (1916, 1923–8) and later investigators have demonstrated, the increase in size of the successive leaves in the sporophyte is attended by a progressive elaboration of their outline and venation—described by Bower (1916) as 'leaf architecture' and by Goebel (1908) as heteroblastic development. The primary phase in the formation of a fern leaf is due to the activity of its distal and subsequently of its marginal meristems; and, without going into details, it may be noted that these regions have many of the organizational and morphogenetic properties of the shoot apical meristem (figs. 2, 3). The leaf apical meristem, and the discrete marginal meristems, however, are of dorsiventral symmetry. Although all fern leaves have much in common, they are nevertheless exceedingly varied morphologically, ranging from simple, undivided organs to highly complex ones, including various interesting dimorphic developments (see below). This range in leaf shape is exemplified in the ontogeny of particular species. There may also be conspicuous differences in leaf morphology in land and water forms of the same species (Allsopp, 1955), these being referable to growth activities in the apical regions under consideration. These varied forms lend themselves to analysis along the lines proposed in this paper.

For convenience, the rate of subapical growth in the direction of elongation of a distal or marginal meristem, i.e. in the direction of the midrib or of a pinna vein, will be referred to as the vertical component (or V); and the rate of growth at right-angles to this in the plane of the lamina will be referred to as the transverse component (or T). Now in a leaf primordium in which a distal meristem and discrete, as distinct from continuous, marginal meristems are present, i.e. as part of the primary pattern, it is evident that if V is considerably greater than T, the lamina will become more or less deeply lobed or pinnate; conversely, if $T > V$, the lamina will tend to remain entire; and if V is only slightly greater than T, the leaf margin may be undulating or notched. In the young sporophyte of *Osmunda regalis*, for example, the earliest formed leaves are of oval outline, or they consist of oval bifurcated lobes, i.e. T tends to exceed V. But as ontogenesis proceeds, V gradually becomes greater than T, and the leaf develops as an elongated rachis with lateral pinnae. Similar relationships also occur in the development of pinnules and pinnulets in species with compound leaves. The ontogenetic elaborations of leaf form typically accompany a progressive increase in size in successive leaves, indicating the importance of nutritional factors among others. The validity of this view is supported by the observation that, in various 'starvation experiments', final leaf size and morphological complexity can both be diminished. Moreover, in organ culture investigations, it has been amply demonstrated that the size and shape of fern leaves can be controlled by varying the composition of the medium (see below), the effects being mainly, though not entirely, on subapical regions.

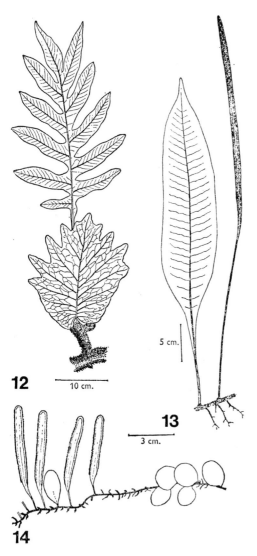

12 *Drynaria sparsisora*. Showing a nest leaf and a foliage leaf. **13** *Leptochilus decurrens*. Showing a normal foliage leaf and a narrowly laminate sporophyll. **14** *Drymoglossum piloselloides*. Showing a piece of rhizome bearing broad sterile and narrower fertile leaves. (Figures 12–14, after R. E. Holttum, *Ferns of Malaya*, 1954.)

Species with dimorphic leaves, e.g. *Drynaria* and *Platycerium* spp., are of particular interest because, as Goebel (1905, 1928) pointed out, their primordia at the shoot apex are essentially alike in size, shape and position of origin. Yet how very different are the adult leaves. The species of *Drynaria* are epiphytes, the elongated rhizome growing upwards on tree-trunks and along the branches. The adult leaves, in two rows on the dorsal side of the rhizome, are of two distinct kinds: (i) there is a sequence of relatively small, somewhat pale, overlapping, entire but notched leaves, with the petiole short or absent: these so-called 'nest-leaves' are closely adpressed to the tree-trunk and constitute reservoirs for humus and water into which the roots penetrate; and (ii) there is a sequence of very large pinnate leaves, full green in colour, bearing the sori, and pendent from the rhizome on long petioles (fig. 12). It is at once evident that both the total growth and the allometric growth patterns in the two kinds of leaf are very different. In the small, sessile, entire but notched leaves, V does not exceed T; but in the large pinnate, assimilating leaves, V greatly exceeds T. For an adequate understanding of these very distinctive leaf formations, environmental, nutritional, and correlative factors, and their interaction with factors in the genetical constitution, must be investigated.

Factors in differential growth
On the substantial evidence now available from culture studies of excised leaf primordia and shoot apices, the variable development in subapical regions is largely determined by hormonal and nutritional factors (Wetmore & Morel, 1949; Wetmore, 1950, 1953, 1954; Allsopp, 1952, 1953a, b; 1954, 1955; Sussex & Steeves, 1953, 1958; Steeves & Sussex, 1957; Sussex, 1958; Steeves, 1959; Sussex & Clutter, 1960). These investigators have shown that, in media in which the other essential nutrients are presumably not limiting, lower sugar concentrations favour the development of the morphologically simple or 'juvenile' type of leaf, whereas higher concentrations tend to promote the development of the more complex 'adult' forms. However, it is pertinent to note that in some so-called 'starvation' experiments (figs. 15–18), abundant starch has been observed very close to apices of greatly diminished growth and morphogenetic activity (Wardlaw, 1945; Cutter, 1955). These variable developments

are largely the result of changes in the amount and distribution of growth in the subapical and maturing regions, though changes in the size and primary pattern of the meristem are also involved. This latter aspect, which was closely studied by morphologists of an earlier period (*see* Bower, 1916, 1923; Goebel, 1900, 1905, 1908, 1930, for full accounts and references), has received rather less attention at the hands of contemporary investigators. Important as sugar demonstrably is in the development of excised fern leaf primordia, the nutrients involved in protein synthesis must also be present in adequate amounts to admit of meristematic activity in a primordium which becomes large and multi-pinnate (Wardlaw, 1945). Heteroblastic development in *Marsilea* is affected not only by sugar but also by variations in the inorganic salts supplied (Allsopp, 1953).

Steeves & Sussex (1957) have reported that whereas cell sizes in excised cultured fern leaves and intact leaves on the plant are approximately the same, the final size of the cultured leaves is usually less, indicating that fewer cells are present. They have accordingly suggested that there is a restriction in mitotic activity in cultured excised leaves, just as there appears to be in the leaves of young sporophytes or in those forced into precocious maturation by experimental treatments, e.g. defoliation. It appears that certain interactions which take place in normal adult leaf development are absent or are precluded; and Sussex & Clutter (1960) have suggested that these probably do not involve carbohydrates but factors of a different kind, not yet closely specified, but with important regulatory effects (Foster, 1928, 1932; Wardlaw, 1954; Ashby, 1948; Ashby & Wangermann, 1950, 1951; Cutter, 1955).

That older leaves affect the morphology of younger ones is well known in flowering plants (Goodwin, 1937); and in the ferns also this aspect has been explored with interesting results. Thus Albaum (1938) showed in the young sporophyte of *Pteris longifolia* that an older leaf, or auxin applied to the stump of an older leaf, regulated, or delayed, the further growth and development of the next younger primordium; and Crotty (1955) has ascribed heteroblastic development to a maturation-delaying effect of older leaves on younger ones in *Acrostichum daneaefolium*. He has shown that in the first leaves of the young sporophyte, the meristematic phase is of brief duration, there being no, or few, older leaves to provide an inhibitory effect on further development. Such leaves mature rapidly and are small and of simple shape. But, as ontogenesis progresses, the successive new leaf primordia are increasingly subjected to the regulatory effects of the surrounding older leaves. The duration of the meristematic phase is thereby extended, the embryonic regions increase in size and complexity and, on maturation, the

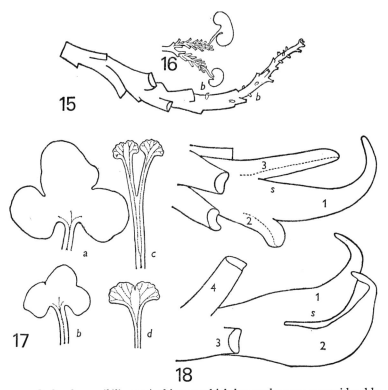

15–18 *Onoclea sensibilis*. **15** A rhizome which has undergone a considerable reduction in diameter as a result of continuous defoliation; *b*, lateral buds. (×¼.) **16** The bifurcated distal region of the same shoot, with small, simple juvenile leaves, 5 months later. (×¼.) **17a–d** Leaves from the rhizome in a 'starvation' experiment showing regression. *a*, *b*, small semi-adult leaves; *c*, *d*, juvenile type of leaf. (×3.) **18** Two specimens illustrating the terminal regions of plantling rhizomes which have been greatly reduced as a result of being grown at a high temperature under feeble illumination. The awl-like leaves are numbered in basipetal sequence; *s*, position of greatly diminished rhizome apical meristem. (×2½.)

adult leaf is an organ of large size and complex morphology. These observations indicate the probable value of fuller investigations of substances which may effect the extent and/or duration of leaf meristem activity (Sussex & Clutter, 1960).

Steeves & Wetmore (1953) have also demonstrated the regulatory effects of older leaves or primordia on the development of the next adjacent inner ones in *Osmunda cinnamomea*. This fern has dimorphic leaves, both fully expanded sterile and fertile fronds and also less developed foliar members known as cataphylls being present (Goebel, 1905; Bower, 1923–8). Special features of the cataphyll are: (i) that, up to a point, its morphological development is identical with that of normal fertile and sterile leaves; (ii) that it shows an arrest of crozier development and concomitant broadening of the marginal regions in the subjacent petiole; and (iii) that prospective cataphylls can be induced to develop as normal leaves, e.g. by the timely, systematic removal of the surrounding older fertile and sterile leaves of the current year. The distinctive morphology of the cataphylls is thus in some way controlled by the adjacent older leaves. Steeves & Wetmore considered that this is probably due to a combination of factors rather than to a single factor, e.g. auxin, though possible auxin effects are not entirely precluded. Other instances of cataphyll formation in the ferns have been recorded (Goebel, 1905). It occurs to the writer that whereas we tend to regard the nest-leaves in dimorphic ferns such as *Drynaria* spp. and the mantle-leaves of *Platycerium* spp. as specialized, adaptive structures, exemplifying division of labour, etc., it may well be that they belong to the same general category of organs as the cataphylls of *Osmunda* and result from the action of similar regulative factors. Cataphylls may readily be observed in *Onoclea sensibilis* (figs. 19–21) and *Matteuccia struthiopteris*.

For the purposes of this paper, emphasis has been laid on growth and development in the apical regions, but other important phases of growth are also involved before a primordium becomes a fully expanded functional leaf. Here the reader may be referred to studies by Steeves & Briggs (1958, 1960), Briggs & Steeves (1958, 1959), Voeller (1960), and Sussex & Clutter (1960), on auxin and other relationships in the expanding and maturing fern frond, including the eventual uncoiling of the crozier. Steeves & Briggs (1958) have shown, for example,

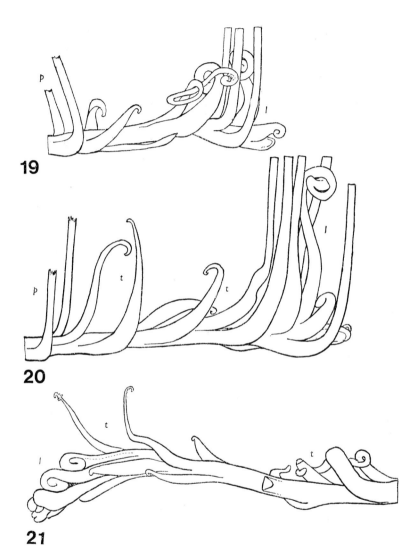

19–21 *Onoclea sensibilis.* Illustrations of the spring condition of rhizomes: *p*, petiole-bases of the previous year's leaves; *t*, transition leaves showing various inhibited cataphyllic conditions and various abnormal curvatures; *l*, the expanded, expanding and still unrolled leaves of the current year. (×⅔.)

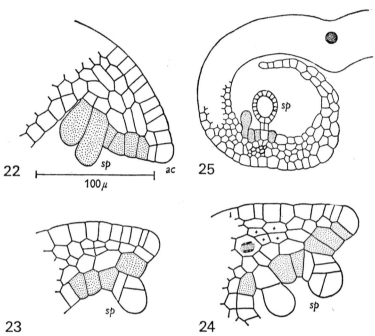

22–25 *Cryptogramme crispa.* Cross-sections of fertile pinnae at different stages in the inception and development of the sorus. **22** The sporogenous meristem, or receptacle (stippled), is initially part of the marginal meristem; two sporangial initials (*sp*) have already become conspicuously enlarged; *ac*, apical cell. **23, 24** Further development of the sporogenous meristem. **25** As a result of sustained marginal growth the sporogenous meristem, with maturing sporangia, now occupies a superficial position.

that there are important and, indeed, striking differences in histological organization as between the basal region of the leaf and the more distal pinna-bearing rachis.

Many factors are clearly operative in controlling differential growth, and these are difficult to dissociate; but work of the kind reviewed above indicates that a beginning has been made.

The distribution of sori

The nature and distribution of the sori and sporangia have long been recognized as being among the most distinctive characters of ferns and as affording important criteria of comparison in phylogenetic studies (Bower, 1923–8). Here again, investigations of the primary pattern laid down in the leaf marginal meristems, and of differential growth in the subjacent regions, enable us to interpret the different kinds of soral distribution. Relevant concepts and facts have been considered elsewhere (Wardlaw, 1958; Wardlaw & Sharma, 1961), but here it may be briefly noted: (i) that the sorus develops from a special kind of meristem, (ii) that all soral meristems originate in the leaf marginal meristem, and (iii) that the ultimate position of the sorus, whether marginal, intramarginal, or superficial, is determined by the distribution of growth in the marginal and submarginal regions. Thus, if marginal growth ceases with the onset of the reproductive phase, the adult sorus will remain marginal; but if growth is maintained in the marginal and submarginal regions of the leaf, the soral meristem will eventually occupy a position on the lower surface of the lamina, in some instances still close to the margin, in others more or less considerably removed from it (figs. 22–5). Here again, the adult morphological development is indicative of the effects and interactions of nutritional and genetical factors. In this connection Sussex & Steeves (1958) have shown, by culturing excised primordia of *Leptopteris hymenophylloides*, *Todea barbara* and *Osmunda cinnamomea* in synthetic culture media, that high concentrations of sugar are among the factors which promote the inception or development of sori, or both, though other metabolic factors must evidently be involved also. The correlative inhibition of growth in the lamina of the sporophyll is exemplified by many ferns (figs. 13, 14).

Discussion

Collectively the ferns afford a wealth of materials for morphogenetic study; and although the examples considered could readily be extended, enough has perhaps been said to validate the hypothesis set out in the Introduction and to indicate where a new concentration of effort on the part of physiologists, geneticists and morphologists, working together or integrating their findings, is most likely to be effective.

While it may be accepted that genetical, ontogenetic, correlative and environmental factors interact to yield the eventual morphological configuration of the leafy shoot, the effects of factors in one or other of these categories may vary widely in different instances. Pervasive as well as specific effects of

x

genetical factors in development are well exemplified by the ferns. Andersson-Kottö (1929, 1938) showed that relatively small genetical differences, by changing the growth pattern, may determine morphological differences of a kind that might well have been regarded as being of critical importance by comparative (phylogenetic) morphologists. In the present paper, it has been shown that, at different stages in ontogenesis, or in relation to particular conditions, very different adult organs may develop from closely comparable primordia. In these instances, the genotype must be such as to admit of the observed developments, but the actual morphological features cannot be directly or closely attributed to the action of specific genetical factors. Thus, although no absolute separation of the several kinds of morphogenetic factors is possible, a more precise evaluation of their specific effects should nevertheless be sought. That the pattern of differential growth is under direct genetical control is often demonstrably true. But, as we have seen in *Dryopteris dilatata* and *Matteuccia struthiopteris*, quite different allometric growth patterns are present in the apices of erect shoots and lateral rhizomes; and species with dimorphic leaves afford further examples of the same general kind. How the primary pattern is established in the apical meristem, and how differential growth, or modifications of it, are determined in the subapical regions, by genetical and other factors, are still very obscure phenomena. The more we study these formative regions as the sites of so many different but integrated reactions, culminating in morphological features with functional significance, the more remarkable and challenging the whole process of development is seen to be. In particular, the subapical region emerges as one which shows great plasticity under the impact of factors of very different kinds. It is, of course, tempting at this point to consider how some of the developments in the subapical region might be interpreted in terms of auxin, auxin-gibberellin, or auxin-kinin activities, as recently discussed by Kefford & Goldacre (1961, where the relevant literature is cited), but what is really needed is a new accession of fact. Recent experimental work indicates that this may well be possible.

The observations discussed here have been mainly concerned with morphological developments. But since the differentiation of tissues is an integral part of the formal development, many of the ideas advanced also apply to the related structural features, e.g. to stelar morphology, leaf venation, etc.

To understand what constitutes both general and specific organization in plants, and to formulate simplified relevant concepts, are among the most important tasks in contemporary botany. It may perhaps be thought that organization must necessarily remain a rather vague and ill-defined notion that does not lend itself to close analysis. This may be so at the present time. But, from the evidence advanced here, it does appear that at least some of the major organizational features in plants can be analysed and subjected to experimental investigation; and they can be defined in terms that are meaningful to morphologists, physiologists and geneticists. Accordingly, it is affirmed that further investigations of the constitution and functional properties of meristems, as yielding the primary morphogenetic pattern, and the control of differential growth in subapical regions, as yielding elaborations and variations of the primary pattern, will advance our knowledge of organization, not only in the ferns but also in flowering plants.

27

Plant embryos as
reaction systems

In that the principal morphological and histological developments in the embryos of all classes of plants have now been
observed with some degree of precision, a legitimate aim during
the next phase of embryological research would be to explain
how these developments are brought about during growth.
This task is made the more attractive by the fact that, as comparative studies show, the embryos from all the major taxonomic groups have many features in common (Wardlaw, 1955).
These homologies of organization may be explained, at least in
part, by envisaging the developing zygote and embryo as a
complex *reaction system* (Wardlaw, 1954). At the present time
what seems to be most needed is a general theory of embryogenesis and a statement of at least some of the basic principles
of embryonic development. The assumption which will be discussed here—that the zygote of any species is a highly complex
and specific organic reaction system which obeys the laws of
physical chemistry—would appear to be basic to any general
theory.

Organismal reaction systems
In referring to developing zygotes as reaction systems, the
author has in mind just such phenomena as those with which
the physical chemist has to deal. But it will at once be apparent
that a developing zygote, with its multifarious metabolism, is
not only a very complex organic reaction system, but also one
which is likely to become increasingly elaborate as development

Recent Advances in the Embryology of Angiosperms; P. Maheshwari. Published
by the International Society of Plant Morphologists Department of Botany,
University of Delhi, Delhi 6, India, 1963.

proceeds. This complexity notwithstanding, the diffusion and reactions of the metabolic substances which surround and enter the ovum and developing zygote must obey the laws of physical chemistry as they apply to organic systems; and already there are ample indications that the general conception now advanced is well supported by the histological facts as observed by embryologists.

It is impossible to give any exact account or definition of a zygotic reaction system at this time, but some of the ingredients and properties of such a system may be tentatively indicated. The importance of trying to do this lies simply in the fact that, however a reaction system may be constituted or defined, it is to its constitution and functioning that the organization of the developing germ is to be referred. The basic factors in a zygotic reaction system are the gene-determined metabolites, together with the physical and chemical properties of their products and, to an extent that cannot yet be defined, the 'fine structure' of the protoplasm and such fixed centres of special metabolism as may be present in it. We may further assume that the newly fertilized ovum is the seat of autocatalytic reactions, as a result of which particular substances important in growth become concentrated in particular positions or loci. In other words, metabolism is accompanied by a patternized distribution of metabolites during the zygotic development, and this precedes and determines the histological and morphological developments. Each visible 'stage' in the ontogeny, in short, will be preceded by its own characteristic biochemical phase. As development proceeds, and under the impact of genetical and environmental factors, and of organismal relationship already established (such as polarity, correlations, nutrient-supplying capacity, etc.), the reaction system will become modified in characteristic ways. This conception could account for the harmonious and epigenetic development of plants. The nutrients (including morphogenetic substances) available to the zygote and young embryo are important determiners of the form and structure of the individual germ. In each instance we may envisage the new substance as entering into the reaction system and exercising its effect there. So also for physical factors such as heat, light, gravity, etc.

General embryological phenomena

In the zygotic development, in species selected from algae to angiosperms, the following phenomena are of general occurrence: (i) In the mature or newly fertilized ovum, the distribution of metabolites is, or quickly becomes, markedly heterogeneous; with or without an attendant elongation of the zygote, an accumulation of different metabolites takes place at diametrically opposite points, the polarity of the new organism being thereby established. (ii) Where the ovum is enclosed, the physiological activity of the surrounding tissue is probably important in determining its polarity. In the free-floating zygotes of algae, factors in the environment may induce the reactions which lead to the establishment of polarity. (iii) In the polarized zygote, the apical or distal pole becomes the principal locus of protein synthesis, growth, and morphogenesis, whereas the basal or proximal pole is characterized by the accumulation of osmotically active substances, its cells becoming vacuolated and distended. (iv) The first division of the zygote is typically by a wall at right angles to the axis, i.e. to the principal direction of growth, cell division being probably stimulated by the increase in size and the instability associated with the drift to cytoplasmic or metabolic heterogeneity. As cell division tends to restore equilibrium in the system, the position of the partition wall will be such that the forces present in the two daughter cells will be balanced. According to the nature and distribution of these forces, the zygote may be more or less equally divided, or it may be divided into a small, densely protoplasmic distal cell and a larger basal cell. The nature of this division is thus ultimately determined by the specific, gene-determined metabolism of the zygote. (v) During the further growth of the embryo, the positions of the successive walls are in general conformity with Errera's law of cell division by walls of minimal area. (vi) As the embryonic development proceeds, the effects of factors in the genetical constitution and in the environment become apparent. Growth is specifically allometric or differential, and the embryo begins to assume a distinctive form. An immense diversity of form is thus possible, but, with some exceptions (e.g. colonial algae), axial development is a general concomitant of the establishment of polarity. (vii) While the embryo is still small, it shows an acropetal gradient of decreasing cell size. With the exception of those algae which grow by means of an intercalary meristem, the distal region of the axis, which may remain perennially embryonic, becomes organized histologically as an apical growing point. (viii) Nutrients are taken up from the environment by the more basal tissues of the embryo and translocated to the apex. Primary growth is in the nature of an accretionary process, the older tissues becoming firm and rigid and showing various characteristic concentric and radiate differentiation patterns.

The determination of biochemical pattern

If the initial distribution of metabolites in the ovum or zygote is homogeneous—it probably is not—a drift toward heterogeneity soon begins. A polarized distribution of metabolites usually follows and this underlies the filamentous or axial development of the zygote. This initial change is evidently of profound importance in all the ensuing embryonic development. In short, from the outset, we are primarily concerned with the patternized distribution of metabolites, and how it may be brought about. The polarized distribution of metabolites, as one example, could be brought about as follows. Let us suppose that in a homogeneous zygote a particular autocatalytic reaction is induced at one end. The stimulus which 'triggers off' the reaction might be due to some external factor such as light or gravity; to a physiological factor such as the presence of a metabolic gradient in the tissue surrounding the zygote; or to some random disturbance such as that caused by the entry of the spermatozoid into the ovum or by some chance accumulation of a particular group of reacting substances in the otherwise homogeneous matrix.

This autocatalytic reaction, once begun, will cause substances to move from other parts of the zygote to the centre of utilization. The drift from the original homogeneous state has now begun. The gradient set up by the initial centre of activity may admit other reactions to take place along its course and after a time it is not difficult to see how, at opposite poles, metabolic activities of rather different kinds might be in progress. Turing (1952) has indicated how an initially homogeneous diffusion reaction system, of the kind that may be present in a developing zygote or apical meristem, may become heterogeneous and give rise to a patternized distribution of metabolites, thus affording a basis for a morphological or histological pattern (*see also* Wardlaw, 1953).

As an embryo enlarges, the underlying reaction system apparently changes in a characteristic manner. The epigenetic development could be referred to the regulated metabolic changes in the morphogenetic regions, and the progressive organization which becomes evident during development could be ascribed to changes in the initial reaction system under the impact of genetical and environmental factors. Any contemporary biochemical theory of embryogenesis will unavoidably be a vast oversimplification of the processes actually involved in it. To know how different metabolites may react and become distributed so as to constitute a particular pattern is to describe only a part of the morphogenetic process. The energy relations within the system and the physical properties of the reacting substances and of the final products must also be known.

Gradients and embryonic organization

Whether we are concerned with algae, bryophytes, pteridophytes, or spermatophytes, their embryos, while still small, show an acropetal gradient of decreasing cell size. The distal region of the axis or filament, which may remain perennially embryonic, becomes organized histologically as a formative apical growing point; the basal or proximal region, on the other hand, is usually of limited growth and tends to develop as a region of enlarged parenchymatous cells. This cellular gradient is evidence of any underlying differential distribution of metabolites in the elongating and enlarging germ. The simple observation of cellular gradients in embryos thus leads to enquiries of a much more fundamental character. These are essentially concerned with the origin of the gradient and with its effect on the further organismal development.

A general survey of plant embryos indicates that the inception of the cellular gradient is in the zygote itself: it may even be in the unfertilized ovum. The histological and cytological evidence is consistent with the view that, in the mature or newly fertilized ovum, the distribution of metabolites in the cytoplasm is, or quickly becomes, markedly heterogeneous, and with or without an attendant elongation of the zygote, an accumulation of different metabolites takes place at diametrically opposite points and the polarity of the new organism is established (*see* previous section). In all embryos the apical pole becomes the seat of protein synthesis, growth, and primary morphogenetic activity. The cells at the basal pole typically become vacuolated and distended and soon lose their capacity for meristematic activity. At this stage the germ not only shows a gradient of cell size but it has acquired the general organizational features characteristic of the adult axial constuction.

Validation of the conception
It is important that the conception briefly outlined above should be tested in all possible ways. That patternized distributions of metabolic substances underlie and precede all visible organogenic and histological developments in embryos can scarcely be doubted. The very young embryos in all classes of plants afford clear evidence of the differential distribution of metabolic materials, and further investigations of these early stages by appropriate techniques, including electron-microscopy, seem likely to be rewarding. In the further elaboration of the embryo, as the various organs—foot, root, leaves, and shoot—begin to be visible, it seems clear that the several distinctive organogenic developments must be referred primarily to the differential distribution and utilization of nutrients. Indeed, no other scientific explanation of these phenomena is in sight.

The most satisfactory kind of evidence, in attempts to validate the concept, will be that based on experiments. In general terms, the essential feature of all such experiments will be that the reaction system is interfered with, or modified, in some characteristic way, preferably under closely controlled conditions. Here the methods of tissue and embryo culture are likely to be of great value. An example may be indicated here. In the normal development of a *Capsella* embryo, two cotyledons are formed. But, as Rijven (1952) has shown, if sufficiently young embryos are placed in certain culture media, they may produce as many as six cotyledons. Van Overbeek *et al.* (1942) obtained similar results with *Datura* embryos. And Buchholz (1946) has shown that the number of cotyledons may vary from 6 to 16 in *Pinus ponderosa*, the cotyledon number being directly correlated with the size of the germ *at the time of cotyledon inception*, and with the size of the seed.

The second experimental approach lies within the sphere of physiological genetics. The embryo in flowering plants is directly dependent on the endosperm for its nutrition. The quality of this nutrition can be varied and controlled by crossing parents of known genetical constitution. In some series of crosses the whole range in embryonic development has been obtained, from embryos which become abortive at an early stage to those which attain to full, normal development. Moreover, embryos which normally become abortive in their particular endospermic matrix can be exercised and successfully grown into adult plants. The already considerable volume of literature on this aspect affords an indication of how the techniques of physiological-genetics and embryo culture, used in conjunction, are likely to advance our knowledge of embryogenesis. They are also likely to afford important evidence relating to the thesis discussed in this chapter.

28
The scope and outlook
for new work
in plant embryology

Relative to the achievements in Zoology, experimental plant embryology has been a sadly though not completely neglected subject. However, it has had its moments: one could not say that it is devoid of growing points, nurtured by competent if rather occasional workers. Thus one may indicate the classical investigations of Whitaker and his collaborators on the free-floating eggs of the Fucales in the 1930's, and a renewal of interest in this group in recent times. One may also refer to the delicate investigations of Wetmore and of Bell and their collaborators on the enclosed zygotes in ferns. Then there are the recent and contemporary studies by Haccius of various changes that can be induced in the embryos of *Eranthis hiemalis in ovulo*. One must also refer to the increasing number of investigations, undertaken with a variable measure of success during the past two decades, on the culture of isolated embryos of flowering plants. This work is now gathering a desirable momentum.* Here, too, one may with propriety include the observations of Steward and his colleagues and others that, in aseptic single cell cultures, pseudo-embryos, or embryoids, of which the further development closely resembles that of the normal embryo *in ovulo*, have been obtained. (Some indication of contemporary work on embryo culture, etc. can be found in *Plant Tissue and Organ Culture—A Symposium*, edited by P. Maheshwari and N. S. Ranga Swamy: and *Recent Advances in the Embryology of*

The Society for Developmental Biology, Inaugural Meeting, Oxford, 20 June 1964.

* Many new experimental observations on the culture of embryos and embryoids have been made since this address was given (see p. 361). C.W.W.

Angiosperms, edited by P. Maheshwari, 1963; both volumes published by the Internat. Soc. Plant Morphologists, Delhi, India).

These have been valuable accessions of information, affording important beginnings. But experimental plant embryology still does not have any considerable place in most standard textbooks. Indeed, it would be a matter of some interest to know if any substantial course of embryology is taught in any British university department of botany. Having commented on the general position regarding experimental studies, one should also note that, on the morphological and anatomical side, the development of embryos, from the zygote onwards to the fully formed embryo, has now been well surveyed in all the major groups of Embryophyta, especially by workers in France, India and the U.S.A. An evident shortcoming is that there have been very few detailed studies of the physico-chemical constitution of the ovum or zygote. Electron-microscopy may be expected to become important in these inquiries. So, as it seems, in common parlance 'we are all set to go', the more so as biology has now entered on a period when intricate and delicate, controlled experimental investigations, of a kind that would have daunted an earlier generation, are becoming the order of the day. So let us consider some of the new possibilities.

The beginning of development

A considerable literature now exists on the morphology and physiology of the embryonic or meristematic regions of the growing plant, i.e. the apices of shoots, leaves, buds, roots and reproductive organs. But, if we are to begin at the beginning, physiological-genetical and morphological investigations of development should properly begin with the ovum, unfertilized or fertilized, or with the single generative cell, from which a new organism may develop. The small size, delicacy and vulnerability of these minute bodies, typically enclosed in archegonia and ovules, have been evident impediments to progress. On the other hand, the advances, which will almost certainly result from intensive investigations, could be of great importance. A general survey of embryogenesis in plants, based mainly on the successive developmental 'stages' described by morphologists, indicates that all embryos have much in common. Such features as the early inception of irreversible polarity, axillary

development, segmentation patterns, gradients, the organization of an apical meristem, and so on, are just as characteristic
of the larger brown algae as they are of angiosperms. In brief,
however different the adult plants in different major systematic
groups may be—and perhaps they are not so different fundamentally as we sometimes assume—their embryos show evident
homologies of organization. Moreover, parallelisms with developments in animal embryos can also be discerned. These homologies of organization suggest that certain physico-chemical
factors or, rather, systems of factors, are of wide generality in
the early stages of development of living things. Accordingly,
for those who entertain the idea that we should work towards
a much closer association of botany and zoology, i.e. that we
should make the most of the common ground of biology, it
would seem that the search for common factors in plant
and animal embryology, as well as being quite fundamental in
itself, is an essential, certainly a desirable, objective. Indeed,
one may go further: the studies indicated will be essential
in any thorough-going attempt to state the 'principles of
biology'.

Embryo culture

To date, investigations of the growth of plant embryos *in vitro*—
chiefly those of flowering plants—have shown that the nurture
of young embryos in artificial media is usually very difficult.
There are, apparently, subtleties in the biochemical and perhaps physical environment of the developing zygote that are
difficult to reproduce in artificial media: the *very young* embryo
is selective and exigent in its nutritional requirements. In the
development of some hybrid embryos, lethal or inhibitory factors become incident, indicating that certain specific metabolic
activities either in the endosperm or in the maternal, i.e. ovular
tissues, are inimical to the developing germ. In some instances
the germ, while still viable, can be excised, successfully grown
in culture and eventually established in soil as a rooted plant.
Such observations indicate that there is an important link
between gene-determined metabolism and embryo nutrition
and growth. In time, the relevant factors will almost certainly
be more fully investigated, with important results, certainly of
academic interest and possibly of practical importance.

A dual approach

The difficulties of nurturing the zygote and young embryo have already been noted. However, new possibilities of overcoming some of these difficulties are beginning to appear. In particular, I refer to the important work of Steward *et al.* and of Reinert on the culture of single cells from carrot phloem tissue, etc. *in vitro*. Steward *et al.* demonstrated that, on further growth, these single cells developed into small cell masses, some of which were closely comparable with the 'stages' observed in normal embryogenesis *in ovulo*. Now, these are remarkably interesting observations. But, after all, is it so strange that single cells, growing in a suitable liquid culture medium, should resemble a zygote which also grows in a liquid medium—the fluid of the embryo sac? These pseudo-embryos eventually developed into normal plants. It thus appears that the physico-chemical and other problems of ontogenesis in plants now admit of a dual approach, from the zygote *in ovulo* or in culture and from the isolated single cell. It is a reasonable expectation that the special techniques and findings of each approach should contribute to the other. The tendency for the small undifferentiated embryos of orchids to develop into more or less bulky masses of callus may also be noted at this point.

Without going into details, it would seem that there is almost unlimited scope for exploring the *potential* developments of single embryonic cells, or embryonic cell masses, under different conditions—a further indication that experimental embryogenesis in plants, which, to some extent, pertains to the contemporary fashionable cellular biology, should not be neglected.

Diversity in embryo development

Because plant embryology receives scanty attention as a 'teaching' subject, there is an understandable tendency for lecturers to demonstrate the structure and development of embryos in lycopods, ferns, gymnosperms and flowering plants by referring to selected 'central types'. This can give a very erroneous impression of the diversity and wealth of materials that await fuller investigation, especially along physiological-genetical lines: the morphological observations, some of which have long been available, point the way. A close examination of the embryos in species of *Lycopodium* and *Selaginella*, for example, reveals a surprising, indeed, fascinating diversity. In some species of *Selaginella*, the long suspensor by which the developing embryo is thrust inwards into the nutritive cells of the prothallus has been replaced, in the course of evolution, by a long tube—a post-fertilization extension of the archegonial cavity—at the inner end of which the embryo is present as a small gemma-like structure, very different indeed from the massive embryo of other species. So, too, reference might be made to the many curious parenchymatous developments in the embryos of some species of *Lycopodium*. Again, in the gymnosperms—a fairly coherent taxonomic group—we encounter a remarkable range in embryonic development, the many extravagances in the elaboration of the suspensor and polyembryony being hard to reconcile with evolutionary theory. In all these developments, the impact of specific genetical factors, from the enlarging zygote onwards, is undeniable; the varied physiological problems relating to the nutrition of the developing germs are as evident as they are untouched.

It is in the flowering plants, however, that the embryonic development reaches its most varied expression. Thus, whereas in many species the zygote undergoes what may, for convenience, be described as a simple and direct course of development, the growing embryo bathed in the nutritive sap of the embryo sac, there are other species in which the embryo, or the embryo-sac, or both, develop more or less extensive outgrowths of a haustorial character. Comparable developments of these kinds have evolved independently in different families. These special structural features evidently afford a wealth of material for further investigation. Along other evolutionary lines, embryos have undergone a more or less marked delay or 'reduction' in their development: in many orchids, for example, and also in other families not closely related, the embryo is a small, undifferentiated, gemma-like body. To be aware of this range of materials, and to reflect on them, is surely to be stirred to curiosity and new endeavour. The relevant investigations should not only contribute to a fuller understanding of the manifold processes that are involved in the crucial early stages in the development of plants: they may also be expected to provide critical new information on major biological phenomena, namely, those of growth, organization, adaptation and evolution.

29
Principles of organization in plants

Definitions

The term *principle* has a considerable range of usage but, following the *Oxford Dictionary*, it can mean 'a fundamental source; a primary element, force or law which produces or determines particular results; the ultimate basis of the existence of something; a fundamental truth, law, or motive force; a highly general or inclusive theorem, or "law"; a general fact or law of nature by virtue of which a machine or instrument operates'. It is in the general sense of these several statements that the term *principle* will be used here.

One of the more evident difficulties in attempting to formulate principles of organization in plants is that, in many instances, all one seems to achieve is the statement of self-evident truths, e.g. that organisms manifest organization. Another difficulty is to state the 'principles' in such a way that the hypotheses relating to them can be validated by observation of their generality, or by analytical and experimental investigations. Some of the 'principles' indicated below admittedly apply more aptly to particular aspects of organogenesis and histogenesis than to the overall organization; for, indeed, it is largely from studies of morphogenesis that the relevant ideas and experience have been gained. It is also hard to avoid the difficulty of stating concepts, i.e. general ideas, rather than well-substantiated *principles*, as defined above, at this stage in the development of the subject.

These and other limitations and difficulties may be duly recognized but, even at the risk of an excessive discharge of aphorisms, a beginning should be made. If this chapter does no

Organization and Evolution in Plants, 1965.

more than provoke someone with greater scholarly equipment and percipience to write a better account of the principles of organization in plants, something will have been achieved.

One of the more general aims of botanical science in the coming decades may well be to prepare a comprehensive classification not only of the morphological and other characters of plants, for that has already been done, but of the *processes* that have been involved in the evolution of both general and specific organization; since, in studying the *evolution of plants,* we are essentially concerned with *the evolution of organizations* (Wardlaw, 1963). This, indeed, may be specified as the central and most comprehensive phenomenon of plant life. It will require a knowledge of the principles of organization, operative at all levels, including the diverse conformations of unicellular organisms and other thallophytes, the organization of the primitive axis, the axis with appendages, and the elaboration of the leafy shoot by various innovations, differential growth, the elaboration of the reproductive organs, etc., all in relation to the general and specific action of factors in the hereditary constitution, to the self-ordering properties of organic and inorganic substances, and to factors in the environment.

Some essential concepts

The interpretation of organization in plants calls for perception, analysis and synthesis of physical, chemical and physiological processes; of the relation of the organism and its environment; of the whole and its parts, and so on. Morphogenesis is sometimes described as being primarily concerned with answering the question: how does the observed form or structure come to be, in terms of the factors that are directly involved? Organization evidently covers much of the same ground; it comprises the same information; but one of its major aims is to understand the development of the organism as an integrated whole.

Among essential concepts,* the following may be indicated.

---

* At this early stage in the development of the subject, some overlapping between 'essential concepts' and 'principles' is difficult to avoid. (A *concept* is defined as a general notion or idea; an idea of a class of objects; the product of the faculty of conception.) The author regards his 'essential concepts' as providing further background for his 'principles'.

Y

i. Organization involves order, stability, flexibility, continuity and specificity.

ii. In the inception of organization in plants, the fundamental processes involve energy and matter, substances being brought together in characteristic spatial relationships, conformations or patterns, which are capable of hereditary perpetuation. Some simple aspects of organization, as in the formation of a crystal, are imminent in physico-chemical processes; but at more complex organismal levels—the result of chemical evolution and selection—organization both manifests and appears to transcend the laws of physical chemistry as presently understood and formulated. But *all* organizational features, whether 'structural' or 'physiological', must be due to innate properties of matter and energy. As molecules (with their associated energy relations) increase in size and complexity, new and often unexpected properties appear. Hence these properties seem to be 'emergent' and have been so described (*see below*).

iii. As already noted, organization in plants is characterized by a high degree of constancy combined with a capacity for change. The hereditary constitution of the zygote or generative cell is relatively very stable and determines the developmental potentiality of the organism, i.e. its range of development and biological activity. But the actual reactivity of the developing organism may change in relation to environmental factors and the stage reached in ontogenesis; and both potency and reactivity may be different in different regions of the organism.

iv. Organization is determined, or affected, by the spatial relationships of parts and their activities; e.g. contiguity is essential for certain reactions to take place. The spatial relationships of parts also affect the distribution of nutrients and therefore, together with other factors, the distribution of growth. In general, the nature of the spatial relationship determines how one part or organ may interact with another.

Physico-chemical factors, organization and 'emergent' properties

While it is assumed that the laws of physics and chemistry apply to all biological activities, the reactions leading to organization typically take place in very complex systems, the bases of which are always transmitted in heredity. The 'hereditary substance' of any species is itself an exceedingly complex physico-chemical system, which has been elaborated as the result of cumulative irreversible changes during great spans of time.

At each successively higher level of evolution, or organization, the new properties or attributes which appear have often been described as being 'emergent', i.e. in the sense used in the concept of emergent evolution (Lloyd Morgan, 1922; Lillie, 1945). Some workers, e.g. Beckner (1959) and Waddington (1962), have denied that any properties are truly emergent: the new, 'emergent' phenomena could equally be regarded as the natural physico-chemical attributes of systems which have attained to a high degree of complexity and organization as a result of hereditary changes and selection. Evolution may be recognized as a process in which chemical materials have become associated into progressively more complex, ordered, persisting systems; and different properties and functional activities will necessarily be associated with each new level of this ordered complexity, or organization. It is only because our ability to predict the properties and activities of complex systems is so limited that we regard them as being 'emergent'. This, of course, is not to deny generally that there can be anything 'new'; for it is undeniable that Mesozoic plants are 'new' relative to Palaeozoic ones. But can there be phenomena which truly transcend the inherent properties and potentialities of matter and energy? Phenomena in plants and animals are, presumably, only 'emergent' in the sense that they are still insufficiently explored by biologists, physicists and chemists. That certain phenomena impress us as being 'emergent' is not really so surprising when we consider how much we still have to learn about 'simple' things. However, as Picken (1960) has noted, the attributes of organisms and their parts and, perhaps, indeed, the very existence of organisms, may be regarded as emergent, in the sense that they are due to properties which are the concomitant of the attainment of a particular level of molecular complexity. But these emergent properties 'can be "explained" as much or as little as we can explain the emergence of molecular from atomic attributes'. Indeed, there appear to be many biological phenomena with which contemporary physical chemistry is not equipped to deal. There are, moreover, many other phenomena, e.g. the reciprocal relationships of parts during development, which are essentially 'organismic' in category.

Other accounts of principles

Although in the section that follows the author has adopted a particular line in setting forth the 'principles of organization', i.e. based on his own working experience and personal convictions, this is by no means the first attempt of its kind. When, in 1868, Hofmeister wrote his *Allgemeine Morphologie der Gewächse* —the general morphology of growing things—he was quite evidently working towards a statement of the general principles relating to form and structure in plants; and in the great classical textbooks of Sachs, Strasburger, de Bary, Haberlandt, etc., we typically find orderly accounts of the Plant Kingdom either beginning with the cell and working upwards to tissues and organs, or beginning with the organs, as the more familiar and evident features, followed by analytical accounts of their tissues and individual cells. By implication, at least, all these authors were working towards comprehensive and coherent accounts of the constitution of plants, i.e. of their organization. In the recent and contemporary periods, writers on philosophical aspects of biology, such as Woodger (1929, 1930–1) and Bertalanffy (1952, where many of his earlier views are summarized) have dealt explicitly with the topic of biological organization and its underlying principles. Bertalanffy, for example, in explaining his *organismic conception*, has affirmed that, fundamentally, organic processes are determined by the mutual interaction of all the factors and conditions in the system, i.e. by dynamic order; and that organisms are not simply reaction systems in the passive sense; i.e. they are not merely machines which operate under some external stimulus or control. An organism, on the contrary, is a basically active system. To quote:

> We can therefore summarize the leading principles of an organismic conception in the following way: *The conception of the system as a whole* as opposed to the *analytical* and *summative* points of view; the *dynamic conception* as opposed to the *static* and *machine-theoretical* conceptions; the consideration of the organism as a *primary activity* as opposed to the conception of its *primary reactivity*.

The following quotations give a fuller impression of Berta-lanffy's point of view (1952, p. 20):

It is not only necessary to carry out analysis in order to know as much as possible about the individual components, but it is equally necessary to know the laws of organization that unite these parts and partial processes and are just the characteristic of vital phenomena. Herein lies the essential and original object of biology. This biological order is specific and surpasses the laws applying in the inanimate world, but we can progressively approach it with continued research. It calls for investigation at all levels: at the level of physico-chemical units, processes, and systems; at the biological level of the cell and the multicellular organism; at the level of supra-individual units of life. At each of these levels we see new properties and new laws. Biological order is, in a wide measure, of a dynamic nature; how this is to be defined we shall see later on.

In this way the autonomy of life, denied in the mechanistic conception, and remaining a metaphysical question mark in vitalism, appears, in the organismic conception, as a problem accessible to science, and, in fact, already under investigation.

The term 'wholeness' has been much misused in past years. Within the organismic conception it means neither a mysterious entity nor a refuge for our ignorance, but a fact that can and must be dealt with by scientific methods.

The organismic conception is not a compromise, a muddling through or mid-course between the mechanistic and vitalistic views. As we have seen, the analytic, summative, and machine-theoretical conceptions have been the common ground of both the classical views. *Organization and wholeness considered as principles of order, immanent to organic systems, and accessible to scientific investigation, involve a basically new attitude* (my italics).

In a chapter on 'Levels of Organization', Bertalanffy (1952) has a section entitled 'General Principles of Organization', in which the 'architecture envisaged in an organism' is perceived as a system of *hierarchical order*. In fact, the principles of hier-archical order had already been defined by Woodger, with the aid of mathematical logic, as early as 1930–1. Let the non-mathematical reader be reassured: this is not nearly as bad as it sounds! In the abstract sense, hierarchical order may be exemplified by a square divided into squares, each of which is also divided into squares; and so on. The hierarchical system is then expounded by Bertalanffy in terms of the *division hierarchy*, i.e. the four-dimensional order of cells which results from the division of a single cell; the *spatial hierarchy*, i.e. the consequences of the relationships of parts at different organizational levels; the *genetic hierarchy*, involving the relationships of successive generations; the *histo-system* of Heidenheim (1923), in which the organism is regarded as being constructed of an ascending order of superordinate systems, including subordinate systems; the morphological and physiological *hierarchy of parts*; and *hierarchical segregation*, as seen in embryonic development, and many other phenomena. Weiss has illustrated his conception of hierarchical order in organisms.

In the section that follows, the author has set out his idea of the principles of organization in plants. These are, as far as possible, 'working principles', i.e. they are intended as a basis for further investigation towards a fuller validation of the 'principles'. Here, two pathways were open: (i) to set out the 'principles' as a coherent statement and (ii) to state each 'principle' with supporting evidence from observation and experiment. After due reflection the first of these choices has been selected, the possibility of the experimental investigation of organization being more fully considered in the following chapter.

Principles of organization in plants

The 'principles' which follow should be regarded as no more than a first approximation, to be improved as our understanding of organization advances. (The 'principles' are in *italics*, but additional statements, in brackets, are added in the interests of fullness of expression.)

i. *Organization in plants involves energy and substances so interrelated and structurally evolved as to constitute viable functional systems, capable of self-maintenance and of being transmitted in heredity.*

ii. *The embryonic cell, an integrated entity comprising the nucleus, the organelles and other structural components (or their precursors), is the unit of biological organization and heredity.* (Many organizational features, in addition to the nucleus, are transmitted during both sexual and asexual reproduction; i.e. all life from life, all cells from cells.)

iii. *Both the 'chemical' and 'physical' attributes of metabolic processes make distinctive and important contributions to organization.* ('Chemical' processes may, purely as a matter of convenience for some purposes, be regarded as primary, and the related 'physical' processes as secondary and consequential; but the two aspects are inseparable. The statement could, for example, be reversed, since no movement, exchange, or recombination of substances can take place without energy; and the physical properties of organic substances, by contributing to the structural organization of the developing individual or biological entity, may profoundly affect all its activities.)

iv. *Biological patterns and other manifestations of organization involve, and are an expression of, the attainment of dynamic equilibrium, or steady state, in complex individual (or unitary) physico-chemical systems, described as organismal reaction systems; or, for convenience, as reaction systems.* (Such reaction systems may consist of single cells, as in unicellular organisms; or they may comprise an associated group of embryonic cells, as in the apices of higher plants. Accordingly, all normal developments in plants, whether regular or seemingly irregular in their morphological or histological conformation, afford evidence of developmental harmony and holistic development, i.e. integrated wholeness. This principle could also account for the capacity of organisms for reorganization after injury or disturbance.)

v. *The way in which a particular cell becomes differentiated is not only related to its position in the plant, but also to its functioning as a component of a coherent tissue pattern which is characteristically determined as a whole. Organs such as leaves also originate as parts of a holistic pattern.* (Holistic tissue patterns are present in cross-sections of roots, petioles, etc. This pattern-forming function, or property, of reaction systems in embryonic regions is a basic phenomenon in organization; i.e. a characteristic distribution, or pattern, of metabolic substances precedes and determines the inception of organs and of tissue systems. This principle not only comprises the spatial aspects of organ and cell differentiation but also the changes in pattern which are correlated with changes in size.)

vi. *However essential, or limiting, in morphogenesis and organization a particular substance may be, it can only produce its effects when it acts as a component of the integrated reaction system in an embryonic cell or region, or in a cell or region still capable of growth.* (By effecting some characteristic change in a reaction system, at a particular stage in ontogenesis, certain substances may, with appropriate qualifications, be regarded as the proximate cause of important and distinctive organizational developments; e.g. certain gene-determined substances are known to act in this way. An organismal system comprises many complex reactions and inter-actions. In the reaction system in a shoot apex, and in the growth centres to which it gives rise, the components include general metabolites, specific gene products, environmental factors and organismal factors (e.g. gradients, correlation factors, etc.). Different genetical factors have different times of action: some are active as long as growth continues: others are evoked at particular stages in ontogenesis, often in characteristic sequence and in relation to the impact of environmental factors, e.g. light, heat, supply of water, nutrients, etc.)

vii. *Differently constituted reaction systems may yield comparable primary patterns; and the same reaction system may yield different primary patterns at different stages in ontogenesis.* (In vascular plants, the reaction system in the vegetative shoot apex typically yields a *primary pattern* of regularly-spaced, discrete *growth centres* which usually develop as leaves during the vegetative phase; but vegetative buds, inflorescences or flower buds may originate from these and from other growth centres. As this primary pattern is usually elaborated in characteristic ways by differential growth in the subapical and maturing regions, similar primary patterns may give rise to very different adult forms.)

viii. *The progressive organization which becomes manifest during ontogenesis is the result of many interrelated serial, or sequential, processes; genetical, organismal and environmental stimuli being involved in the induction and regulation of the successive phases of development.* (The harmonious development of the leafy and flowering shoot of an angiosperm can be analysed in terms of (1) pervasive, serial and specific genic activity; (2) the establishment of polarity in the zygote and young embryo, and the inception of acropetal and basipetal gradients; (3) sustained meristematic activity in the shoot apex and the inception of a primary pattern of organ primordia; and (4) the effects of differential growth and of correlative and other factors in the subapical regions. The serial, or sequential, evocation of gene-determined substances, as components of the reaction system, by progressively changing the reaction system in characteristic ways, makes a major contribution to specific organization, e.g. in the formation of the distinctive and characteristic organs of a flower.)

ix. *The physiological factors which determine reciprocal relationships* *of various kinds, e.g. between the whole and its parts,* between distal and proximal regions, between organ and cell, between contiguous cells of different kinds, etc., contribute significantly to the organization of the individual plant.* (While it is true that 'organs build cells, not cells organs', the substances which determine differential growth, and therefore the size and shape of an organ, proceed from the genetically- and environmentally-controlled metabolism of particular cells, or groups of cells. Although a biological relationship is not an agent, as Beckner (1959) pointed out, nevertheless the positional or spatial relationships of contiguous differentiating cells and nascent organs may affect organizational developments. This is seen (1) in various autonomous or self-ordering morphogenetic processes (*see* Chapter 7); (2) in the translocation and distribution of metabolic substances; (3) in the competition for nutrients, and in mutual stimulation, inhibition, etc.)

x. *Integration is an essential condition for the endurance of organismal reaction systems and for the development of adaptive features in organisms.* (Hence any viable gene mutation must be such that the specific substance which it determines must be capable of acting as a component of the system. A lethal gene is one which does not react compatibly with the other components of the system; a neutral gene is one which can persist in the system without specific activity other than self-reproduction. The adaptive characters of plants are equilibrium or steady state resultant effects of the interaction of hereditary and environmental factors, primarily in the reaction system.)

* Beckner (1959), in discussing biological statements such as 'the whole determines the part', has suggested that, on critical philosophical grounds, it would be more acceptable to say that 'the concept of the whole determines the concept of the part'.

30
Problems of organization in plants: a practical approach

Today there is so much activity in the several branches of botanical science that any major integrating theme, idea or concept is to be welcomed. Such a theme is the *organization* that we perceive in every aspect and phase of development in plants. Moreover, to the extent that we can specify the factors in such organization, the better we are equipped to interpret the phenomena of disorganization, of which there is also abundant evidence. Accordingly, in this colloquium, in which the two aspects will be considered, it is hoped that we may be able to shed new light on the important subject of plant morphogenesis.

The idea or concept of organization in plants, the 'inner order, which normally pervades, and sometimes seems to guide, the whole course of ontogenesis, is in no sense new. Yet it is always new! At one and the same time, the relevant phenomena seem to be self-evident, i.e. that plant species are indeed organized entities, and yet a close scientific account of what is fundamentally involved always seems to elude us. One may still sympathize with the difficulties and efforts of biologists of an earlier period to express their thoughts on this tantalizing theme; for their problems are still our problems! Thus Driesch (1908), an experimentalist of profound experience, was so impressed with the orderly developments of plants and animals, and with their powers of orderly reconstitution after injury or experimental treatments, that he ascribed to each individual an 'ordering inner principle', or *entelechy*. And, later, Smuts (1922), in a more general, philosophical vein, perceiving that development in plants and animals is characterized by a pervasive, integrating unity propounded his thesis of Holism. By this he meant that, at all organizational levels in the Universe, and in the evolu-

Colloque de Morphologie Expérimentale, Grenoble, 28 March 1966.

tionary process, there is a fundamental tendency towards the creation of wholes. Holism, he stated categorically, is a causal factor with a real existence. Both Driesch and Smuts, as also other thinkers, found it impossible to accept that there could be an adequate interpretation of the orderly development of living organisms along purely mechanistic lines.

But let us be frank. As general ideas, entelechy, holism, and other ideas in this category, have their value, their interest and, indeed, their intellectual charm. But, for us, as active experimentalists, have they any working value? Our scientific aims are to understand the phenomena of plant life in terms of their component factors, and to be able to state in the language of science how these factors, working together in characteristic ways and sequences, yield the functional integrated organism, with its specific and distinctive morphology, anatomy and histology. Alas, the general notions of entelechy and holism do not help us on our way; and, perhaps more important, they are unlikely to stimulate the interest of younger workers. Yet, we all want to associate our scholarly work with some major theme. I propose to argue the case that the general theme of organization, including its more particular aspect, i.e. that of *specific organization*—the distinctive organization of the individual species as a unique physical entity—is central to all our biological studies and that, properly developed, this study could impart to our science a high philosophic and scholarly distinction.

Organization is, admittedly, a very complex and mysterious phenomenon; but, if we are to progress, we must discover practical ways of dealing with this complexity and we must state our findings in language that is not mystical or intangibly philosophical. Assuredly, we are only at the beginning; but the possibility of making substantial progress is within our grasp.

Studies of organization should conform to some simple criteria, including the formulation of *working hypotheses*. These should be of such a nature that (i) they can accommodate the results of morphological, physiological, cytogenetical and ecological studies, treated integratively; and (ii) they can incorporate the ideas and information from relevant physical and chemical investigations. In brief, I envisage the possibility not only of a fully integrated botanical science but of a modern botanical science well supported by biochemistry, biophysics and mathematics. The attractiveness of bringing the new 'molecular biology' into a substantial relationship with the general corpus of botanical knowledge needs no emphasis.

Four basic phenomena

In an exercise such as this, much must be assumed, i.e. taken *as given*. For my purposes, I propose to take as given (i) that the individual embryonic cell possesses specific organization; and (ii) that phenomena of organization manifest themselves at all the levels from the 'molecular' to that seen in the fully developed individual plant; and that each level requires its own data and criteria.

According to my present views, a knowledge of four organizational phenomena affords a working basis for interpreting, in comprehensive and integrative terms, the morphology of vascular plants. These are: (1) polarity; (2) the inception, or initiation, or primary patterns of organs and tissues in embryonic regions; (3) the elaboration and modification of primary patterns by differential, or allometric, growth; and (4) the serial, or sequential, evocation and action of genes.

Polarity

Polarity, a phenomenon for which an adequate biophysico-chemical explanation is still awaited, is irreversibly established by the time the zygote has undergone its first division. In other generative cells, and in bud formation, polarity is also evident from an early stage. As Bünning noted, this is the first stage in the process of differentiation. In the embryos of all classes of plants, the establishment of polarity is typically accompanied by a filamentous or axial development, with detectable metabolic and organogenic differences at the apical and basal poles. Without going into the details, polarity, viewed as a phenomenon for investigation, provides materials and interests for experimental morphologists, physiologists and geneticists and affords scope for the skills of biochemists and biophysicists.

The inceptions of primary patterns

During ontogenesis, the shoot apex becomes enlarged and develops a characteristic histological organization. Some seven or more different 'types' of shoot apical organization have now been recognized. This histological organization of shoot apices, already much investigated by morphologists, has also important physiological and cytogenetical aspects. Moreover, like all structural conformations, it has fundamental biophysical and mathematical aspects, even though we have not yet made much progress in these directions.

However, notwithstanding all the differences in size and shape, and the great cellular diversity of shoot apices, they all have a function in common: they give rise to a characteristic pattern of lateral primordia (leaves and buds) and concomitantly to a tissue pattern within. Similarly, although the cellular construction of root apices is also very varied, they all give rise to a tissue pattern that is closely comparable in ferns, dicotyledons and monocotyledons.

Thus, as a major phenomenon, we recognize that in organized embryonic regions there is an active system of some kind which gives rise to organogenic and histogenic patterns. Indeed, we may go further and say that such an embryonic region is the locus of a physicochemical reaction system which has the function or property of yielding a patternized distribution of metabolic substances; and that this pattern, established at the chemical level, is the basis for the ensuing visible morphological and anatomical developments. In his chemical theory of morphogenesis, Turing (1952) indicated how, in an initially homogeneous embryonic tissue, a regular distribution, or pattern, of different metabolic substances could be brought about by ordinary physico-chemical processes. The alternating pattern of the substances determining the differentiation of the protophloem and protoxylem in a nascent root stele affords a commonplace and simple illustration of Turing's theory. So also does the formation of 6–16 cotyledons at the shoot apex, according to its size, in the embryos of some species of *Pinus*.

The substance in the system
Of course, all this is still very hypothetical, although there is no lack of supporting general evidence. We know very little about the nature and mode of action of these organismal reaction systems; but Turing's theory perhaps gives us a clue to the kind of explanation that, at some future time, may be validated by

further work at the hands of biologists and physical chemists, working together. Meanwhile, if we do not understand the physiological mechanism, we have the positive fact that meristematic tissues, organized and supplied with particular metabolic substances, yield specific, *unitary* patterns of organs and tissues, i.e. *the systems are typically holistic in their mode of action.* Thus the pattern which we see in a differentiating root stele is formed as a whole. So we may agree with Vöchting that a cell differentiates as it does because of the position which it occupies in the plant. But we may now go further and say that *a cell differentiates as it does because of its position in a holistic pattern.*

At an earlier stage in the study of morphogenesis in plants, botanists hoped to solve their problems by discovering special 'morphogenetic substances', e.g. root-forming, bud-forming, leaf-forming, flower-forming substances, and so on. During recent years, very important growth-regulating substances have indeed been discovered, in particular, the auxins, the gibberellins and the kinins. But no single substance, however essential it has been shown to be, is uniquely and specifically organ-forming. As it now seems to me, our task is to understand *the action of the substance in the system,* e.g. in a pattern-forming reaction system in a shoot or root apex. A point that is becoming daily more evident is that it is not only the spectacular physiologically-active substances that are important in these organismal reaction systems: relatively simple metabolic substances, such as sugar at appropriate concentrations, have been shown to be highly important, for example, in the differentiation of phloem and xylem, in heteroblastic development, etc.

The concept of the organismal reaction system—an *open system* always tending towards a dynamic equilibrium or steady state—has many advantages. For one thing, there must be a physico-chemical basis for pattern-inception in embryonic regions: some kind of reaction system concept now seems to me to be a *sine qua non.* In the inception of a bud in an undifferentiated tissue culture involving the presence, in suitable concentrations, of IAA, kinetin and sugar, how can these substances react with one another, to yield a holistic end-product, in a state of near-equilibrium, unless they do so in some kind of physico-chemical system? During ontogenesis, with its attendant changes in leaf size, shape and phyllotaxis, we may think of new or increased metabolic components entering the open reaction system and effecting harmonious changes in the morphogenetic developments. And similarly, at the transition to flowering, or during flower formation, it is convenient to think of special, or gene-determined, substances becoming active components of the system, with consequential changes in morphogenesis. Furthermore, as we know, morphogenetic developments are affected in characteristic ways by factors in the environment, e.g. heat, light, water supply, etc. Here we may envisage the reaction system as the locus of the inter-actions of external and internal factors. And lastly, for Neo-Lamarckians and others who consider that external factors may be the cause of certain genetical changes, one has to think of a locus where such factors could make their impact. Where other than in the reaction system in an embryonic region?

Differential growth
The subapical region of the vegetative shoot has long been recognized as *the* important region of growth and expansion. So it is. But, from the organizational standpoint, it is much more than that: it is a region of differential growth. According to the species and the stage reached in ontogenesis, the vertical and transverse components of growth in the subapical region may be closely comparable. More usually, however, one component greatly exceeds the other. In tall species, the vertical com-ponent greatly exceeds the transverse component and the inter-nodes are long. On the other hand, in many ferns with a squat rhizome, in flowering plants with the rosette habit, and in the vegetative phase of many monocotyledons, the vertical growth of the axis is small whereas transverse growth is a more or less conspicuous feature. Many other examples of a differential distribution and utilization of the materials of growth could be cited, e.g. in the development of leaves. In all of them, the primary pattern or organs and tissues, established in the apical meristem, can be greatly modified and diversified according to the nature of differential growth.

Now differential growth, as a major feature in the organiza-tion of plants, is a multi-aspect phenomenon, capable of being investigated in quite different ways by different observers. Taxonomists and experimental morphologists have their own evident interests and opportunities; physiologists have found a special interest in substances such as the auxins and gibberellins which are active in promoting growth in length, with related phenomena concerned with the planes and rates of cell divisions, etc.; geneticists have long had an interest in the hereditary factors at work in determining the tallness and dwarfness of varieties; and ecologists make observations on the effects of habitat factors, e.g. sun or shade on stem length and leaf form. So, in this aspect of organization, which lends itself to analytical and experimental investigations, one can hardly fail to perceive a convergence of many interests, even although we may not yet have made as much progress as we might have in the positive integration of the available information. Indeed, we are still very much at the beginning: the basic problem still remains namely, to ascertain the cause, or the fundamental nature, of differential growth. Since differential growth affords evidence of the movements of metabolic materials and their utilization in the formation of cells, tissues and organs, with resulting characteristic conformations, it is apparent that factors of a physicochemical and biophysical kind must be involved. Here, we may recognize that it is a task for students of organization to bring together, in appropriate ways, the facts from both biological and physical investigations.

Serial, or sequential, developments

When a morphologist examines the development of a flower, e.g. of a species of *Ranunculus*, *Solanum*, *Rosa*, etc., he is aware that bracts, sepals, petals, stamens and carpels are typically and normally formed in a certain sequence, each kind of organ being of the expected size, shape and number for the species; i.e. *they appear with the expected regularity and fidelity of hereditary characters*. On the metabolic aspect, it is evident that some of the materials utilized in the formation of the coloured petals are very different from those utilized earlier in the formation of the sepals, or later in the formation of the stamens. So, whether we view the formation of the flower from the morphological, physiological or genetical standpoint, *we are concerned with serial, or sequential, developments—another major facet of the organizing process*. In broad outline, the hypothesis is that the sequence of specific morphological developments, seen in the formation of a flower, are referable to gene-controlled metabolic processes. These are the result of the serial evocation of genes, consequent on changes in substrates present in the reaction system.

As a result of a stimulus from an environmental factor, e.g. a suitable photoperiodic exposure, the reaction system in the vegetative shoot apical meristem begins to undergo a change from its previous state; the new substrate which is formed evokes a particular gene-determined enzyme. The substrate undergoes further changes and the process of serial or sequential evocation continues. The result is an orderly sequence of biochemical phases, with attendant morphogenetic developments, till eventually the formation of the flower is completed.

All this is admittedly very hypothetical; but, to be realistic, can we *begin* to give *any* account of the orderly formation of a flower without some such hypothesis? There is the further point that, if this kind of interpretation is valid for flower formation, may it not apply to other morphogenetic developments, e.g. the orderly and characteristic differentiation of the primary vascular tissues? Moreover, the hypothesis of the serial evocation of genes, and of sequential chemical phases, in floral development is not without factual support. An increasing body of evidence, derived from surgical experiments and from the application of growth-regulating substances to plants approaching, or in, the flowering phase, supports the general idea now advanced. The hypothesis lends itself to validation by various means. As 'molecular biology' advances, this is one aspect of organization to which it may, in time, make a substantial and critical contribution.

Conclusion

We are still very much at the beginning of our exploration of a major theme—perhaps the major theme—in biology. But, in my view, at least the right kind of beginning is in sight. Unless I have failed in this essay, the theme or organization in plants need no longer be viewed with distrust by scientists because it can only be talked about in a vague and rather general way: on the contrary, the problems can be envisaged in such a way that their fuller exploration can utilize the skills—observational, analytical and experimental—of *all* the botanical disciplines, and include the physical sciences also.

Part three
Perspectives in morphogenesis

The wind bloweth where it listeth.

I have been tempted to use the word *vistas* in the title of this concluding essay, but a botanist who uses it may well be deemed rash, foolish, or simply imprudent. When I consult that infallible source of entertaining erudition *The Oxford Dictionary*, I find that a *vista*—evidently a view or prospect—is further defined, rather pertinently for my present purpose, as a view 'esp. one seen through an avenue of trees or other long or narrow openings'. It is further expounded as 'a long narrow opening in a wood, etc., through which a view may be obtained'. Not least—and very challenging it is too—it may be used figuratively as 'a mental view or vision of a far-reaching nature (1673)'. However, I have chosen to use the word *perspectives*, like my readers, hoping for the best! After all, to gaze into the crystal ball and attempt to predict future developments in botany or, for that matter in any other branch of science, is a hazardous undertaking. If one is successful, it will be accepted as having been self-evident all along! The expectation is that one will mostly be wrong, or fall far short. For example, who in 1915 could have foretold the output of papers on pH, the auxins, the gibberellins, the kinins, the DNAs and RNAs, or the numerous new, simplified chemical techniques; and so on; each new discovery being optimistically regarded by some as a kind of master-key that would unlock many doors.

But should not older scholars occasionally venture on a little prediction, *pour encourager les autres?* During the past year, when I have been seeing a volume on *Morphogenesis in Plants—A Contemporary Study* (to which my readers are directed for factual details not included here) through the press, I can tell you that from my general reading I could have added a new page to the book virtually every day. So, perspectives are in order: they may even be useful. I have confidence in my theme. Some understanding and perception of morphogenesis is a great enrichment in life. It is there by the wayside wherever you go. (Of course, in these days of a prevalence of artificial flowers, the exuberant protagonist has occasionally to be alert and cautious!)

General outlook
One thing can be prophesied without fear of contradiction: botanists are never going to find themselves short of materials. On the one hand, taxonomists are still hard at it, trying to bring order into about a quarter of a million species of flowering plants—not to mention other great and evolving groups; on the other, adherents of molecular biology would have to be blind if they did not see a vast and colourful spectrum of new investigational work spread out before them; for there is virtually no end to the diversified antics of carbon molecules. But the aims, outlook and methods of adherents of these two aspects are in many ways far apart. Tolerance as well as scholarship will be needed if we are to occupy effectively the intermediate territory.

In general, I think that certain major pervasive themes will continue to be of active interest, e.g. the origin of life, evolution, organization, the abundant specificity and homology which are characteristic of most major groups, the 'inwardness', or manifold properties, of large organic molecules and, not least, some reasonably coherent philosophy of organism.

Biochemistry and biophysics
Some kinds of prediction regarding future developments in botany, i.e. as they specially affect morphogenesis, can probably be made with reasonable assurance. For example, plant physiologists, organic chemists, biochemists (or biological chemists) have, for many decades past, been more or less deeply preoccupied with the detection, analysis and paths of synthesis of both general and 'special' substances involved in metabolic processes. This investigational work, in its many branches and aspects, has now gathered a truly astonishing momentum. Just

look over some of the literature if you entertain a contrary view! So it can hardly be doubted that the next two or three decades will see great advances in our knowledge of special substances, their part in cellular activities, growth and development, etc. The accumulated information in its totality cannot be other than formidable: for it is already encyclopaedic. It will also be highly specialized. Indeed, except for those who have had a sound basic training in organic chemistry, and an aptitude for it, much of this new discovery will remain a closed book. No doubt useful popular expositions will become available but, at the scholarly level, we are not concerned with a handy veneer: our aim must be the advancement of botanical science in depth and in coherence. I think what perhaps bothers me most is the sheer complexity of the reaction systems that are involved. For example, organic chemists, interested in plant and animal materials, have been able to demonstrate, indicate, or suggest, the pathway of synthesis of some particular growth-regulating substance from a fairly simple initial substance. Now, the chemical statement, or formulation, of this *one* process is often quite elaborate, i.e. it may occupy a page, or a half-page, of a text-book. Yet, in any morphogenetic process, the synthesis of the substance in question is only one of a whole system of regulated, simultaneous or serial reactions, as well as other physiological activities, all contributing to the eventual morpho-genetic development. Even if we could tabulate the separate pieces of information—the chains of syntheses, etc.—we still have to find verbal expressions for these complex dynamic, growing, differentiating systems. It is true that computers can be used to process the multitudinous accumulation of records. But, in the end, we come back to ourselves: the botanist is an individual, striving to reason inductively and intuitively, e.g. about specific organization, or a particular organogenic de-velopment, etc.; and he can only advance his thinking if he has clear and sufficiently simplified propositions on which to cogitate. So, in this matter, I can see the difficulty, as others have before me; but I cannot present my readers with a far-reaching vision. But, in time, both the vision and relatively simple expressions for at least some of the complex systems in biology will be achieved.

In the foregoing very brief commentary on metabolism and

morphogenesis, I have said nothing about the cognate bio-physical aspects. At the present time, although comparatively little is being done about the teaching of biophysics in university botanical laboratories—there are exceptions—the need for such studies is being increasingly appreciated and developments are on the way. Thus, several university schools of biology have already included, or are making preparations to include, both biochemistry and biophysics. That biophysics will undergo big developments as the century wears on is, I think, almost certain. But *how* the subject will develop can be no more than the wildest guesswork. The number of research workers is still very small, physicists being deeply immersed in their own manifold prob-lems, both pure and applied. Indeed, some considerable time may elapse before effective numbers of physicists with an insight into the problems of biology, or biologists with an adequate competence in physics, become available. As we know, the really big advances in a science are made by rather occasional, exceptionally gifted individuals. We need some of these in biophysics! But who can predict when or where such individuals will make their impact on biology? What one can perhaps say is that the time is close at hand when botanical science could profit greatly from the ideas and practical skills of both physicists and physical chemists.

In the foregoing, I most emphatically do not wish to sug-gest that there has been no progress in the introduction of the physical sciences into biology. The *Bulletin of Mathematical Biophysics* already runs to many volumes. From time to time useful texts have appeared on the general topic of physics for biologists, but I do not know to what extent such books have gained a real foothold in biology laboratories. Mention, too, must be made of such works as the now considerable series *Progress in Biophysics and Molecular Biology*,* the articles being intended to bridge the gap between the physical and the biological sciences. In these volumes there is undoubtedly a wealth of information for students of morphogenesis, once they can see how to use it! On my assessment, each of the volumes so far published contains at least three or four articles, written by specialists, which could have an application, direct or indirect, to morphogenetic situations. But it will be heavy going, for

* There are, of course, other important journals of biophysics.

teachers and students alike, unless the latter come to the university reasonably well read in the physical sciences. So, in the next ten to twenty years, as the impact of physics on botany gains in force, the nature and quality of the student intake in botany, in some universities at least, may be expected to undergo extensive changes. It may well happen that advances in biophysics will be not unlike those in physics. Here one recalls that Sir Isaac Newton, at the height of his powers, stated that all his life he had been but 'a child gathering pebbles on the seashore'. His success had come from seizing with zest on whatever new knowledge or ideas came his way: the systematic treatment of this knowledge came later!

Plant embryology and related developments

Relative to the zoologists' achievements, *experimental* plant embryology has lagged far behind. There are, however, adequate accounts of the morphological development of embryos in all classes of plants. Thus, when I was working on my book *Embryogenesis in Plants* (1955), from about 1950 onwards, there was an abundance of descriptive morphological literature available, much of it agreeably illustrated; but there was comparatively little on experimental aspects, such as embryo culture. The past ten years have witnessed some astonishing changes. In 1955, among dicotyledonous species, only older embryos, e.g. at the 'heart-shaped' or 'torpedo' stages, had been grown to full size in culture; younger embryos at the globular proembryo stage had still not been successfully cultured. A few years later, however, as a concomitant of work on tissue culture, revolutionary changes began to take place in what may be regarded as the general field of plant embryology; and the whole subject is now advancing with gathering impetus. In fact, the classical aim of Haberlandt, 1902, to grow single cells in pure culture has been achieved. Observers in different parts of the world have now reported the growth of isolated cells in pure culture and the formation of embryoids or pseudo-embryos, these more or less closely resembling stages in the normal embryogenesis *in ovulo*. Some of these embryoids have grown into normal adult plants.

The successful culture of single cells and of simple filaments and cell clusters encouraged embryologists to renew their attempts to grow very young excised embryos in various media. In fact, this has now been done in a number of species. From a perusal of contemporary botanical journals from different parts of the world, one can hardly avoid the conclusion that embryo and embryoid culture is becoming a very competitive subject! If to these achievements one adds such novel observations as the formation, in *Ranunculus sceleratus*, of very embryo-like embryoids, capable of developing into adult plants, from epidermal cells of cultured stem pieces, it will be seen that experimental plant embryology has at length 'got itself right off the ground'.

So, what of the future? In my view, there is wonderful scope here for young investigators, especially if they come to the subject equipped with new techniques in physiology, cyto-chemistry and biochemistry, and, I would add, with a liberal botanical education and an open mind. Many kinds of new investigational work, some of which could have applied aspects of the greatest importance in horticulture and agriculture, can be done either on excised embryos or on embryoids. In paren-thesis, for the time being, some suspension of judgment regard-ing the similarities and dissimilarities between true embryos and embryoids seems desirable until further critical information has been obtained. In the new phase of work, metabolic studies, together with cytochemistry and electron microscopy will un-doubtedly have an important place. One of the earliest and most significant morphogenetic events in the normal develop-ment of the zygote—the establishment of polarity—is still not understood. Polarity is also established at an early stage in em-bryoids. But the underlying cause (or causes), whether related to physico-chemical, i.e. metabolic processes, or to some struc-tural feature in the cytoplasm, or to a combination of both, has still to be ascertained.

Now that individual cells can be grown under controlled conditions and, if necessary, under continuous observation, with photographic recording of the changes that take place in them, many new investigations become possible. So far, only a very small number of species has been examined experimentally. As the range of species examined widens, new and surprising developments are to be expected; for, as in the past, the dis-covery of specially favourable materials for the application of the new techniques will almost certainly have its impact on future progress. The incorporation of various substances into the nucleus and cytoplasm has already been attempted with interesting results. Indeed, the amount of new observation is already considerable, e.g. on the effects of applying various tritiated substances which may be incorporated in the chromosomes, organelles, etc., or of substances which inhibit, or disorganize, some particular reaction in the nucleus or the cytoplasm. It is, therefore, reasonable to assume that the application of these and other techniques to single cells, embryoids and young embryos, growing under controlled conditions and under sustained observation must, in time, add greatly to our knowledge of the processes involved in cellular growth, differentiation and organization. There are some contemporary botanists who consider that, eventually, by special chemical or physical treatments of cells, or a combination of both, we shall be able to induce new, viable entities—in fact, mutants—at will. The production of neomorphs in aquatic species of Umbelliferae, e.g. in *Oenanthe* spp., as studied by Waris and his co-workers, may perhaps be regarded as preliminary indications of much more far-reaching new developments to be achieved in the future. Obviously, if we are thinking of the new discoveries that may be made in the course of decades, we must indeed keep a very open mind. Sometimes, in the light of recent technological innovations and successes, almost any new development seems possible. But Nature, like Art, is long. The specific organization of an embryonic cell of any species is a very special thing: it is, in fact, a unique, complex physical system, the result of a *very* long evolutionary process. We know that, by interfering with developmental processes, e.g. by chemical or radiation treatments of individual embryonic cells or regions, a very considerable range of monstrosities, teratologies or morphological disorganizations, can be induced. Some of these may have little or no survival value in Nature; but, as in evolutionary processes, many mutants, hybrids, polyploids, etc. can, and do, survive and some of those induced by experimental treatments can also be maintained from generation to generation. For example, by breeding and selection, many of our major crop plants have been so changed as to be quite unlike their wild ancestral forms—where these, indeed, are known! These phenomena have important morphogenetic aspects and for younger research

workers who are interested there are major opportunities both
in the fundamental and applied fields. Personally, I expect that
there will be many exciting new developments, of a kind that
we are not yet in a position to predict; but, subject to the
proviso that small differences in the genetic constitution are
sometimes attended by very striking morphological differences,
e.g. the peloric to the zygomorphic flower and *vice-versa*, I
would be surprised if any neomorphs that may be induced
greatly transcend the morphological range inherent in the
slowly evolved specific organization.

Floral studies

For more than a hundred years, botanists have tackled problems of floral morphology either by observations on floral
construction in the interests of taxonomy, or by morphological
and anatomical studies, typically descriptive, to illustrate the
course of development. Teratologies were of interest because
they showed what could happen when departures from the
normal had been induced, either by genetical factors, as in
horticultural species, or by external agents. The overall effect
of this period of study was a very considerable accumulation of
descriptive information—the basis of systematic studies. Plant
breeding and genetical studies over the past decades have
greatly enlarged our understanding of the extent to which the
floral morphology of the parental wild type can be modified.
During more recent years, new impetus has been given to floral
studies by observation of the onset of the reproductive phase
under various controlled light and other conditions, and by the
induction of flowering in tissue cultures—a notable achievement
of recent date and of great potential value. But, notwithstanding the considerable efforts of physiologists, over some thirty
years, we still have not isolated and identified a specific 'florigenic' substance, though there are indications that, in some
species, certain substances, such as the gibberellins, are very
closely involved. Meanwhile, rightly or wrongly, it is my
impression that, in these days of specialization, studies of floral
morphogenesis have had a rather restricted appeal and that
there are contemporary botanists who have never had an
adequate introduction to the subject. Unwittingly, or wilfully,
they turn their backs on the most numerous and highly evolved

materials in the Plant Kingdom. One can, however, detect a dawning recognition of the vast potentialities of this field of endeavour, and already there are confirmed experimental observations on the conditions that determine the change from the vegetative to the reproductive phase, observations of differential growth in the developing flower and its organs, some histochemical studies of the floral meristem, and comparative observations on various related materials in which the genetical constitutions have been explored.

Of course, I do not pretend to predict what will take place in the future, but, sooner or later, botanists must see that, in the development of the flower or inflorescence in a quarter of a million species, there is a vast reservoir of materials for many kinds of new investigations. Perhaps it is the sheer wealth of materials that is so 'off-putting'! Let us try to meet the difficulty noting, in passing, that floristic studies are still being undertaken in a massive way by groups of taxonomists in different major regions. It is desirable, indeed essential, that we should have these basic surveys, but one may doubt if they will be regarded as the pulsating heart-blood of the science by a new generation of botanists, especially if their university courses have equipped them with all kinds of elegant experimental techniques in biochemistry, histochemistry, cytochemistry, electron microscopy, and so on.

Let us accept that, professional taxonomists apart, the great mass of morphological-anatomical detail relating to the reproductive phase of Angiosperms will be unacceptable to a majority of botanists if it is presented in an encyclopaedic fashion. There is vastly too much of it. But it seems imperative to do something about this extensive and wonderful range of morphogenetic materials. So, to begin at the beginning, it has been established that, typically (though not absolutely), at a certain stage in ontogenesis, and under particular environmental conditions—cold, duration of light, etc.—critical changes are induced in the vegetative shoot apex and the transition to the flowering phase has been stimulated or effected. The subsequent growth developments in the inflorescence and in individual flowers exemplify specificity, differential growth, the sequential formation of the several groups of organs, correlation and a holistic overall harmony of development which, at maturity, endows the flower with its several special functional characteristics. (*See* C. W. Wardlaw's *Morphogenesis in Plants*, 1968, for details.)

Against this background of superabundantly varied, complex and still evolving materials, which lend themselves to many kinds of experimental treatments, it is difficult to predict what new discoveries will be made. At the scholarly level, I think what is most needed, to set our house in order and to encourage potential research workers in the future, is *an adequate general theory of floral development*. This theory must attempt to deal effectively, i.e. biochemically, biophysically and organizationally, with at least three major aspects: (i) the harmonious transition from the vegetative to the reproductive phase; (ii) the specific and orderly inception and formation of the several groups of floral organs; (iii) the very great diversity and large numbers of angiosperm species, on the one hand; and the many parallelisms of evolutionary development, in different families and orders, on the other.

I have already expressed my thoughts on some of the constituent aspects of a general theory and how they may be approached in a practical way, (*See Organization and Evolution in Plants*, 1965, and some of the essays in this volume). It remains to be seen what the future brings. But at least a major integrating theme, however inadequate, has been proposed. Any theory, of course, has to be judged on its merits and, in the end, by its usefulness. What are the merits of the theory proposed? Apart from being a means of extracting the scientific essence from a vast and important assemblage of materials, I think they also include: (i) the necessary association and integration of morphological, physiological, environmental, genetical and organismal (ontogenetical) factors during the development of individual plants of selected species; (ii) a basis for understanding (*a*) the great diversity and specificity of flowering plants and (*b*) their many similarities; (iii) avoidance of the insoluble difficulties associated with earlier work and ideas on comparative morphology, e.g. how one adult organ could be *transformed* into another: now, in the views which I have presented (Essay No. 18, etc.), we need only think in terms of growth centres, each yielding a characteristic organ according to the metabolism of its reaction system; (iv) making available, without encyclopaedic treatment, a working insight into the most numerous, highly evolved, elaborate and prevalent morphological materials in the Plant Kingdom.

But this is only *one* possible beginning!

Special adaptations

Even if I had the knowledge and the capacity, it is not my intention in this brief concluding essay to comment on all the possible developments in morphogenesis. My aim, rather, is to show by some selected examples that the subject can be extended in many different directions. What will be achieved in the future will, I am sure, greatly surpass anything so far ascertained. However, I do wish to touch on an aspect of morphogenesis which, so far, has perhaps received too little attention, but in which there are unquestionably major opportunities for new thought and work. I refer to the numerous examples of special adaptation in plants, e.g. as seen in insect-pollinated flowers, in the trapping mechanisms of insectivorous plants, the many curious morphological developments in species with irregular nutrition, the special features in aquatic and xeromorphic plants, and so on. It is probable, though one can never be sure, that there are many botanists who now take these 'special adaptions' for granted: they appear to be self-evident and indisputable, like the eye for seeing, wings for flying, and so on. But there are others, like myself, who also subscribe to the theory of evolution and who are not lacking in an appreciation of the ingenious and compelling arguments of Darwin and of the neo-Mendelians or neo-Darwinians, yet who are considerably less sure that our understanding of these biological developments is in any sense adequate. It is not so much that they are opposed to a general acceptance of evolutionary theory but rather that they have a kind of deep unbelief. This is not, I think, simply due to a 'suspension of judgment', as T. H. Huxley enjoined on his contemporaries. Rather it is a critical, cautious and detached attitude to a region of biological scholarship in which much fuller information is needed. In my experience this attitude appears to be the result of long cogitation on the facts of development, i.e. on an insight into morphogenetic processes. One becomes acutely and critically aware of great and enigmatic gaps in our knowledge. As several eminent botanists have said to me, or have explicitly stated in their writings, each in his own way: how does it happen that, during the development of a specially adapted organ, or indeed of any part of a plant, the 'right thing always seems to happen at the right time in the right place'? Another aspect of the same general thought takes the form: how does it come about that the *initial pattern of growth and differentiation* of an organ or a tissue in an embryonic region, which must be determined by the nexus of genetical-organismal-environmental factors *at work at that time in that region*, is such that, in the *adult stage*, the organ or tissue has the highly functional, i.e. specially adaptive, properties essential to the life and survival of the species?

There are evidently some very considerable gaps in our knowledge, and thinking, not to be filled by arguments alone, but by new facts or a new insight. Some of the gaps may, I think, be at least partly closed by comprehensive morphogenetic investigations. Wherever such investigations have been undertaken, no matter how fully the materials had been examined by earlier generations of morphologists, new and often surprising facts, inferences and conclusions have emerged. Witness our contemporary information and views on the differentiation of vascular tissues as compared with those held no more than thirty years ago!

In conclusion

There is no finality in biology: our scholarship and research are always in continuation and there are always new and interesting things to do. Contemporary botany keeps moving on, usually in the form of increasingly specialized investigations, and there is, accordingly, the ever-present problem of incorporating adequately the new information into the general corpus of knowledge; for a new discovery, however important it may be, still gains much in value when it is brought within the general ambiance of botanical scholarship. So, if we maintain the ideal that botany is the science of plants, and not simply of selected biological processes, studies in morphogenesis must also include the investigation of 'whole plant morphogenesis'. This cannot be other than an exercise of great complexity. To make the mass of relevant information manageable, we may well have to resort to symbols and formulae and to new, preferably simplified, methods of statement. If this work is to move forward, both enlightened leadership and co-operative effort will be essential. For me, the *beau ideal* of the botanist of the future would be one who contributed to the science of plants both by the quality of his individual research and by the liberality of his scholarly outlook.

Notes and references

Essay 3

AVERY, G. S. (1940). *Growth*, Suppl. for 1940.

BARY, A. DE (1884). *Comparative Anatomy of the Phanerogams and Ferns* (English trans. by F. O. Bower and D. H. Scott). Oxford.

BOWER, F. O. (1889). *Ann. Bot. Lond.*, **3,** 305.—(1890–1). **5,** 109.—(1908). *Origin of a Land Flora.* London.—(1921). *Proc. Roy. Soc. Edin.* **41,** 1.—(1922). **43,** 1.—(1923). *The Ferns*, Vol. 1. Camb. Univ. Press.—(1930). *Size and Form in Plants.* London.—(1935). *Primitive Land Plants.* London.—(1937). *Bot. Mag. Tokyo*, **51,** 183.

BRUCHMANN, H. (1909). *Flora*, **99.**—(1909). **99.**—(1910). **100.**

BUDER, J. (1928). *Ber. dtsch. bot. Ges.* **46.**

CAMPBELL, D. H. (1890). *Bot. Gaz.* **15,** 1.—(1921). *Amer. J. Bot.* **8,** 303.—(1940). *The Evolution of Land Plants.* Stanford Univ. Press.

CHILD, C. M. (1915). *Individuality in Organisms.* Chicago.—(1941). *Patterns and Problems of Development.* Chicago Univ. Press.

CONARD, H. (1908). Carnegie Inst. Washington.

DINGLER, H. (1882). *Über das Scheitelwachsthum des Gymnospermen-Stammes.*—(1886). *Ber. dtsch. bot. Ges.* **4,** 18.

DOULIOT, H. (1890). *Ann. Sci. Nat.* (*Bot.*), Sér. **7,** 11, 283.—(1891). Sér. **7,** 12, 93.

ERRERA, L. (1886). *C.R. Soc. Biol. Paris*, **103,** 822; *Bull. Soc. belge Micr.* **13.**

FARMER, J. B. (1891). *Ann. Bot. Lond.* **5,** 37.

FOSTER, A. S. (1939). *Bot. Rev.* **5,** 454.

GOEBEL, K. (1880–1). *Bot. Z.*—(1900). *Organography of Plants* (Engl. Trans. by I. B. Balfour). Oxford.—(1908). *Einleitung in die experimentelle Morphologie der Pflanzen.* Leipzig.—(1926). *Wilhelm Hofmeister* (Engl. Trans. by H. M. Bower). Ray Soc. London.—(1928). *Organographie der Pflanzen*, **3,** Aufl. Jena.

HABERLANDT, G. (1914). *Physiological Plant Anatomy* (Engl. trans. by J. M. F. Drummond). London.

HANSTEIN, J. (1868). *Festschr. niederrhein Ges. Natur.-u. Heilkunde usw.*, Bonn, pp. 109–43.

HARTEL, K. (1938). *Beitr. Biol. Pfl.* **25,** 125.

AA

HEGELMAIER, F. (1872). *Bot. Z.* **30.**—(1874). *Bot. Z.* **32.**

HELM, J. (1923). *Planta*, **15,** 105.

HOFMEISTER, W. (1857). *Abh. Kön. sächs. Ges. Wiss.* **3,** 603.—(1862). *Higher Cryptogamia.* Ray Soc., London.—(1863). *Jb. wiss. Bot.* **3,** 272; *Handb. Physiol. Bot.* **1,** 129.—(1868). *Allgemeine Morphologie.* Leipzig.

HOLLOWAY, J. E. (1939). *Ann. Bot. Lond.,* N.S., **111,** 324.

JOST, L. (1891). *Bot. Z.* **36,** 593.

KAPLAN, R. (1937). *Planta*, **27,** 224.

KLEIN, L. (1884). *Bot. Z.,* p. 577.

KNY, L. (1875). *Nova Acta Leop. Carol.* **37,** 1.

KOCH, L. (1891). *Jb. wiss. Bot.* **22,** 491.

KORSCHELT, P. (1884). *Jb. wiss. Bot.* **15,** 642.

LANG, W. H. (1913–15). *Ann. Bot. Lond.,* **27–9.**—(1915*a*). Pres. Address Brit. Ass. Adv. Sci. Sect. K, Manchester.—(1915*b*). *Mem. Proc. Manchr. Lit. Phil. Soc.* **59.** (1924).

LUDWIGS, K. (1911). *Flora,* N.F., **3,** 385.

MEKEL, J. C. (1933). *Réc. trav. bot. neérl.* **30,** 627.

MILDE, J. (1867). *Nova Acta Leop. Carol.* **32.**

NAEGELI (& SCHLEIDEN) (1844–6). *Z. wiss. Bot.*

NAEGELI, C. (1878). *Bot. Z.* **36,** 124.

PRAEGER, R. L. (1934). *J. Bot. Lond.* **72.**

PRIESTLEY, J. H. (1928). *Biol. Rev.* **3,** 1.—(1929). *New Phytol.* **28,** 54.

PRIESTLEY, J. H. & SCOTT, L. I. (1933). *Biol. Rev.* **8,** 241.

PRINGSHEIM, N. (1869). *Mber. K. preuss. Akad. Wiss.* pp. 92–115.

ROSTOWZEW, S. (1892). *Beiträge zur Kenntniss d. Ophioglosseen,* **1,** Moscow.

SACHS, J. (1878). *Arb. bot. Inst. Würzburg,* **2,** 46.—(1879). *Arb. bot. Inst. Würzburg,* **2,** 2.—(1887). *Lectures on the Physiology of Plants* (English trans. by H. Marshall Ward). Oxford.—(1890). *A History of Botany* (Engl. trans. by H. E. F. Garnsey). Oxford.

SAHNI, N. (1917). *New Phytol.* **16,** 1.

SCHLEIDEN, M. J. (1842). *Grundzüge der wissenschaftlichen Botanik.* Leipzig.

SCHMIDT, A. (1924). *Bot. Arch.* **8,** 345.

SCHOUTE, J. C. (1902). *Die Stelar Theorie.* Jena.

SCHÜEPP, O. (1926). Meristeme. Linsbauer's *Handbuch der Pflanzenanatomie,* **4.** Berlin.—(1938). *Biol. Rev.* **13,** 59.

SCHWENDENER, S. (1879). *S.B. Ges. Naturf. Freunde.* (Also in Schwendener's *Gesamm. Bot. Mitt.* **2,** 47, 1898.)—(1880). *Mber. K. preuss. Akad. Wiss.* **3**A. Berlin.—(1885). *S.B. preuss. Akad. Wiss.*

SINNOTT, E. W. (1938). *Bot. Gaz.* **99,** 803.

SOLMS-LAUBACH, H. (1902). *Bot. Z.* p. 179.

STENZEL, K. G. (1861). *Nova Acta Leop. Carol,* **28.**

STRASBURGER, E. (1872). *Die Coniferen und die Gnetaceen.* Leipzig.—(1873). *Bot. Z.* (Summary in De Bary's *Comparative Anatomy,* p. 21.)

TANSLEY, A. G. (1907). *New Phytol.* **6,** 150.

THIMANN, K. V. & SKOOG, F. (1934). *Proc. Roy. Soc.* B, **114.**

THOMPSON, D'ARCY W. (1942). *On Growth and Form.* Camb. Univ. Press.

WAND, A. (1914). *Flora,* **106,** 237.

WARDLAW, C. W. (1925). *Phil. Trans. Roy. Soc. Edin.* **54,** 281.—(1943).

Ann. Bot. Lond., N.S. **7**, 171.—(1943*a*). **7**, 357.—(1944). **8.**—(1944*a*). **8.**—(1944*b*). *Nature, Lond.* **153**, 588.—(1945). *Ann. Bot. Lond.*, N.S., **9.**—(1945*a*). **9.**

WHITE, P. R. (1944). *A Handbook of Plant Tissue Culture.* Lancaster, Penn.

WILLIAMS, S. (1931). *Trans. Roy. Soc. Edin.* **57**, 1.—(1933). **57**, 29.—(1937). *Nature, Lond.* **139**, 966.—(1938). Experimental Morphology (in Verdoorn's *Manual of Pteridology*, pp. 105–40, The Hague).

WOLFF, K. F. (1759). *Theoria Generationis.* Halle.

Essay 6

[1] SNOW, M. & SNOW, R. (1931). *Phil. Trans.* B, **221**, 1.—(1933). **222**, 353.—(1935). **225**, 63.—(1947), *New Phytol.* **46**, 15.

[2] WARDLAW, C. W. (1948). *Ann. Bot.*, N.S., **21**, 97.—(1949). **13**, 163.

[3] WARDLAW, C. W. (1949). *Phil. Trans.* B, **233**, 415.

[4] SCHOUTE, J. C. (1913). *Réc. trav. bot. néerl.* **10**, 153.

[5] RICHARDS, F. J. (1948). Symposium on Growth, S.E.B., No. 2. Camb. Univ. Press.

[6] BÜNNING, E. (1948). Entwicklungs- und Bewegungs-Physiologie der Pflanze. Springer, Berlin.

[7] WARDLAW, C. W. (1943). *Ann. Bot.*, N.S., **7**, 171.

Essay 8

ARBER, A. (1950). *The Natural Philosophy of Plant Form.* Camb. Univ. Press.

BOWER, F. O. (1935). *Primitive Land Plants.* London.

CHILD, C. M. (1941). *Patterns and Problems of Development.* Chicago Univ. Press.

DRIESCH, H. (1908). *The Science and Philosophy of the Organism* (Gifford Lectures). London.

GOEBEL, K. (1900). *Organography of Plants.* (Engl. trans.) Clarendon Press. Oxford.

HERSCH, A. H. (1941). Allometric growth: the ontogenetic and phylogenetic significance of differential rates of growth. *Growth*, 3rd Suppl., 113–45.

LANG, W. H. (1915). *Causal and phyletic morphology.* Presidential Address, British Assoc. Advancement of Sci. Sect. K. (Manchester).

MATHER, K. (1948). Nucleus and cytoplasm in differentiation. Symposium, Soc. Exper. Biol. (*Growth*) **2**, 196–216.

NEEDHAM, J. (1942). *Biochemistry and Morphogenesis.* Camb. Univ. Press.

NORTHROP, F. S. C. & BURR, H. S. (1937). Experimental findings concerning the electrodynamic theory of life and an analysis of their physical meaning. *Growth*, **1**, 78–88.

PRAT, H. (1945). Les gradients histo-physiologiques et l'organogenèse végétale. Contrib. l'Inst. Bot. Univ. Montreal, Canada **58**, 1–151.

SINNOTT, E. W. (1946). Substance or system: the riddle of morphogenesis. *Amer. Nat.* **80**, 497–505.

SMUTS, J. (1921). *Holism and Evolution.* London.

THOMPSON, D'ARCY W. (1917, 1942). *Growth and Form*. Camb. Univ. Press.

WADDINGTON, C. H. (1939). *An Introduction to Modern Genetics*. London.

WADDINGTON, C. H. (1948). The genic control of development: Symposium, Soc. Exper. Biol. (*Growth*) **2,** 145–176.

WARDLAW, C. W. (1949). Experiments on organogenesis in ferns. *Growth* (Suppl.), **13,** 93–131.

WEISS, P. (1939). *Principles of Development*. New York.

WRIGHT, S. (1945). Genes as physiological agents. *Amer. Nat.* **79,** 289–303.

ZIMMERMANN, W. (1938). *Die Telomtheorie*. Der Biologe, 7th year, 12, 385.

Essay 9

BALL, E. (1948). *Soc. Exp. Biol. Symp.* Camb. Univ. Press.

CROSS, G. L. & JOHNSON, T. J. (1941). *Bull. Torrey Bot. Cl.* **68,** 618.

ESAU, K. (1943). *Bot. Rev.* **9,** 125.

FLEET, D. S. VAN (1948). *Amer. J. Bot.* **35,** 219.

FOSTER, A. S. (1949). *Practical Plant Anatomy*. New York: Van Nostrand.

OVERBEEK, J. VAN (1935). *Proc. Nat. Acad. Sci. Wash.* **21,** 292.—(1938). *Plant Physiol.* **13,** 587.

RANDOLPH, L. F., ABBÉ, E. C. & EINSET, J. (1944). *J. Agric. Res.* **69,** 47.

SINNOTT, E. W., HOUGHTALING, H. & BLAKESLEE, A. F. (1934). *Publ. Carneg. Instn*, no. 451, p. 1.

SKOOG, F. & TSUI, C. (1948). *Amer. J. Bot.* **35,** 782.

SNOW, M. & SNOW, R. (1931–5). *Phil. Trans.* B, **221,** 1; **222,** 353; **225,** 63.—(1948). *Soc. Exp. Biol. Symp.* **2.**

WARDLAW, C. W. (1944). *Ann. Bot. Lond.*, N.S., **8,** 387.—(1947). *Phil. Trans.* B, **232,** 343.—(1950). *Phil. Trans.* B, **234,** 583.

WOODGER, J. H. (1946). *Essays on Growth and Form*. Camb. Univ. Press.

Essay 10

ARBER, A. (1950). *Natural Philosophy of Plant Form*. Camb. Univ. Press.

DE BEER, G. R. (1951). *Embryos and Ancestors*. Oxford Univ. Press.

GOEBEL, K. (1922). Gesetzmässigkeiten in Blattaufbau. *Bot. Abh.* **1,** Jena.

HARLAND, S. C. (1936). The genetical conception of the species. *Biol. Rev.* **11,** 83.

THODAY, D. (1939). The interpretation of plant structure. Pres. Addr. (Sect. K), Brit. Assoc. Adv. Sci. (also *Nature, Lond.* **144,** 571).

TURING, A. M. (1952). The chemical basis of morphogenesis. *Phil. Trans.* B, **237,** 37.

WARDLAW, C. W. (1951). Organization in plants. *Phytomorphology*, **1,** 22.

Essay 11

BALL, E. (1949). The shoot apex and normal plant of *Lupinus albus* L., bases for experimental morphology. *Amer. J. Bot.* **36,** 440.

BLAKESLEE, A. F., BERGNER, A. D., SATINA, S. & SINNOTT, E. W. (1939). Induction of periclinal chimaeras in *Datura stramonium* by colchicine treatment. *Science*, **89,** 402.

BOWER, F. O. (1948). *Botany of the Living Plant* (4th edn.). London. Macmillan.

CROSS, G. L. & JOHNSON, T. J. (1941). Structural features of the shoot apices of diploid and colchicine-induced tetraploid strains of *Vinca rosea* L. *Bull. Torrey Bot. Cl.* **68,** 618.

FOSTER, A. S. (1943). Zonal structure and growth of the shoot apex in *Microcycas calocoma. Amer. J. Bot.* **30,** 56.

FRITSCH, F. E. (1945). *The Structure and Reproduction of the Algae.* II. Camb. Univ. Press.—(1945). Observations on the anatomical structure of the *Fucales.* I. *New Phytol.* **44,** 1.

GOEBEL, K. (1897). *Organographie der Pflanzen,* 1st edn., 2 vols. Jena (Engl. edn. by I. B. Balfour as *Organography of Plants, especially of the Archegoniatae and Spermaphyta.* Oxford Univ. Press, 1900.)

JOHNSON, M. A. (1951). The shoot apex in gymnosperms. *Phytomorph.* **1,** 188.

LANG, W. H. (1915). Phyletic and causal morphology. Pres. Addr. (Sect. K.). *Brit. Assoc. Adv. Sci.*

LANKESTER, E. R. (1870). On the use of the term homology in modern zoology and the distinction between homogenetic and homoplastic agreements. *Ann. Mag. Nat. Hist.* **4,** ser. 6.

MANTON, I. (1950). *Problems of Cytology and Evolution in the Pteridophyta.* Camb. Univ. Press.

OVERBEEK, J. VAN (1935). The growth hormone and the dwarf type of growth in corn. *Proc. Nat. Acad. Sci. Wash.* **21,** 292.

OVERBEEK, J. VAN (1938). 'Laziness' in maize due to abnormal distribution of growth hormone. *J. Hered.* **29,** 339.—(1938). Auxin production in seedlings of dwarf maize. *Plant Physiol.* **13,** 587.

PRIESTLEY, J. H. & SCOTT, L. I. (1933). Phyllotaxis in the dicotyledon from the standpoint of developmental anatomy. *Biol. Rev.* **8,** 241.

RANDOLPH, L. F., ABBÉ, E. C. & EINSET, J. (1944). Comparison of shoot apex and leaf development and structure in diploid and tetraploid maize. *J. Agric. Res.* **69,** 47.

SATINA, S., BLAKESLEE, A. F. & AVERY, A. G. (1940). Demonstration of the three germ layers in the shoot apex of *Datura* by means of induced polyploidy in periclinal chimeras. *Amer. J. Bot.* **27,** 895.

SCHUEPP, O. (1914). Wachstum und Formwechsel des Sprossvegetationspunktes der Angiospermen. *Ber. dtsch. Bot. Ges.* **32,** 328.

SNOW, M. & SNOW, R. (1931). Experiments on phyllotaxis. I. The effect of isolating a primordium. *Phil. Trans.* B, **221,** 1.— (1933). Experiments on phyllotaxis II. The effect of displacing a primordium. *Phil. Trans.* B, **222,** 353.—(1935). Experiments on phyllotaxis. III. Diagonal splits through decussate apices. *Phil. Trans.* B, **225,** 63.—(1947). On the determination of leaves. *New Phytol.* **46,** 15.

STANT, M. Y. (1952). The shoot apex of some monocotyledons. 1. Structure and development. *Ann. Bot.,* N.S. **16,** 115.

SUSSEX, I. M. (1951). Experiments on the cause of dorsiventrality in leaves. *Nature, Lond.* **167,** 651.

THOMPSON, D'ARCY W. (1917, 1942). *On Growth and Form.* (1st and 2nd edns.) Camb. Univ. Press.

TURING, A. M. (1952). The chemical basis of morphogenesis. *Phil. Trans.* B, **237,** 37.

WARDLAW, C. W. (1925). Size in relation to internal morphology. 2. The vascular system of *Selaginella*. *Trans. Roy. Soc. Edin.* **54,** 281.—(1948). Experimental and analytical studies of pteridophytes. 13. The shoot apex of a tree fern, *Cyathea Manniana. Ann. Bot.*, N.S., **12,** 371.—(1949). Further experimental observations on the shoot apex of *Dryopteris aristata* Druce. *Phil. Trans.* B, **233,** 415.—(1949). Experiments on organogenesis in ferns. *Growth* (Suppl.), **13,** 93.—(1949). Experimental and analytical studies of pteridophytes. 14. Leaf formation and phyllotaxis in *Dryopteris aristata* Druce. *Ann. Bot.*, N.S., **13,** 163.—(1950). The comparative investigations of apices of vascular plants by experimental methods. *Phil. Trans.* B, **234,** 583.—(1950). Comparative morphogenesis in pteridophytes and spermatophytes by experimental methods. *Rep. Proc. Int. Bot. Congr.* Stockholm; also *New Phytol.* **50,** 127 (1951).—(1950). Experimental and analytical studies of pteridophytes. 16. The induction of leaves and buds in *Dryopteris aristata* Druce. *Ann. Bot.*, N.S., **14,** 435.—(1952). Experimental and analytical studies of pteridophytes. 18. The nutritional status of the apex and morphogenesis. *Ann. Bot.*, N.S., **16,** 207.

Essay 12

BUCHHOLZ, J. T. (1946). Volumetric studies of seeds, endosperms, and embryos in *Pinus ponderosa* during embryonic differentiation. *Bot. Gaz.* **108,** 232.

CHAMBERLAIN, C. J. (1935). *Gymnosperms.* Chicago Univ. Press.

FAHN, A. (1953). The origin of the banana inflorescence. *Kew Bull.* no. 3, 299.

GARRETT, S. D. (1953). Rhizomorph behaviour in *Armillaria mellea* (Vahl) Quel. 1. Factors controlling rhizomorph initiation by *A. mellea* in pure culture. *Ann. Bot. Lond.*, N.S., **17,** 63.

GORTER, C. J. (1949). The influence of 2,3,5-triiodobenzoic acid on the growing points of tomatoes. *Proc. Acad. Sci. Amst.* **52,** 1185.

GORTER, C. J. (1951). The initiation of ring-fasciations. *Proc. Acad. Sci. Amst.* **54,** 181.

HARRISON, Y. HESLOP. (1953). *Nuphar intermedia* Ledeb., a presumed relict hybrid. *Watsonia,* **3,** 7.

HEIJNINGEN, W. VAN, BLAAU, A. H. & HARTSEMA, A. M. (1928). Gevolgen van Röntgenbestralingen bij Tulpenbollen. *Verh. Akad. Wet. Amst.* **26,** 1.

MARSDEN-JONES, E. M. (1933). *Proc. Linn. Soc. (Bot.),* **50,** 39.

MOSS, E. H. (1924). Fasciated roots of *Caltha palustris. Ann. Bot. Lond.* **38,** 789.

PERJE, A. M. (1952). Some causes of variation in *Ranunculus ficaria* L. *Arkiv. Bot.* **2,** 251.

RIJVEN, A. H. G. C. (1950). *In vitro* studies on the embryo of *Capsella bursa-pastoris.* N. Holland Publ, Co. Amsterdam.

SANDERS, M. E. (1950). Development of self and hybrid embryos in artificial culture. *Amer. J. Bot.* **37,** 6.

TURING, A. M. (1952). The chemical basis of morphogenesis. *Phil. Trans.* B, **237,** 37.

DE WAARD, J. & ROODENBURG, J. W. M. (1948). *Proc. Acad. Sci. Amst.* **51,** 248.

WARDLAW, C. W. (1928). Size in relation to internal morphology. 3. The vascular system of roots. *Trans. Roy. Soc. Edin.* **54,** 281.—(1952). *Phylogeny and Morphogenesis.* London: Macmillan.—(1953*a*). A commentary on Turing's diffusion reaction theory of morphogenesis. *New Phytol.* **52,** 40.—(1953*b*). Action of triiodobenzoic and trichlorobenzoic acids in morphogenesis. *New Phytol.* **52,** 210.

Essay 13

[1] WARDLAW, C. W. (1949). *Growth* (Supp.), **13,** 93.
[3] CUTTER, E. G. (1954). *Nature,* **173,** 440.
[3] WARDLAW, C. W. & CUTTER, E. G. (1954). *Nature,* **174,** 734.
[4] SKOOG, F. (1954). Brookhaven Symp. Biology, **6,** 1.

Essay 14

ALLSOPP, A. (1953). Experimental and analytical studies of pteridophytes. XXI. Investigations on *Marsilea*. 3. The effect of various sugars on development and morphology. *Ann. Bot. Lond.,* N.S., **17,** 447.

BALL, N. G. (1953). The effects of certain growth-regulating substances on the rhizomes of *Aegopodium podograria*. *J. Exp. Bot.* **4,** 349.

BENNET-CLARK, T. A. & BALL, N. G. (1951). The diageotropic behaviour of rhizomes. *J. Exp. Bot.* **2,** 169.

BONNER, J. T. (1952). *Morphogenesis; an Essay on Development.* Princeton Univ. Press.

GOEBEL, K. (1922). Gesetzmässigkeiten in Blattaufbau. *Bot. Abh.* **1,** Jena.

HERSCH, A. H. (1941). Allometric growth: the ontogenetic and phylogenetic significance of differential rates of growth. *Growth* (Suppl.), p. 113.

NEEDHAM, J. (1942). *Biochemistry and Morphogenesis.* Camb. Univ. Press.

SCHOUTE, J. C. (1936). Fasciation and dichotomy. *Réc. trav. bot. néerl.* **33,** 649.

THOMPSON, D'ARCY W. (1917, 1942). *On Growth and Form* (1st and 2nd edns.). Camb. Univ. Press.

TURING, A. M. (1952). The chemical basis of morphogenesis. *Phil. Trans.* B, **237,** 37.

WARDLAW, C. W. (1952). *Phylogeny and Morphogenesis.* London: Macmillan. —(1953). Experimental and analytical studies of pteridophytes. XXIII. The induction of buds in *Ophioglossum vulgatum*. L. *Ann. Bot. Lond.,* N.S., **17,** 513.—(1953). A commentary on Turing's diffusion reaction theory of morphogenesis. *New Phytol.* **52,** 40.—(1955). *Embryogenesis in Plants.* London. Methuen.

WETMORE, R. H. (1950). *Rep. Proc. 7th Int. Bot. Congr.* Stockholm.

WOODGER, J. H. (1945). On biological transformations. *Essays on Growth and Form.* Oxford Univ. Press.—(1948). Observations on the present stage of embryology. *Symp. Soc. Exp. Biol.* **2,** 351. Camb. Univ. Press.

Essay 15

[1] WARDLAW, C. W. (1951). *New Phytol.* **50,** 127.
[2] PLANTEFOL, L. (1946). *Ann. Sci. Nat. Bot.,* 11th Ser., **7,** 15, 229;—(1947), **8,** 1, 66;—(1947). *Rev. Gen. Bot.* **54,** 49.
[3] BUVAT, R. (1952). *Ann. Sci. Nat. Bot.,* 11th Ser., **13,** 199.
[4] BUVAT, R. (1955). *Ann. Biol.* **31,** 595.
[5] LOISEAU, J. E. (1954). *C.R. Acad. Sci.* **238,** 149.
[6] LOISEAU, J. E. (1954). *C.R. Acad. Sci.* **238,** 385.
[7] SNOW, R. (1948). *Nature,* **162,** 798.
[8] SNOW, R. (1949). *Nature,* **163,** 332.
[9] RICHARDS, F. J. (1951). *Phil. Trans.* B, **235,** 509.
[10] CAMEFORT, H. (1956). *Ann. Sci. Nat. Bot.,* 11th Ser., **17,** 1.
[11] WARDLAW, C. W. (1953). *New Phytol.* **52,** 195.
[12] SCHOUTE, J. E. (1936). *Réc. trav. bot. Néer,* **33,** 649.
[13] WARDLAW, C. W. (1955). *New Phytol.* **54,** 302.
[14] GIFFORD, E. M., JUN. (1954). *Bot. Rev.* **20,** 477.
[15] NEWMAN, I. V. (1956). *Phytomorph.* **6,** 1.
[16] WARDLAW, C. W. (in the press).
[17] TURING, A. M. (1952). *Phil. Trans.* B, **237,** 37.
[18] WARDLAW, C. W. (1953). *New Phytol.* **52,** 40.
[19] WARDLAW, C. W. (1955). *New Phytol.* **54,** 39.
[20] SEWARD, A. C. (1933). 'Plant Life through the Ages' (Camb. Univ. Press).
[21] MARTENS, P. (1950). *Bull. Sci. Acad. Roy. Belge,* **36,** 811.
[22] PLANTEFOL, L. (1948). *Ann. Sci. Nat. Bot.,* 11th Ser., **9,** 35.
[23] SNOW, M. & R. (1947). *New Phytol.* **46,** 15.
[24] SNOW, R. (1950). *Endeavour,* **14,** 190.
[25] PLANTEFOL, L. (1949). 'L'Ontogénie de la Fleur' (Paris);—(1948). *Ann. Sci. Nat. Bot.,* 11th Ser., **9,** 35.
[26] WARDLAW, C. W. (1947). *Phil. Trans.* B, **232,** 343.
[27] BALL, E. (1948). *Symp. Soc. Exper. Biol.* **2,** 246 (1948).
[28] WARDLAW, C. W. (1949). *Growth Symp.* **9,** 93.

Essay 16

BÜNNING, E. (1952). *Surv. Biol. Progr.* **2,** 105–40.
ESAU, K. (1953). *Plant Anatomy.* New York: Wiley.
RICHARDS, F. J. (1951). *Phil. Trans.* B, **235,** 509–64.
SNOW, M. & R. (1955). *Proc. Roy. Soc.* B, **144,** 222–9.
SUSSEX, I. (1955). *Phytomorph.* **5,** 286–350.
TURING, A. M. (1952). *Phil. Trans.* B, **237,** 37–72.
WARDLAW, C. W. (1952). *Phylogeny and Morphogenesis.* London: Macmillan, 166.

Essay 17

ABBÉ, E. C. & STEIN, O. L. (1954). The growth of the shoot apex in maize: embryogeny. *Amer. J. Bot.* **41,** 285–93.

BAIN, H. F. & DERMEN, H. (1944). Sectorial polyploidy and phyllotaxy in the cranberry (*Vaccinium macro-carpon* Ait.). *Amer. J. Bot.* **31**, 581–7.

BERSILLON, G. (1955). Recherches sur les Papavéracées. Contribution à l'étude du développment des Dicotylédones herbacées. *Ann. Sci. Nat. Bot.* XI. sér. **16**, 225–443.

BÜNNING, E. (1948). Entwicklungs- und Bewegungsphysiologie der Pflanze. Springer, Berlin.—(1952). Morphogenesis in plants. Survey of biological progress. **2**, 105–40. New York.

BUVAT, R. (1952). Structure, évolution et fonctionnement du méristeme apical de quelques Dicotylédones. *Ann. Sci. Nat. Bot.* XI. sér. **13**, 199–300.—(1955). Le méristeme apical de la tige. *Ann. Biol.* **31**, 595–656.

CROSS, G. L. & JOHNSON, T. J. (1941). Structural features of the shoot apices of diploid and colchicine-induced tetraploid strains of *Vinca rosea* L. *Bull. Torrey Bot. Cl.* **68**, 618–35.

CUTTER, E. G. (1956). Experimental and analytical studies of Pterido- phytes. XXXIII. The experimental induction of buds from leaf primordia in *Dryopteris aristata* Druce. *Ann. Bot.*, N.S., **20**, 143–65.

DERMEN, H. (1945). The mechanism of colchicine-induced cytohisto- logical changes in cranberry. *Amer. J. Bot.* **32**, 387–94.—(1947). Periclinal cytochimeras and histogenesis in cranberry. *Amer. J. Bot.* **34**, 32–43.—(1951). Ontogeny of tissues in stem and leaf of cytochimeral apples. *Amer. J. Bot.* **38**, 753–60.—(1953). Periclinal cytochimeras and origin of tissues in stem and leaf of peach. *Amer. J. Bot.* **40**, 154–68.

GIFFORD, E. M., JR. (1954). The shoot apex in angiosperms. *Bot. Rev.* **20**, 477–529.

HACCIUS, B. (1955). Experimentally induced twinning in plants. *Nature*, **176**, 355.

NEWMAN, I. V. (1956). Pattern in meristem of vascular plants. I. Cell partition in living apices and in the cambial zone in relation to the con- cepts of initial cells and apical cells. *Phytomorph.* **6**, 1–19.

PLANTEFOL, L. (1947). Hélices foliaires, point végétatif et stele chez les Dicotylédones. La notion d'anneau initial. *Rev. Gén. Bot.* **54**, 49–80.—(1951). La phyllotaxie. Colloques internationaux. Morphogènese **28**, 447–60.

RANDOLPH, L. F., ABBÉ, E. C. & EINSET, J. (1944). Comparison of shoot apex and leaf development and structure in dipoid and tetraploid maize. *J. Agr. Res.* **69**, 47–76.

RICHARDS, F. J. (1948). The geometry of phyllotaxis and its origin. *Symp. Soc. Exper. Biol.* **2**, 217–45.—(1951). Phyllotaxis: its quantitative expression and relation to growth in the apex. *Phil. Trans.* B, **235**, 509–64.

SATINA, S., BLAKESLEE, A. F. & AVERY, A. G. (1940). Demonstration of the three germ layers in the shoot apex of *Datura* by means of induced poly- ploidy in periclinal chimeras. *Amer. J. Bot.* **27**, 895–905.—(1941). Periclinal chimeras in *Datura stramonium* in relation to development of leaf and flowers. *Amer. J. Bot.* **28**, 862–71.—(1943). Periclinal chimeras in *Datura* in relation to the development of the carpel. *Amer. J. Bot.* **30**, 453–62.

SCHOUTE, J. C. (1913). Beiträge zur Blattstellungslehre. I. Die Theorie. *Rec. Trav. Bot. Néerlandais*, **10**, 153–339.—(1936). Fasciation and dichotomy. *Réc. Trav. bot. néerl.* **33**, 649–69.

SKOOG, F. (1954). Substances involved in normal growth and differentiation. Abnormal and pathological plant growth. *Brookhaven Symp. Biol.* **6,** 1–21.

STEIN, O. L. (1955). Rates of leaf initiation in two mutants of *Zea mays,* Dwarf-1 and Brachytic-2. *Amer. J. Bot.* **42,** 885–92.—& KROMAN, R. A. (1954). An analysis of growth rates in substage A of plastochron nine in *Zea mays* L. *Minnesota Acad. Sci.* **22,** 104–8.

SUSSEX, I. M. (1955). Morphogenesis in *Solanum tuberosum* L: Experimental investigation of leaf dorsiventrality and orientation in the juvenile shoot. *Phytomorph.* **5,** 286–300.

TURING, A. M. (1952). The chemical basis of morphogenesis. *Phil. Trans.* B, **237,** 37–72.

WARDLAW, C. W. (1948). Experimental and analytical studies of pteridophytes. XIII. On the shoot apex in a tree fern, *Cyathea manniana* Hooker. *Ann. Bot.,* N.S. **12,** 371–84.—(1949). Experiments on organogenesis in ferns. *Growth* (Suppl.), **9,** 93–131.—(1952). *Phylogeny and morphogenesis,* p. 536, Macmillan. London.—(1953). A commentary on Turing's diffusion reaction theory of morphogenesis. *New Phytol.* **52,** 40–7.—(1953). Comparative observations on the shoot apices of vascular plants. *New Phytol.* **52,** 195–208.—(1955). Evidence relating to the diffusion-reaction theory of morphogenesis. *New Phytol.* **54,** 39–48.—(1955). The chemical concept of organization in plants. *New Phytol.* **54,** 302–10.—(1955). Experimental and analytical studies of pteridophytes. XXVIII. Leaf symmetry and orientation in ferns. *Ann. Bot.,* N.S., **19,** 389–99.—(1955). Responses of a fern apex to direct chemical treatments. *Nature,* **176,** 1098–1100.—(1956). Experimental and analytical studies of pteridophytes. XXXIV. On the shoot apex of the bird's nest fern, *Asplenium nidus* L. *Ann. Bot.,* N.S., **20,** 363–74. —(1957). Experimental and analytical studies of pteridophytes. XXXVII. A note on the inception of microphylls and macrophylls. *Ann. Bot.,* N.S., **21,** 427–37.—(1957). The reactivity of the apical meristem as ascertained by cytological and other techniques. *New Phytol.* **56,** 221–9.—(1957). The inception of leaf primordia. *Symp. on Leaf Growth,* Nottingham (1956), 53–65.—(1957). Experimental and analytical studies of pteridophytes. XXXV. The effects of direct applications of various substances to the shoot apex of *Dryopteris austriaca* (*D. aristata*). *Ann. Bot.,* N.S., **21,** 85–120.

WEBER, A. V. & STEIN, O. L. (1954). A comparison of the rates of leaf initiation in seedlings of *Zea mays* L. under field and growth chamber conditions. *Minnesota Acad. Sci.* **22,** 94–8.

WETMORE, R. H. (1954). The use of *in vitro* cultures in the investigation of growth and differentiation in vascular plants. Abnormal and pathological plant growth. *Brookhaven Symp. Biol.* **6,** 22–40.

Essay 18

BAILEY, I. W. (1954). *Contributions to Plant Anatomy.* Waltham, Mass.

BÜNNING, E. (1948). *Entwicklungs- und Bewegungsphysiologie der Pflanze.* Berlin.—(1952). Morphogenesis in plants. *Surv. Biol. Progr.* **2,** 105.

BUVAT, R. (1952). Structure, évolution et fonctionnement du méristème

apical de quelques dicotylédones. *Ann. Sci. Nat. (Bot.)* **13**, 199–300.—
(1955). Le méristème apical de la tige. *Ann. Biol.* **31**, 596–656.

CUSICK, F. (1956). Some experimental studies of floral development.
Trans. Roy. Soc. Edin. **63**, 153–66.

CUTTER, E. G. (1955). Observations on some abnormal fruits of the tomato,
Lycopersicon esculentum Mill. *Phytomorph.* **5**, 274–86.

DOUGLAS, G. E. (1944). The inferior ovary. *Bot. Rev.* **10**, 125–86.

EAMES, A. J. (1931). The vascular anatomy of the flower with refutation of
the theory of carpel polymorphism. *Amer. J. Bot.* **18**, 147–88.

ENGARD, C. J. (1944). Organogenesis in *Rubus*. *Res. Publ. Univ. Hawaii*, **21**,
1–234.

ESAU, K. (1953). *Plant Anatomy*. New York.

FOSTER, A. S. (1929). Investigations on the morphology and comparative
history of development of foliar organs. I. The foliage leaves and cata-
phyllary structures in the horse-chestnut (*Aesculus hippocastanum* L.). *Amer.
J. Bot.* **18**, 243–9.

FOSTER, A. S. (1935). A histogenic study of foliar development in *Carya
Buckleyi* var. *Arkansana*. *Amer. J. Bot.* **22**, 88–147.

GOEBEL, K. (1922). Gesetzmässigkeiten in Blattaufbau. *Bot. Abh.* **1**.

GRÉGOIRE, V. (1938). La morphogenèse et l'autonomie morphologique
de l'apparail floral. I. Le carpelle. *Cellule*, **47**, 285–452.

GUNCKEL, J. E. & SPARROW, A. H. (1954). Aberrant growth in plants
induced by ionizing radiation. *Brookhaven Symp. Biol.* **6**, 252–79.

GUNCKEL, J. E. (1956). Morphological effects of ionizing radiations on
plants. *Amer. J. Bot.*

HARRISON, Y. HESLOP. (1953). *Nuphar intermedia* Ledeb., a presumed relict
hybrid. *Watsonia*, **3**, 7.

MARSDEN-JONES, E. M. (1933). *Proc. Linn. Soc. (Bot.)* **50**, 39.

MATHER, K. (1944). Genetical control of incompatibility in angiosperms
and fungi. *Nature*, **153**, 392–4.—(1948). Nucleus and cytoplasm in differentiation. *Soc. Exp. Biol. Symp.* **2**, 195–216.

MURNEEK, A. F. (1927). Physiology of reproduction in horticultural plants.
II. The physiological basis of intermittent sterility with special reference to
the spider flower. *Res. Bull. Univ. Mo. Agric. Exp. Sta.* **106**, 1–30.

PERJE, A. M. (1952). Some causes of variation in *Ranunculus ficaria* L. *Ark.
Bot.* **2**, 251–64.

PLANTEFOL, L. (1938). *L'ontogénie de la Fleur*. Paris.

SCHOUTE, J. C. (1936). Fasciation and dichotomy. *Réc. trav. bot. néerl.* **33**, 649.

SCUÜEPP, O. (1929). Untersuchungen zur Beschreibenden und experi-
mentellen Entwicklungsgeschichte von. *Acer pseudoplantanus*. *Jb. wiss. Bot.*
70, 743–804.

SIRONVAL, C. (1956). *La photoperiode et la sexualisation du fraisier des quatre-
saisons (metabolisme chlorophyllien et hormone florigène)*. Thesis: Univ. Liege.

STEBBINS, G. L. JR. (1950). *Variation and Evolution in Plants*. New York.

TEPFER, S. S. (1953). Floral anatomy and ontogeny in *Aquilegia formosa*
var. *truncata* and *Ranunculus repens*. *Univ. Calif. Publ. Bot.* **25**, 513–648.

THOMPSON, J. MCL. (1937). On the place of ontogeny in floral inquiry.
Publ. Hartley Bot. Labs. L'pool. Univ. **17**, 3–20.

TURING, A. M. (1952). The chemical basis of morphogenesis. *Phil. Trans.* B, **237,** 37–72.

WARDLAW, C. W. (1953). A commentary on Turing's diffusion reaction theory of morphogenesis. *New Phytol.* **52,** 40–7.—(1955). Evidence relating to the diffusion-reaction theory of morphogenesis. *New Phytol.* **54,** 39–48.— (1955). The chemical concept of organization in plants. *New Phytol.* **54,** 302–10.—(1955). Responses of a fern apex to direct chemical treatments. *Nature,* **176,** 1098–1100.—(1957). Experimental and analytical studies of pteridophytes. XXXV. The effects of direct applications of various substances to the shoot apex of *Dryopteris austriaca* (*D. aristata*). *Ann. Bot.* **21,**— (1957). On the organization and reactivity of the shoot apex in vascular plants. *Amer. J. Bot.* **44,** 176–85.

WORSDELL, W. C. (1916). *The Principles of Plant-Teratology.* London.

Essay 19

ANDERSSON-KOTTÖ, I. (1929). A genetical investigation in *Scolopendrium vulgare. Hereditas,* **12,** 109–78.—(1938). *Verdoorn's Manual of Pteridology,* Chap. IX, 284–302.

BOWER, F. O. (1923–8). *The Ferns,* Vols. I, II, III, Camb. Univ. Press.

COPELAND, E. B. (1947). *Genera Filicum.* Waltham, Mass.

GOEBEL, K. (1928–30). *Organographie der Pflanzen.*

HOLTTUM, R. E. (1954). *Flora of Malaya.* Vol. II. *Ferns.*

MANTON, I. (1950). *Problems of Cytology and Evolution in the Pteridophytes.* Camb. Univ. Press.

WARDLAW, C. W. (1952). *Phylogeny and Morphogenesis.* London.—(1957). On the organization and reactivity of the shoot apex in vascular plants. *Amer. J. Bot.* **44,** 176–85.—(1962). The sporogenous meristems of ferns. *Phytomorphology,* **12,** 394–408.

Essay 22

BALL, E. (1946). Development in sterile culture of stem tips and subjacent regions of *Tropaeolum majus* L., and *Lupinus albus* L. *Amer. J. Bot.* **33,** 301.

BOWER, F. O. (1919). *Botany of the Living Plant.* Macmillan: London.— (1935). *Primitive Land Plants.* Macmillan: London.

BROWN, R. & ROBINSON, E. (1955). Cellular differentiation and the development of enzyme proteins in plants. *Biological Specificity and Growth,* 93, Princeton Univ. Press.

BUCHHOLZ, J. T. (1946). Volumetric studies of seeds, endosperms, and embryos in *Pinus ponderosa* during embryonic differentiation. *Bot. Gaz.* **108,** 232.

CHAMBERLAIN, J. C. (1935). *Gymnosperms.* Univ. Chicago Press.

CUTTER, E. G. & VOELLER, B. R. (1959). Changes in leaf arrangement in individual fern apices. *Jour. Linn. Soc. Lond.* (*Bot.*) **56,** 225.

ENGARD, C. J. (1944). Organogenesis in *Rubus. Res. Publ. Univ. Hawaii,* **21,** 1.

FOSTER, A. S. (1929). Investigations on the morphology and comparative history of development of foliar organs. I. The foliage leaves and cata-phyllary structures in the horse-chestnut (*Aesculus hippocastanum* L.). *Amer. J. Bot.* **18,** 243.—(1935). A histogenic study of foliar development in *Carya buckleyi* var. *arkansana*. *Amer. J. Bot.* **22,** 88.

GOEBEL, K. (1900). *Organography of Plants*. (English edn., I. B. Balfour.) Clarendon Press, Oxford.—(1913). *Organographie der Pflanzen*. (2nd edn.) Gustav Fischer, Jena, Pt. I, p. 181.

HESLOP-HARRISON, J. (1959). Growth substances and flower morpho-genesis. *Jour. Linn. Soc. Lond.* (*Bot.*), **56,** 269.—& HESLOP-HARRISON, Y. (1958). Long-day and auxin induced male sterility in *Silene pendula* L. *Portug. Acta Biol.* 5, 79.

HESLOP-HARRISON, Y. & WOODS, I. (1959). Temperature-induced meristic and other variation in *Cannabis sativa*. *Jour. Linn. Soc. Lond.* (*Bot.*) **56,** 290.

HEYES, J. K. & BROWN, R. (1956). Growth and cellular differentiation. In *The Growth of Leaves*, edited by F. L. Milthorpe. Butterworth's Scientific Publs., 31.

KIDSTON, R. & LANG, W. H. (1917–21). Old red sandstone plants showing structure; from the Rhynie chert bed, Aberdeenshire, Pts. 1–4, *Trans. Roy. Soc. Edin*, **51–2.**

LAM, H. J. (1959). Some fundamental considerations on the 'New Morpho-logy'. *Trans. Bot. Soc. Edin.* **38,** 100.

LANG, W. H. (1915). Phyletic and causal morphology. Pres. Addr. (Sect. K.) *Brit. Assoc. Adv. Sci.*

LANG, W. H. & COOKSON, I. C. (1935). On a flora including vascular land plants, associated with *Monograptus*, in rocks of Silurian age, from Victoria, Australia. *Phil. Trans.* B, **224,** 421.

LANKESTER, E. R. (1870). On the use of the term homology in modern zoology and the distinction between homogenetic and homoplastic agree-ments. *Ann. Mag. Nat. Hist.* **4,** ser. 6.

MANTON, I. (1950). *Problems of Cytology and Evolution in the Pteridophyta*. Camb. Univ. Press.

PAYER, J. B. (1857). *Traité D'organogénie Comparée de la Fleur*. Paris.

RICHARDS, F. J. (1951). Phyllotaxis: its quantitative expression and rela-tion to growth in the apex. *Phil. Trans.* B, **221,** 1.

ROBINSON, E. & BROWN, R. 1952. The development of the enzyme com-plement in growing root cells. *Jour. Exp. Bot.* **3,** 356–74.

ROBINSON, E. & BROWN, R. (1954). Enzyme changes in relation to cell growth in excised root tissues. *Jour. Exp. Bot.* **5,** 71.

SCHOUTE, J. C. (1913). Beiträge zur Blattstellungslehre. *Réc trav. bot. Néerl.* **10,** 153.

SKOOG, F. (1954). Substances involved in normal growth and differentia-tion. *Brookhaven Symp. Biol.* **6,** 1.—(1954). Chemical regulation of growth in plants. In *Dynamics of Growth Processes*. (E. F. Boell, ed.) Princeton Univ. Press, 148.—& MILLER, C. O. (1957). Chemical regulation of growth and organ formation in plant tissues cultured *in vitro*. *Symp. Soc. exp. Biol.* **11,** 118.—& TSUI, C. (1948). Chemical control of growth and bud formation in tobacco stem segments and callus cultured *in vitro*. *Amer. J. Bot.* **35,** 782.—& TSUI, C. (1951). Growth substances and the formation of buds in plant tissues. In *Plant Growth Substances* (F. Skoog, ed.). Madison: Univ. Wisconsin Press, 263.

SNOW, M. & R. (1931). Experiments on phyllotaxis. I. The effect of isolating a primordium. *Phil. Trans.* B, **221**, 1.—(1933). Experiments on phyllotaxis. II. The effect of displacing a primordium. *Phil. Trans.* B, **222**, 353.—(1935). Experiments on phyllotaxis. III. Diagonal splits through decussate apices. *Phil. Trans.* B, **225**, 63.

STEEVES, T. A. & SUSSEX, I. M. (1957). Studies in the development of excised leaves in sterile culture. *Amer. J. Bot.* **44**, 665.

STEWARD, F. C., MAPES, MARION O. & SMITH, J. (1958). Growth and division of freely suspended cells. *Amer. J. Bot.* **45**, 693.

SUNDERLAND, N. & BROWN, R. (1956). Distribution of growth in the apical region of the shoot of *Lupinus albus*. *Jour. Exp. Bot.* **7**, 127.—& HEYES, J. K. (1956). Growth and metabolism in the shoot apex of *Lupinus albus*. In *The Growth of Leaves* (ed. F. L. Milthorpe). Butterworth's Scientific Publs., 77.—& HEYES, J. K. (1957). Protein and respiration in the apical region of the shoot of *Lupinus albus*. *Jour. Exp. Bot.* **8**, 55.

TANSLEY, A. G. (1923). Some aspects of the present position of botany. Pres. Addr. (Sect. K.) *Brit. Assoc. Adv. Sci.*

TEPFER, S. S. (1953). Floral anatomy and ontogeny in *Aquilegia formosa* var. *truncata* and *Ranunculus repens*. *Univ. Calif. Publ. Bot.* **25**, 513.

THOMAS, H. HAMSHAW, (1932). The old morphology and the new. *Proc. Linn. Soc. Lond.* **145**, 17.

TURING, A. M. (1952). The chemical basis of morphogenesis. *Phil. Trans.* B, **237**, 37.

VAN FLEET, D. S. (1959). Analysis of the histochemical localization of peroxidase related to the differentiation of plant tissues. *Canad. J. Bot.* **37**, 449.

WARDLAW, C. W. (1952). *Phylogeny and Morphogenesis*. Macmillan: London. —(1953). Experimental and analytical studies of pteridophytes. XXIII. The induction of buds in *Ophioglossum vulgatum* L. *Ann. Bot. Lond.*, N.S., **17**, 513.—(1953). A commentary on Turing's diffusion reaction theory of morphogenesis. *New Phytol.* **52**, 201.—(1954). Experimental and analytical studies of pteridophytes, XXVI. *Ophioglossum vulgatum*: comparative morphogenesis in embryos and induced buds. *Ann. Bot. Lond.*, N.S., **18**, 397.—(1955). *Embryogenesis in Plants*. Wiley: New York; Methuen: London.—(1955). Experimental and analytical studies of pteridophytes. XXVIII. Leaf symmetry and orientation in ferns. *Ann. Bot.*, *Lond.*, N.S., **19**, 389.—(1957). Experimental and analytical studies of pteridophytes. XXXVII. A note on the inception of microphylls and macrophylls. *Ann. Bot. Lond.*, N.S., **21**, 427.—(1957). On the organization and reactivity of the shoot apex in vascular plants. *Amer. J. Bot.* **44**, 176.—(1957). The floral meristem as a reaction system. *Proc. Roy. Soc. Edin.* B, **66**, 394.—(1959). Methods in plant morphogenesis. *Jour. Linn. Soc. Lond.* **56**, 154.—(1960). The inception of shoot organization. *Phytomorph.* **10**, 107.—(1961). Growth and development of the inflorescence and the flower (in *Growth in Living Systems*, 491–523, ed. M. X. Zarrow. Internat. Symp., Purdue University, 1960).

WETMORE, R. H. (1953). Tissue and organ culture as a tool for studies in development. *Proc. VII Internat. Bot. Congr. Stockholm,* 369.—(1954). The use of *in vitro* cultures in the investigation of growth and differentiation in vascular plants. *Abnormal and Pathological Plant Growth. Brookhaven Symp. Biol.* **6,** 22.—(1955). Differentiation of xylem in plants. *Science,* **121,** 626. —(1956). Growth and development in the shoot system of plants. *Cellular Mechanisms in Differentiation and Growth.* Princeton Univ. Press, 173.—(1959). Morphogenesis in plants—a new approach. *American Scientist,* **47,** 326.— & MOREL, G. (1949). Growth and development of *Adiantum pedatum* L. on nutrient agar. *Amer. J. Bot.* **36,** 805. & SOROKIN, S. (1955). On the differentiation of xylem. *Jour. Arnold Arboretum,* **36,** 305.—& WARDLAW, C. W. (1951). Experimental morphogenesis in vascular plants. *Ann. Rev. Plant Physiol.* **2,** 269.

ZIMMERMANN, W. (1930). *Phylogenie der Pflanzen.* Jena.—(1938). Die Telomtheorie. *Der Biologe,* 7th year, **12,** 385.

Essay 24

HESLOP-HARRISON, J. (1953). *New Concepts in Flowering Plant Taxonomy.* Heinemann: London.

HOLTTUM, R. E. (1949). The classification of ferns. *Biol. Rev.* **24,** 267–96.

HUTCHINSON, J. (1926). *The Families of Flowering Plants.* Macmillan, London.

HUXLEY, J. S. (ed.). (1941). *The New Systematics.* Oxford Univ. Press.

LINSLEY, E. G. & USINGER, R. L. (1961). Taxonomy. In *The Encyclopaedia of the Biological Sciences.* Ed. P. Gray. 992–6. Reinhold Publ. Corp. New York.

MATTHEWS, J. R. (1957). Obituary: Sir William Wright Smith, F.R.S., D. ès Sc., LL.D., F.L.S., F.R.S.E. *Trans. & Proc., Bot. Soc. Edin.* **37,** 142–5.

STEBBINS, G. L. (1950). *Variation and Evolution in Plants.* Oxford Univ. Press.

TURRILL, W. B. (1938). The expansion of taxonomy with special reference to the spermatophyta. *Biol. Rev.* **13,** 342.

Essay 26

ALBAUM, H. G. (1938). Inhibitions due to growth hormones in fern prothallia and sporophytes. *Amer. J. Bot.* **25,** 124–33.

ALLSOPP, A. (1952). Experimental and analytical studies of pteridophytes. 17. The effect of various physiologically active substances on the development of *Marsilea* in sterile culture. *Ann. Bot. Lond.,* N.S., **16,** 165–83.— (1953*a*). Experimental and analytical studies of pteridophytes. 19. Investigations on *Marsilea.* 2. Induced reversion to juvenile stages. *Ann. Bot. Lond.,* N.S., **17,** 37–55.—(1953*b*). Experimental and analytical studies of pteridophytes. 20. Investigations on *Marsilea,* 3. The effect of various sugars on development and morphology. *Ann. Bot. Lond.,* N.S., **17,** 447–63.— (1954). Experimental and analytical studies of pteridophytes. 24. Investigations on *Marsilea.* 4. Anatomical effects of changes in sugar concentration. *Ann. Bot. Lond.,* N.S., **18,** 449–61.—(1955). Experimental and analytical studies of pteridophytes. 27. Investigations on *Marsilea.* 5. Cultural conditions and morphogenesis, with special reference to the origin of land and water forms. *Ann. Bot. Lond.*, N.S., **19**, 247–64.

ANDERSSON-KOTTÖ, I. (1929). A genetical investigation in *Scolopendrium vulgare*. *Hereditas*, **12**, 109–78.—(1938). Genetics, in Verdoorn's *Manual of Pteridology*, Chap. IX, 284–302.

ASHBY, E. (1948). Studies in the morphogenesis of leaves. 2. The area, cell size and cell number of leaves of *Ipomoea* in relation to their position on the shoot. *New Phytol.* **47**, 177–95.—& WANGERMANN, E. (1950). Studies in the morphogenesis of leaves. 4. Further observations on area, cell size and cell number of leaves of *Ipomoea* in relation to their position on the shoot. *New Phytol.* **49**, 23–35.—WANGERMANN, E. (1951). Studies in the morphogenesis of leaves. 7. Part 2. Correlative effect of fronds in *Lemma minor*. *New Phytol.* **50**, 200–9.

BOWER, F. O. (1916). On leaf-architecture as illuminated by a study of Pteridophyta. *Trans. Roy. Soc. Edin.* **51**, 657–788.—(1923–8). *The Ferns* (Filicales). I (1923). Analytical examination of the criteria of comparison. II (1926). The eusporangiate and other primitive ferns. III (1928). The leptosporangiate ferns. Cambridge.

BRIGGS, W. R. & STEEVES, T. A. (1958). Morphogenetic studies on *Osmunda cinnamomea* L.—The expansion and maturation of vegetative fronds. *Phytomorphology*, **8**, 234–48.—(1959). Morphogenetic studies on *Osmunda cinnamomea* L.—The mechanism of crozier uncoiling. *Phytomorphology*, **9**, 134–47.

BROWN, R. (1958). Cellular basis for the induction of morphological structures. *Nature, Lond.* **181**, 1546–7.

CROTTY, W. J. (1955). Trends in the pattern of primordial development with age in the fern *Acrostichum daneaefolium*. *Amer. J. Bot.* **42**, 627–36.

CUTTER, E. G. (1955). Experimental and analytical studies of Pteridophytes. 29. The effect of progressive starvation on the growth and organization of the shoot apex of *Dryopteris aristata* Druce. *Ann. Bot. Lond.*, N.S., **19**, 485–99.

CUTTER, E. G. & VOELLER, B. R. (1959). Changes in leaf arrangement in individual fern apices. *J. Linn. Soc. (Bot.)* **56**, 225–36.

DASANAYAKE, M. D. (1960). Aspects of morphogenesis in a dorsiventral fern, *Pteridium aquilinum* (L.) Kuhn. *Ann. Bot. Lond.*, N.S., **24**, 317–29.

FOSTER, A. S. (1928). Salient features of the problem of bud-scale morphology. *Biol. Rev.* **3**, 123–64.—(1932). Investigations on the morphology and comparative history of development of foliar organs III. Cataphyll and foliage-leaf ontogeny in the black hickory (*Carya buckleyi* var. *arkansana*). *Amer. J. Bot.* **19**, 75–99.

GOEBEL, K. (1900). *Organography of Plants*. Part I. Oxford: Clarendon Press.—(1905). *Organography of Plants*. Part II. Oxford: Clarendon Press.—(1908). *Einleitung in die experimentelle Morphologie der Pflanzen*. Leipzig.—(1928). Morphologische und biologische Studien. XIII. Weitere Untersuchungen über die Gruppe der Drynariaceae. *Ann. Jard. bot. Buitenz.* **39**, 117–26.—(1930). *Organographie der Pflanzen*, 3rd edn. Jena.

GOODWIN, R. H. (1937). The role of auxin in leaf development in *Solidago*. *Amer. J. Bot.* **24**, 43–51.

GOTTLIEB, J. E. (1958). Development of the bracken fern, *Pteridium aquilinum* (L.) Kuhn. I. General morphology of the sporeling. *Phytomorphology*, **8**, 184–94.

HOFMEISTER, W. (1862). *Higher Cryptogamia*. Engl. ed. London.

KEFFORD, N. P. & GOLDACRE, P. L. (1961). The changing concept of auxin. *Amer. J. Bot.* **48**, 643–50.

KLEIN, L. (1884). Vergleichende Untersuchungen über Organbildung und Wachstum am Vegetationspunkt dorsiventraler Farne. *Bot. Ztg.*, **42**, 577–87.

RICHARDS, F. J. (1948). The geometry of phyllotaxis and its origin. *Symp. Soc. Exp. Biol.* **2**, 217–45.—(1951). Phyllotaxis: its quantitative expression and relation to growth in the apex. *Phil. Trans.* B, **235**, 509–64.

SCHOUTE, J. C. (1936). Fasciation and dichotomy. *Réc. trav. bot. néerl.* **33**, 649–69.

STEEVES, T. A. (1959). The development of leaves in sterile nutrient culture. *IX Int. Bot. Congr. Proc.*, **2**, 380.—& BRIGGS, W. R. (1958). Morphogenetic studies on *Osmunda cinnamomea* L.: the origin and early development of vegetative fronds. *Phytomorphology*, **8**, 60–72.—& BRIGGS, W. A. (1960). Morphogenetic studies on *Osmunda cinnamomea* L. The auxin relationships of expanding fronds. *J. exp. Bot.* **11**, 45–67.—& SUSSEX, I. M. (1957). Studies on the development of excised leaves in sterile culture. *Amer. J. Bot.* **44**, 665–73.—& WETMORE, R. H. (1953). Morphogenetic studies on *Osmunda cinnamomea* L.: Some aspects of the general morphology. *Phytomorphology*, **3**, 339–54.

SUSSEX, I. M. (1958). A morphological and experimental study of leaf development in *Leptopteris hymenophylloides* (A. Rich.) Presl. *Phytomorphology*, **8**, 97–107.—& CLUTTER, M. E. (1960). A study of the effect of externally supplied sucrose on the morphology of excised fern leaves *in vitro*. *Phytomorphology*, **10**, 87–99.—& STEEVES, T. A. (1953). Growth of excised leaves in sterile culture. *Nature, Lond.*, **172**, 624.— & STEEVES, T. A. (1958). Experiments on the control of fertility of fern leaves in sterile culture. *Bot. Gaz.*, **119**, 203–8.

TURING, A. M. (1952). The chemical basis of morphogenesis. *Phil. Trans.* B, **237**, 37–72.

VOELLER, B. R. (1960). Regulation of 'fiddlehead' uncoiling in ferns. *Naturwissenschaften*, **47**, 70–1.

WARDLAW, C. W. (1945). Experimental and analytical studies of pteridophytes. VI. Stelar morphology: the occurrence of reduced and discontinuous vascular systems in the rhizome of *Onoclea sensibilis*. *Ann. Bot. Lond.*, N.S., **9**, 383–97.—(1948). Experimental and analytical studies of pteridophytes. XII. On the shoot apex in a tree fern, *Cyathea manniana* Hooker. *Ann. Bot. Lond.*, N.S., **12**, 372–84.—(1952). Experimental and analytical studies of pteridophytes. XVIII. The nutritional status of the apex and morphogenesis. *Ann. Bot. Lond.*, N.S., **16**, 207–17.—(1953). A commentary on Turing's diffusion reaction theory of morphogenesis. *New Phytol.*, **52**, 40–7.—(1955). Experimental and analytical studies of pteridophytes. XXVIII. Leaf symmetry and orientation in ferns. *Ann. Bot. Lond.*, N.S., **19**, 389–99.—(1956). Experimental and analytical studies of pteridophytes. XXXIV,

On the shoot apex of the Bird's Nest Fern. *Asplenium nidus* L. *Ann. Bot. Lond.*, N.S., **20**, 363–74.—(1957). On the organization and reactivity of the shoot apex in vascular plants. *Amer. J. Bot.*, **44**, 176–85.—(1957). Experimental and analytical studies of pteridophytes. 35. The effects of direct applications of various substances to the shoot apex of *Dryopteris austriaca* (*D. aristata*). *Ann. Bot. Lond.*, N.S., **21**, 85–120.—(1958). Reflections on the unity of the embryonic tissues in ferns. *Phytomorphology*, **8**, 323–7.—(1960). The inception of shoot organization. *Phytomorphology*, **10**, 107–10.—& SHARMA, D. N. (1961). Experimental and analytical studies of pteridophytes. XXXIX. Morphogenetic investigations of sori in leptosporangiate ferns. *Ann. Bot. Lond.*, N.S., **25**, 477–90.

WEBSTER, B. D. & STEEVES, T. A. (1958). Morphogenesis in *Pteridium aquilinum* (L.) Kuhn.—General morphology and growth habit. *Phytomorphology*, **8**, 30–41.

WETMORE, R. H. (1950). Tissue and organ culture as a tool for studies in development. *Proc. 7th Int. Bot. Congr.* (*Stockholm*), 369–70.—(1953). Carbohydrate supply and leaf development in sporeling ferns. *Science*, **118**, 578.—(1954). The use of *in vitro* cultures in the investigation of growth and differentiation in vascular plants. Abnormal and pathological plant growth. *Brookhaven Symp. Biol.* **6**, 22–40.—& MOREL, G. (1949). Growth and development of *Adiantum pedatum* L. on nutrient agar (Abstr.). *Amer. J. Bot.* **36**, 805–6.

WETTER, R. & C. (1954). Studien über das Erstarkungswachstum und das primäre Dickenwachstum bei Leptosporangiaten Farnen. *Flora*, **141**, 598.

Essay 27

BUCHHOLZ, J. T. (1946). Volumetric studies of seeds, endosperms and embryos in *Pinus ponderosa*, during embryonic differentiation. *Bot. Gaz.* **108**, 232–44.

RIJVEN, A. H. G. C. (1952). *In vitro* studies on the embryo of *Capsella bursa-pastoris*. *Acta Bot. Néerl.* **1**, 158–200.

TURING, A. M. (1952). The chemical basis of morphogenesis. *Phil. Trans.* B, **237**, 37–72.

VAN OVERBEEK, J., CONKLIN, M. E. & BLAKESLEE, A. F. (1942). Cultivation *in vitro* of small *Datura* embryos. *Amer. J. Bot.* **29**, 472–7.

WARDLAW, C. W. (1953). A commentary on Turing's diffusion-reaction theory of morphogenesis. *New Phytol.* **52**, 40–7.—(1954). The interpretation of embryos as reaction systems. *Proc. 8th Internat. Bot. Congr.* (*Paris*), Sec. 8, pp. 257–9.—(1955). *Embryogenesis in Plants*. London.

Index